INTERMEDIATE
PHYSICAL
CHEMISTRY

INTERMEDIATE PHYSICAL CHEMISTRY

STATIONARY PROPERTIES OF CHEMICAL SYSTEMS

Joseph B. Dence
St. Louis University

Dennis J. Diestler
Purdue University

A WILEY-INTERSCIENCE PUBLICATION
JOHN WILEY & SONS
NEW YORK • CHICHESTER • BRISBANE • TORONTO • SINGAPORE

Library of Congress Cataloging in Publication Data:

Dence, Joseph B.
    Intermediate physical chemistry.

    "A Wiley-Interscience publication."
    Includes bibliographies and indexes.
    1. Chemistry, Physical and theoretical.   I. Diestler,
Dennis J., joint author   II. Title.

QD453.2.D54   1987   541.3   86-26638
ISBN 0-471-81243-9

Printed in the United States of America

10 9 8 7 6 5 4 3 2 1

To W. Heinlen Hall and William Sly
who first taught us physical chemistry

Preface

This book was written to fill the need for a text in a one semester course of physical chemistry at an intermediate level. It is intended to bridge the gap between the first undergraduate course in physical chemistry and the strictly graduate level courses that commence at an advanced level, with the aim of introducing the student to the primary literature of current research. At institutions with graduate programs, this book would be suitable for a one semester course for seniors and first year graduate students.

Modern physical chemistry is a borderline discipline between chemistry and physics. Depending on the emphasis, it is often called chemical physics. In any case, most of the fundamental work being done by physical chemists is aimed at understanding chemical phenomena from a microscopic (i.e., molecular) viewpoint.

Chemical systems in general can be divided into two classes, according to whether their properties are time-dependent or time-independent. For example, the properties of an *isolated* molecule in a stationary quantum state do not change with time, whereas a molecule interacting with light undergoes transitions between stationary states, with its properties depending on the particular mix of those states induced by the light. On a macroscopic scale, the properties of a mixture of reacting substances change with time as the system evolves toward a condition of equilibrium, in which the (macroscopic) properties become stationary.

Since we cannot treat all the diverse areas of physical chemistry in a reasonable number of pages commensurate with a one semester course at an intermediate level, we have chosen to restrict the book primarily to time-independent chemical systems. The main theme is the *quantitative microscopic understanding of the stationary properties of chemical systems*. The essential tools are quantum and statistical mechanics. The general principles of quantum mechanics (nonrelativistic version) are developed in Chapter 2 and then applied to atoms and molecules in Chapters 3 to 5. Similarly, the basic principles of statistical mechanics are presented in Chapter 6. They are then applied to systems of noninteracting particles (ideal gases and harmonic solids) in Chapter 7. The scope of statistical mechanics is expanded to cover systems of interacting particles in Chapter 8, where it is applied to fluids.

Consistent with our emphasis on quantitative description, we have included special sections called "Computer Highlights" in the chapters on applications. These promote the use of computers for handling calculations more involved than the "paper-and-pencil" variety encountered in the standard first course in physical chemistry. The computer has become so prominent in physical science today that no serious student should remain unexposed to its use. The computational problems (designated by a superscript C) have been designed so that they can be carried out on contemporary microcomputers now commonly accessible to students at most institutions. In fact, it has been argued that even useful research in physics (and by inference, chemistry and other sciences as well) can be done on a microcomputer.*

We do not attempt to teach the fundamentals of computer programming. It is assumed that readers have some familiarity with a computer language. A number of good introductory manuals are available.[†] In the book, programs have been displayed, arbitrarily, in BASIC. For background in physics, only the standard two semester, sophomore level course is assumed. For mathematics, we assume varying degrees of familiarity with calculus of functions of more than one variable, vector algebra, vector calculus, orthogonal coordinate systems, ordinary differential equations, linear algebra, complex numbers, and series. Throughout the text numerous references are made to general and specialized works in physics, mathematics, and physical chemistry; these are listed, with annotations, at the ends of the chapters. The reader is strongly encouraged to consult as much of this literature as possible.

We are grateful to Dolores Naylor for patiently typing many rough drafts, and to our wives for their understanding during the course of this project.

JOSEPH B. DENCE
DENNIS J. DIESTLER

Saint Louis, Missouri
West Lafayette, Indiana
January 1987

*P. Bak, "Doing Physics with Microcomputers," *Phys. Today*, **36** (12), 25 (1983).
[†] The following are representative: B. S. Gottfried, *Programming with* FORTRAN IV, Prentice-Hall, Englewood Cliffs, 1972; S. Lipschutz and A. Poe, *Programming with* FORTRAN, Schaum's Outline Series, McGraw-Hill, New York, 1978; F. Scheid, *Computers and Programming*, Schaum's Outline Series, McGraw-Hill, New York, 1982; L. Graff and L. J. Goldstein, *Applesoft* BASIC *for the Apple II and IIe*, Robert S. Brady, Bowie, Maryland, 1984; D. D. McCracken, *Computing for Engineers and Scientists with* FORTRAN 77, Wiley, New York, 1984; J. H. Noggle, *Physical Chemistry on a Microcomputer*, Little, Brown, Boston, 1985.

Contents

INTERMEDIATE
PHYSICAL
CHEMISTRY

CHAPTER 1

Classical Background

By the end of the 19th century it was generally accepted that Newton's laws of motion for material bodies, Maxwell's equations for electromagnetic radiation, and thermodynamics were sufficient to explain phenomena in the physical world. Most of our everyday experience can be accounted for in terms of these classical principles. While classical physics ultimately breaks down when one attempts to explain the behavior of microscopic objects such as atoms, it is still possible to give a satisfactory classical molecular description of the macroscopic behavior of certain chemical systems. For example, simple liquids such as argon can be accurately simulated by numerically solving Newton's equations of motion on a computer for a collection of a few hundred atoms.

The current theory of matter and radiation—quantum mechanics—has its roots in classical physics. Moreover, the interpretation of the results of quantum mechanics is usually attempted in the more familiar classical terms. For these reasons it is appropriate at the outset to summarize the basic concepts of classical mechanics.

1.1   THE NEWTONIAN FORMULATION

To describe the motion of a *single* particle, we need only give its vector position $\mathbf{r}$ as a function of time t. According to **Newton's Second Law** (after the English mathematical physicist Isaac Newton, 1642–1727)

$$\mathbf{f} = m\ddot{\mathbf{r}} = m\dot{\mathbf{v}} = \dot{\mathbf{p}} \tag{1.1-1}$$

where m is the particle's (fixed) mass, $\mathbf{v}$ is the velocity, $\mathbf{p}$ is the momentum,

1

and $\mathbf{f}$ is the force acting on the particle. By convention, a dot over a symbol denotes the derivative with respect to time, that is, $\dot{\mathbf{r}} = d\mathbf{r}/dt$ and $\ddot{\mathbf{r}} = d^2\mathbf{r}/dt^2$. The vector relation (1.1-1) can be rewritten

$$m\ddot{x} = f_x \qquad m\ddot{y} = f_y \qquad m\ddot{z} = f_z \qquad (1.1\text{-}2)$$

in terms of the Cartesian components. We assume that the force $\mathbf{f}$ is a known function of $\mathbf{r}$ and its derivatives, and possibly also of time explicitly. Thus, in general, Newton's equations comprise a set of coupled, nonlinear, second-order, ordinary differential equations. To solve eqs. (1.1-2) uniquely, we must specify the initial position [$x(0)$, $y(0)$, and $z(0)$] and the initial velocity [$\dot{x}(0)$, $\dot{y}(0)$, and $\dot{z}(0)$]. We are then able to predict the position for all future and past times, that is, we know the trajectory of the particle. Note that we are implicitly regarding the particle as a *point mass* whose position is precisely known at any particular instant.

If time is "reversed," that is, if t is replaced by $-t$, then $\mathbf{r}(t)$ is simply replaced by $\mathbf{r}(-t)$; velocity and acceleration transform as follows:

$$\dot{\mathbf{r}}(t) = \frac{d\mathbf{r}(t)}{dt} \xrightarrow{t \to -t} \frac{d\mathbf{r}(-t)}{d(-t)} = \frac{-d\mathbf{r}(-t)}{dt} = -\dot{\mathbf{r}}(-t)$$

$$\ddot{\mathbf{r}}(t) = \frac{d\dot{\mathbf{r}}(t)}{dt} \xrightarrow{t \to -t} \frac{d\dot{\mathbf{r}}(-t)}{d(-t)} = \frac{d\dot{\mathbf{r}}(-t)}{dt} = \ddot{\mathbf{r}}(-t) \qquad (1.1\text{-}3)$$

Newton's equation [eq. (1.1-1)] then becomes

$$m\ddot{\mathbf{r}}(-t) = f[\mathbf{r}(-t)]$$

where we have assumed that the force depends only on the coordinates. Thus, if $\mathbf{r}(t)$ is a possible trajectory, then so is $\mathbf{r}(-t)$. Newton's equation is said to be **invariant under time reversal**. Some caution is required in the interpretation of the time-reversed trajectory $\mathbf{r}(-t)$. One does not actually "go backward in time." Rather, eq. (1.1-3) indicates that the velocity is reversed. If we imagine the particle's motion to be recorded on movie film, time reversal corresponds to running the film backward.

In most cases of interest to us for which the trajectory of a particle is a closed path, no net work is done on the particle, that is,

$$\oint \mathbf{f} \cdot d\mathbf{r} = 0$$

where the circle through the integral sign denotes integration around a closed path. Then, it can be shown that the force $\mathbf{f}$ is necessarily expressible as the negative of the gradient of a function $V(\mathbf{r})$:[1]

$$\mathbf{f} = -\nabla V(\mathbf{r})$$

The function $V(\mathbf{r})$ is called the **potential energy** of the particle.

Newton's equation (1.1-1) can now be written

$$-\nabla V = m\dot{\mathbf{v}} \qquad (1.1\text{-}4)$$

The work done by $\mathbf{f}$ on the particle as it moves along a differential element $d\mathbf{r} = \mathbf{v}(t)\, dt$ of its trajectory is given by

$$\mathbf{f} \cdot d\mathbf{r} = -\nabla V \cdot d\mathbf{r} = m\dot{\mathbf{v}} \cdot \mathbf{v}\, dt = d\left(\tfrac{1}{2}mv^2\right) = dT \qquad (1.1\text{-}5)$$

where the quantity $T = \tfrac{1}{2}mv^2$ is defined as the **kinetic energy**. On the other hand, from the definition of the gradient we can write

$$-\nabla V \cdot d\mathbf{r} = -\left[\left(\frac{\partial V}{dx}\right) dx + \left(\frac{\partial V}{\partial y}\right) dy + \left(\frac{\partial V}{dz}\right) dz\right]$$

$$= -dV \qquad (1.1\text{-}6)$$

Combination of eqs. (1.1-5) and (1.1-6) yields

$$d(T + V) = 0 \qquad (1.1\text{-}7)$$

We conclude that the *total* energy of the particle, $T + V$, is constant. A one-particle system having this property is said to be **conservative**.

The results for a one-particle system can be extended to a collection of N interacting particles. The force acting on any particle i is expressed as

$$\mathbf{F}_i = \mathbf{F}_i^{\text{ext}} + \sum_{j \neq i}^{N} \mathbf{f}_{ij}$$

where $\mathbf{f}_{ij}$ is the force on i due to particle j, and the external force $\mathbf{F}_i^{\text{ext}}$ arises from interaction with the rest of the universe. Newton's equations, eqs. (1.1-2), for the N-particle system can be cast in the compact form

$$m_i\ddot{x}_i = F_i \qquad (i = 1, 2, \ldots, 3N) \qquad (1.1\text{-}8)$$

where the index i now runs over Cartesian components of the successive

---

[1]We recall from vector calculus that the **nabla operator** is defined as $\nabla = \mathbf{e}_x(\partial/\partial x) + \mathbf{e}_y(\partial/\partial y) + \mathbf{e}_z(\partial/\partial z)$, where $\mathbf{e}_x, \mathbf{e}_y, \mathbf{e}_z$ are unit vectors in the positive x, y, and z directions. When ∇ operates on a scalar-valued function f, the result is called the **gradient** of f. The gradient is itself a vector.

vectors, that is, $\mathbf{r}_1 = (x_1, x_2, x_3)$, $\mathbf{r}_2 = (x_4, x_5, x_6)$, and so on, $\mathbf{F}_1 = (F_1, F_2, F_3)$, $\mathbf{F}_2 = (F_4, F_5, F_6)$, and so forth, and $m_1 = m_2 = m_3$, $m_4 = m_5 = m_6$, and so on. According to **Newton's Third Law**, if two bodies exert forces on each other, then these forces are equal in magnitude and opposite in direction:

$$\mathbf{f}_{ij} = -\mathbf{f}_{ji} \tag{1.1-9}$$

If all forces are conservative, then we can write

$$\mathbf{F}_i^{\text{ext}} = -\nabla_i\left(V_i^{\text{ext}}\right) \qquad \mathbf{f}_{ij} = -\nabla_i V_{ij}$$

where ∇_i denotes the gradient with respect to the position of particle i. Moreover, if the potential energy V_{ij} is taken to be a function only of the distance $|\mathbf{r}_i - \mathbf{r}_j|$ between particles i and j, then eq. (1.1-9) holds automatically. It can then be shown that the total energy

$$E = T + V$$
$$= \sum_{i=1}^{N} \tfrac{1}{2} m_i v_i^2 + \sum_{i=1}^{N} V_i^{\text{ext}}(\mathbf{r}_i) + \sum_{i=1}^{N} \sum_{j>i} V_{ij}\left(|\mathbf{r}_i - \mathbf{r}_j|\right) \tag{1.1-10}$$

is constant, that is, the system is conservative. Equation (1.1-10) is the many-particle analog of eq. (1.1-7).

1.2   ALTERNATIVE FORMULATIONS

If the potential energy function exhibits spatial symmetry, it may be more convenient to describe the motion in terms of non-Cartesian coordinates. For example, for an isolated system of just two particles, $\mathbf{F}_i^{\text{ext}} = 0$. If the potential energy is a function of only the distance between the particles, one intuits that it should be possible to reduce the original Newtonian equations, eqs. (1.1-8) with $N = 2$, to perhaps just one nontrivial equation involving $r = |\mathbf{r}_1 - \mathbf{r}_2|$. We shall see that this can be accomplished by reformulating Newton's equations.

In the first such formulation, one defines the **Lagrangian** L (after the French mathematician and physicist Joseph-Louis Lagrange, 1736–1813) by

$$L = T - V = \frac{1}{2} \sum_{i=1}^{3N} m_i \dot{x}_i^2 - V(\{x_i\}) \tag{1.2-1}$$

where for a system of N particles we regard L as a function of $6N$ independent variables, the $3N$ coordinates $\{x_i\}$ and the $3N$ velocities $\{\dot{x}_i\}$, and we assume that V depends only upon the coordinates. We can then recast

Newton's equations, eqs. (1.1-8), as

$$\frac{d}{dt}\left(\frac{\partial L}{\partial \dot{x}_i}\right) = \frac{\partial L}{\partial x_i} \qquad (i = 1, 2, \ldots, 3N) \qquad (1.2\text{-}2)$$

Now introduce a new set of coordinates, termed **generalized coordinates**, $\{q_k\}$.

$$q_k = q_k(x_1, x_2, \ldots, x_{3N}) \qquad (k = 1, 2, \ldots, 3N) \qquad (1.2\text{-}3)$$

Equations (1.2-3) can be viewed simply as a transformation from one set of independent variables, $\{x_i\}$, to another set, $\{q_k\}$. The *inverse* transformation is then

$$x_i = x_i(q_1, q_2, \ldots, q_{3N}) \qquad (i = 1, 2, \ldots, 3N) \qquad (1.2\text{-}4)$$

Differentiating eq. (1.2-4) with respect to time, we obtain

$$\dot{x}_i = \sum_{k=1}^{3N} \left(\frac{\partial x_i}{\partial q_k}\right) \dot{q}_k \qquad (1.2\text{-}5)$$

It follows from eqs. (1.2-1) and (1.2-5) that the transformed kinetic energy generally depends upon both the generalized coordinates and the **generalized velocities**, $\{\dot{q}_k\}$. The transformed potential energy, however, depends only on the coordinates $\{q_k\}$. As we demonstrate presently, the form of eqs. (1.2-2) is invariant under the transformation eq. (1.2-3), that is,

$$\frac{d}{dt}\left(\frac{\partial L}{\partial \dot{q}_k}\right) = \frac{\partial L}{\partial q_k} \qquad (k = 1, 2, \ldots, 3N) \qquad (1.2\text{-}6)$$

These are **Lagrange's equations**; they are equivalent to Newton's equations, eqs. (1.1-8).

To prove eqs. (1.2-6), it is convenient to develop some auxiliary relations. Upon differentiating both sides of eq. (1.2-5) with respect to $\dot{q}_j$, we obtain

$$\frac{\partial \dot{x}_i}{\partial \dot{q}_j} = \frac{\partial x_i}{\partial q_j} \qquad (1.2\text{-}7)$$

Since $\partial x_i / \partial q_j$ is a function only of the q_k's, we can write according to the chain rule

$$\frac{d}{dt}\left(\frac{\partial x_i}{\partial q_j}\right) = \sum_{k=1}^{3N} \frac{\partial}{\partial q_k}\left(\frac{\partial x_i}{\partial q_j}\right) \dot{q}_k \qquad (1.2\text{-}8)$$

Likewise, from eq. (1.2-5) we have

$$\frac{\partial \dot{x}_i}{\partial q_j} = \sum_{k=1}^{3N} \frac{\partial}{\partial q_j} \left(\frac{\partial x_i}{\partial q_k} \right) \dot{q}_k \tag{1.2-9}$$

Comparing eqs. (1.2-8) and (1.2-9) and noting that the order of differentiation is immaterial, we deduce

$$\frac{d}{dt} \left(\frac{\partial x_i}{\partial q_j} \right) = \frac{d}{dt} \left(\frac{\partial \dot{x}_i}{\partial \dot{q}_j} \right) = \frac{\partial \dot{x}_i}{\partial q_j} \tag{1.2-10}$$

Since V depends only upon the $\{x_i\}$, we can write from the chain rule

$$\frac{\partial L}{\partial q_k} = \frac{\partial T}{\partial q_k} - \frac{\partial V}{\partial q_k} = \frac{\partial T}{\partial q_k} - \sum_{i=1}^{3N} \frac{\partial V}{\partial x_i} \frac{\partial x_i}{\partial q_k} \tag{1.2-11a}$$

$$\frac{\partial L}{\partial \dot{q}_k} = \frac{\partial T}{\partial \dot{q}_k} = \sum_{i=1}^{3N} \frac{\partial T}{\partial \dot{x}_i} \frac{\partial \dot{x}_i}{\partial \dot{q}_k} \tag{1.2-11b}$$

Differentiating eq. (1.2-11b) with respect to time gives

$$\frac{d}{dt} \left(\frac{\partial L}{\partial \dot{q}_k} \right) = \sum_{i=1}^{3N} \left[\frac{d}{dt} \left(\frac{\partial T}{\partial \dot{x}_i} \right) \frac{\partial \dot{x}_i}{\partial \dot{q}_k} + \frac{\partial T}{\partial \dot{x}_i} \frac{d}{dt} \left(\frac{\partial \dot{x}_i}{\partial \dot{q}_k} \right) \right]$$

$$= \sum_{i=1}^{3N} \frac{d}{dt} \left(\frac{\partial T}{\partial \dot{x}_i} \right) \frac{\partial x_i}{\partial q_k} + \frac{\partial T}{\partial q_k} \tag{1.2-12}$$

where the second line follows from eqs. (1.2-7) and (1.2-10). Now subtract eq. (1.2-11a) from eq. (1.2-12) to obtain

$$\frac{d}{dt} \left(\frac{\partial L}{\partial \dot{q}_k} \right) - \frac{\partial L}{\partial q_k} = \sum_{i=1}^{3N} \left[\frac{d}{dt} \left(\frac{\partial T}{\partial \dot{x}_i} \right) + \frac{\partial V}{\partial x_i} \right] \frac{\partial x_i}{\partial q_k}$$

$$= \sum_{i=1}^{3N} \left[\frac{d}{dt} \left(\frac{\partial L}{\partial \dot{x}_i} \right) - \frac{\partial L}{\partial x_i} \right] \frac{\partial x_i}{\partial q_k}$$

$$= 0 \quad (k = 1, 2, \ldots, 3N) \tag{1.2-13}$$

which is identical to eqs. (1.2-6). The second line of eqs. (1.2-13) follows from the facts that T does not depend upon the $\{x_i\}$ and V does not depend upon the $\{\dot{x}_i\}$; the third line follows from eqs. (1.2-2).

Considerable manipulation has been required to derive eqs. (1.2-13).[2] Their importance is that they allow us to write the equations of motion for a system described in any coordinate system; in general, the $\{q_k\}$ need not have dimensions of length, and the $\{\dot{q}_k\}$ need not have dimensions of length time^{-1}.

Still another formulation of the equations of motion can be given in terms of the **generalized momenta conjugate** to the generalized coordinates, defined by

$$p_k = \frac{\partial L}{\partial \dot{q}_k} \qquad (k = 1, 2, \ldots, 3N) \qquad (1.2\text{-}14)$$

Let us now introduce the **Hamiltonian** H,

$$H = \sum_k p_k \dot{q}_k - L$$

In differential form this is

$$
\begin{aligned}
dH &= \sum_k (\dot{q}_k \, dp_k + p_k \, d\dot{q}_k) - \sum_k \left[\left(\frac{\partial L}{\partial q_k} \right) dq_k + \left(\frac{\partial L}{\partial \dot{q}_k} \right) d\dot{q}_k \right] \\
&= \sum_k (\dot{q}_k \, dp_k + p_k \, d\dot{q}_k - \dot{p}_k \, dq_k - p_k \, d\dot{q}_k) \\
&= \sum_k (\dot{q}_k \, dp_k - \dot{p}_k \, dq_k) \qquad\qquad (1.2\text{-}15)
\end{aligned}
$$

from eqs. (1.2-6) and (1.2-14). Now regarding H as a function of the independent variables $\{q_k\}$ and $\{p_k\}$, we can express dH alternatively as

$$dH = \sum_k \left[\left(\frac{\partial H}{\partial p_k} \right) dp_k + \left(\frac{\partial H}{\partial q_k} \right) dq_k \right] \qquad (1.2\text{-}16)$$

Hence, making the identification of corresponding terms in eqs. (1.2-15) and (1.2-16), we arrive at **Hamilton's equations** (after the Irish mathematical physicist William R. Hamilton, 1805–1865).

$$\dot{q}_k = \frac{\partial H}{\partial p_k} \qquad \dot{p}_k = -\frac{\partial H}{\partial q_k} \qquad (k = 1, 2, \ldots, 3N) \qquad (1.2\text{-}17)$$

---

[2]In standard treatments of classical mechanics, Lagrange's equations are more conveniently derived from **Hamilton's Principle**. This principle, which is taken as a postulate, states that a system pursues that path during a specified time interval $[t_1, t_2]$ such that the time integral of the Lagrangian is an extremum, that is, the integral $\int_{t_1}^{t_2} L \, dt$ is a minimum (usually the case) or a maximum. Hamilton's principle is not physically transparent, but is nevertheless fundamental because it reduces all of classical mechanics to a single statement (Marion, 1965; Feynman et al., 1964).

We have in eqs. (1.2-17) yet a third version of the classical equations of motion. Hamilton's equations replace a single second-order differential equation for the coordinate (i.e., Newton's or Lagrange's equation) by two first-order equations, one for the coordinate and one for the conjugate momentum. In eqs. (1.2-17), q_k and p_k are on an equal footing. Note that the concept of force is not needed (for conservative systems) in Hamilton's formulation of mechanics, whereas in the Newtonian version force is of central importance.

An important corollary of Hamilton's equations is that for conservative systems, H is equal to the total energy. To show this, we combine the definitions of H and L.

$$H = \sum_k p_k \dot{q}_k - T + V$$

Specializing to Cartesian coordinates, we recast this as

$$H = \sum_k (m_k \dot{x}_k)\dot{x}_k - T + V$$
$$= 2T - T + V$$
$$= T + V \tag{1.2-18}$$

This result holds, however, for any coordinate system. For conservative systems $T + V$ is constant [see eq. (1.1-10)], and hence so is H.

Exercises

1.1. The Coulomb potential for the interaction of charges q_1 and q_2 located at positions $\mathbf{r}_1$ and $\mathbf{r}_2$ is given in SI units by the expression $V(\mathbf{r}_1, \mathbf{r}_2) = q_1 q_2/[4\pi\varepsilon_0|\mathbf{r}_1 - \mathbf{r}_2|]$. Compute the force on q_1 due to q_2, and on q_2 due to q_1, and thus show that they are equal in magnitude but oppositely directed along the line of centers.

1.2. A harmonic oscillator consists of a body of mass m moving in one dimension under the influence of a force, $\mathbf{f} = -kx\mathbf{e}_x$. Solve the classical equations of motion in closed form. Show explicitly that the system is conservative. The harmonic oscillator is very important in physical chemistry, and is mentioned several times later in the book.

1.3. Consider an electrolytic solution contained between the plane parallel plates of a conductance cell. A uniform electric field $\mathbf{E}$ exists between the plates. Solve Newton's equation to determine the trajectory of an ion of charge q and mass m; assume that in addition to the force due to the field, the ion also experiences a frictional force that is directly proportional to its velocity (coefficient of friction $= \beta$). What is the terminal speed of the ion?

1.4. A function f is a **homogeneous function of order** n in the variables $x_1, x_2, \ldots, x_m$ if for any nonzero λ

$$f(\lambda x_1, \lambda x_2, \ldots, \lambda x_m) = \lambda^n f(x_1, x_2, \ldots, x_m)$$

By writing the derivative $df/d\lambda$ in two different ways, prove **Euler's theorem on homogeneous functions**:

$$\sum_{i=1}^{m} x_i \left(\frac{\partial f}{\partial x_i} \right) = nf$$

1.5. The kinetic energy of a system of particles is homogeneous of order two in the Cartesian velocities because T is given by $\sum_i \frac{1}{2} m_i \dot{x}_i^2$. Prove that T is also homogeneous of order two in the generalized velocities. Assume that time does not appear explicitly in the transformation equations (1.2-3). [*Hint:* Use eq. (1.2-5).]

1.6. Refer once again to eq. (1.2-18), which we obtained by considering the $\{p_k\}$ and $\{\dot{q}_k\}$ in terms of Cartesian coordinates. Use the results of Exercises 1.4 and 1.5, however, to show that eq. (1.2-18) is true in generalized coordinates. Assume (as is usually the case) that the potential energy is a function only of the $\{q_k\}$.

1.7. Confirm that eq. (1.2-12) follows from eq. (1.2-11b).

1.8. A water molecule is oriented with the oxygen atom at the origin, atom H_a at $(-R, 0)$, and atom H_b in the upper right quadrant of the xy plane (R is the equilibrium H–O bond length). Atom H_b moves in the xy plane in such a way as to vary θ, the H–O–H bond angle, without changing the H–O bond lengths. Let θ_0 be the equilibrium bond angle, and assume the potential energy of the molecule is given roughly by $V(\theta) = \frac{1}{2} k(\theta - \theta_0)^2$. Write Lagrange's equation in the generalized coordinate θ. Solve the equation according to suitable initial conditions on θ and $\dot{\theta}$ of your own choosing.

1.3 MOTION IN A CENTRAL FORCE FIELD

To demonstrate the utility of Lagrange's or Hamilton's equations, we now consider the *isolated two-particle* system, for which

$$T = \tfrac{1}{2} m_1 \dot{\mathbf{r}}_1^2 + \tfrac{1}{2} m_2 \dot{\mathbf{r}}_2^2 \tag{1.3-1}$$

We define the **center of mass** ($\mathbf{R}$) and the **relative** ($\mathbf{r}$) **coordinates** by

$$\mathbf{R} = \frac{m_1 \mathbf{r}_1 + m_2 \mathbf{r}_2}{M} = X\mathbf{e}_x + Y\mathbf{e}_y + Z\mathbf{e}_z \tag{1.3-2a}$$

$$\mathbf{r} = \mathbf{r}_1 - \mathbf{r}_2 = x\mathbf{e}_x + y\mathbf{e}_y + z\mathbf{e}_z \tag{1.3-2b}$$

$$M = m_1 + m_2$$

Solving eqs. (1.3-2), we obtain

$$\mathbf{r}_1 = \mathbf{R} + \left(\frac{m_2}{M}\right)\mathbf{r}$$

$$\mathbf{r}_2 = \mathbf{R} - \left(\frac{m_1}{M}\right)\mathbf{r}$$

If these are differentiated with respect to time and the expressions for $\dot{\mathbf{r}}_1$ and $\dot{\mathbf{r}}_2$ are substituted into eq. (1.3-1), the result is

$$T = \tfrac{1}{2}M\dot{\mathbf{R}} \cdot \dot{\mathbf{R}} + \tfrac{1}{2}\mu\dot{\mathbf{r}} \cdot \dot{\mathbf{r}}$$

where $\mu = m_1 m_2/M$ is the **reduced mass**. The Lagrangian is then

$$L = \tfrac{1}{2}M\dot{\mathbf{R}} \cdot \dot{\mathbf{R}} + \tfrac{1}{2}\mu\dot{\mathbf{r}} \cdot \dot{\mathbf{r}} - V(\mathbf{r})$$

We observe that L can be partitioned as $L = L_{\text{cm}} + L_{\text{rel}}$, where

$$L_{\text{cm}} = \tfrac{1}{2}M\dot{\mathbf{R}} \cdot \dot{\mathbf{R}} \qquad (1.3\text{-}3a)$$

$$L_{\text{rel}} = \tfrac{1}{2}\mu\dot{\mathbf{r}} \cdot \dot{\mathbf{r}} - V(\mathbf{r}) \qquad (1.3\text{-}3b)$$

It is clear that since V depends only on the relative coordinate, Lagrange's equations do not couple the center of mass (cm) and relative (rel) motions. Hence, the entire motion can be separated into the *free* motion of the center of mass plus the motion of a *fictitious* particle of mass μ relative to a *fixed* center of force. Application of Lagrange's equation, eq. (1.2-6), gives for the center-of-mass trajectory

$$\mathbf{R}(t) = \frac{t}{M}\mathbf{P}(0) + \mathbf{R}(0)$$

This equation describes rectilinear motion and is a quantitative expression of **Newton's First Law**: A body moves in a straight line at constant velocity unless acted upon by an external force.

If the potential energy depends only upon the *magnitude* of $\mathbf{r}$, then the force acting on the fictitious particle is directed along $\mathbf{r}$. For such a **central force**, the trajectory for the relative motion is more conveniently described in terms of spherical polar coordinates, which are given by

$$x = r \sin \theta \cos \phi$$
$$y = r \sin \theta \sin \phi$$
$$z = r \cos \theta$$

Using the chain rule for differentiation, we compute the time derivatives of the

Cartesian coordinates. For example,

$$\dot{x} = \dot{r}\sin\theta\cos\phi + r\dot{\theta}\cos\theta\cos\phi - r\dot{\phi}\sin\theta\sin\phi$$

Similar expressions obtain for $\dot{y}$ and $\dot{z}$. Substitution of these into eq. (1.3-3b) yields

$$L_{\text{rel}} = \tfrac{1}{2}\mu\left[\dot{r}^2 + r^2(\dot{\theta}^2 + \dot{\phi}^2\sin^2\theta)\right] - V(r) \qquad (1.3\text{-}4)$$

We observe now that the **relative orbital angular momentum**, defined as the cross product

$$\boldsymbol{\ell} = \mathbf{r} \times \mathbf{p} \qquad (1.3\text{-}5)$$

is *constant*. This can be seen easily by differentiating both sides of eq. (1.3-5) with respect to time to get

$$\dot{\boldsymbol{\ell}} = \dot{\mathbf{r}} \times \mathbf{p} + \mathbf{r} \times \dot{\mathbf{p}}$$

Since $\mathbf{p} = \mu\dot{\mathbf{r}}$, the first term on the right-hand side vanishes. Likewise, the second term is zero because $\dot{\mathbf{p}}$ is the force, and by definition a central force is directed along $\mathbf{r}$. Therefore, $\boldsymbol{\ell}$ is constant in time. Since $\boldsymbol{\ell}$ is perpendicular to both $\mathbf{r}$ and $\mathbf{p}$, the constancy of $\boldsymbol{\ell}$ implies that the trajectory associated with the relative motion lies in a plane, which for convenience we may take to be the xy plane (see Fig. 1.1). Thus, we can describe the relative motion in terms of plane polar coordinates r and ϕ, as given by $x = r\cos\phi$ and $y = r\sin\phi$, and which result from the spherical polar coordinates by fixing $\theta = \pi/2$. The Lagrangian (1.3-4) for the relative motion then simplifies to

$$L_{\text{rel}} = \tfrac{1}{2}\mu\left[\dot{r}^2 + r^2\dot{\phi}^2\right] - V(r)$$

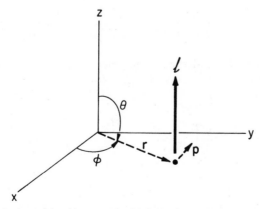

FIGURE 1.1. A particle with constant orbital angular momentum moves in a plane.

A second application of Lagrange's equation, eq. (1.2-6), generates the equations for the relative motion:

$$\frac{d}{dt}(\mu\dot{r}) = -\frac{\partial V}{\partial r} + \mu r\dot{\phi}^2 \tag{1.3-6a}$$

$$\frac{d}{dt}(\mu r^2\dot{\phi}) = 0 \tag{1.3-6b}$$

The solution of eq. (1.3-6b) is

$$\mu r^2\dot{\phi} = \text{constant} \tag{1.3-7}$$

This constant can be shown to be the orbital angular momentum ℓ. Combining eqs. (1.3-6a) and (1.3-7), we obtain

$$\mu\ddot{r} = -\frac{\partial V}{\partial r} + \frac{\ell^2}{\mu r^3} \tag{1.3-8}$$

This is the last remaining equation of motion that needs to be solved. Therefore, as desired, the original six Newtonian equations (one equation for each of the coordinates of the two particles) have been reduced to a single equation that describes the motion of a particle in one dimension under the influence of an *effective* potential energy $V_{\text{eff}} = V + \ell^2/2\mu r^2$. In general, it is necessary to employ numerical methods to solve eq. (1.3-8), although there are a number of special cases that admit closed-form solutions (Spiegel, 1967).

1.4   THE TIME RATE OF CHANGE OF DYNAMICAL VARIABLES

The time rate of change of any dynamical variable A can be expressed generally as

$$\dot{A} = \sum_k \left[\frac{\partial A}{\partial q_k}\right]\dot{q}_k + \sum_k \left[\frac{\partial A}{\partial p_k}\right]\dot{p}_k + \frac{\partial A}{\partial t} \tag{1.4-1}$$

where A is regarded as a function of the generalized coordinates and momenta, as well as of time explicitly. Using Hamilton's equations (1.2-17), we can rewrite this

$$\dot{A} = \sum_k \left(\frac{\partial A}{\partial q_k}\frac{\partial H}{\partial p_k} - \frac{\partial A}{\partial p_k}\frac{\partial H}{\partial q_k}\right) + \frac{\partial A}{\partial t}$$

$$= \{A, H\} + \frac{\partial A}{\partial t} \tag{1.4-2}$$

where the second line defines the **Poisson bracket** of A and H. It is convenient to introduce **Liouville's operator** $\mathscr{L}$ (after the French mathematician Joseph

Liouville, 1809–1882),

$$\mathscr{L}A = -i\{A, H\} \qquad (1.4\text{-}3)$$

where $i = \sqrt{-1}$, as usual.[3] Combining eqs. (1.4-2) and (1.4-3), we obtain

$$\dot{A}(t) = i\mathscr{L}A(t) + \frac{\partial A}{\partial t} \qquad (1.4\text{-}4)$$

In case A does not depend *explicitly* upon the time, that is, $\partial A/\partial t = 0$, then the formal solution of eq. (1.4-4) is

$$A(t) = \exp(it\mathscr{L})A(0) \qquad (1.4\text{-}5)$$

where the exponential is to be interpreted as

$$\exp(it\mathscr{L}) = 1 + it\mathscr{L} - \tfrac{1}{2}t^2\mathscr{L}^2 + \cdots + \frac{(it)^n}{n!}\mathscr{L}^n + \cdots \qquad (1.4\text{-}6)$$

That expression (1.4-5) is the solution can be demonstrated simply by differentiating both sides of eq. (1.4-5) with respect to t. Note that since $i^n\mathscr{L}^nA$ is just the nth time derivative of A, eq. (1.4-6) is equivalent to the Maclaurin series expansion of $A(t)$, that is,

$$A(t) = A(0) + \dot{A}(0)t + \tfrac{1}{2}\ddot{A}(0)t^2 + \cdots \qquad (1.4\text{-}7)$$

The concept of **operator** is extremely important, more so in quantum mechanics than in classical mechanics. In order to become accustomed to dealing with this abstract notion, it is useful here to derive a few simple results. For example, if we let $A = q$, then from eq. (1.4-2)

$$\dot{q}_j = \{q_j, H\} = \sum_k \left(\frac{\partial q_j}{\partial q_k} \frac{\partial H}{\partial p_k} - \frac{\partial q_j}{\partial p_k} \frac{\partial H}{\partial q_k} \right)$$
$$= \frac{\partial H}{\partial p_j}$$

which we recognize immediately as one of Hamilton's equations, eqs. (1.2-17). Also, if we let $A = H$ and if H does not depend explicitly on the time, then it follows from eq. (1.4-2) that H is constant in time. This is consistent with the result in eq. (1.2-18). Note that *any* dynamical variable A that does not depend explicitly on the time and whose Poisson bracket with the Hamiltonian

---

[3] The imaginary number i is included in the definition of Liouville's operator on account of a connection with quantum mechanics to be made in Chapter 2. In any case, the use of complex numbers in quantum mechanics is necessary and the reader should be familiar with their basic properties (Boas, 1983; Flanigan, 1983).

vanishes, is time independent. Such variables are referred to as **constants of the motion.** In Section 2.9 we shall see that this concept has its exact counterpart in quantum mechanics.

A somewhat more involved example of the application of the operator equation (1.4-5) is provided by a model system of great importance in physical chemistry—the **harmonic oscillator.** For the one-dimensional case, the Hamiltonian is

$$H = \frac{p^2}{2m} + \frac{1}{2}m\omega^2 q^2 \qquad (1.4\text{-}8)$$

where ω is the fundamental frequency. We compute $q(t)$ by using eq. (1.4-6). From eqs. (1.4-2) and (1.4-3), we obtain successively

$$(i\mathscr{L})^0 q = q$$
$$(i\mathscr{L})q = m^{-1}p$$
$$(i\mathscr{L})^2 q = -\omega^2 q$$
$$(i\mathscr{L})^3 q = -m^{-1}\omega^2 p$$
$$(i\mathscr{L})^4 q = \omega^4 q$$
$$(i\mathscr{L})^5 q = m^{-1}\omega^4 p$$
$$(i\mathscr{L})^6 q = -\omega^6 q$$

and so on. Substituting these explicit expressions into eq. (1.4-6) with $A(t) \equiv q(t)$ and rearranging terms, we have

$$q(t) = \left[1 - \frac{1}{2}\omega^2 t^2 + \frac{1}{4!}\omega^4 t^4 - \frac{1}{6!}\omega^6 t^6 + \cdots - \right] q(0)$$
$$+ \left[\omega t - \frac{1}{3!}\omega^3 t^3 + \frac{1}{5!}\omega^5 t^5 - \cdots + \right] m^{-1}\omega^{-1} p(0)$$

The coefficients multiplying $q(0)$ and $m^{-1}\omega^{-1}p(0)$ are recognizable as the beginning terms of the respective Maclaurin series for $\cos \omega t$ and $\sin \omega t$. Computation of the general term $(i\mathscr{L})^n q$, left as an exercise for the reader, verifies this. Therefore, the complete result is

$$q(t) = q(0)\cos \omega t + m^{-1}\omega^{-1}p(0)\sin \omega t \qquad (1.4\text{-}9)$$

It can be shown directly by differentiation that expression (1.4-9) is a solution of Newton's equation (1.4-10) for the harmonic oscillator:

$$\ddot{q}(t) + \omega^2 q(t) = 0 \qquad (1.4\text{-}10)$$

1.5 COMPUTER HIGHLIGHT: NUMERICAL SOLUTION OF THE CLASSICAL EQUATIONS OF MOTION

The harmonic oscillator is a very special system in that the trajectories $q(t)$ can be expressed in terms of familiar functions, namely, sines and cosines. These can be easily evaluated using a pocket calculator. For most mechanical systems one must resort to numerical methods of solving the classical equations of motion. Our first nontrivial application of the computer is to implement a simple scheme for solving Lagrange's equation (1.3-8) for the motion of a particle in a central field.

Before beginning any actual programming, we must devise a numerical method that yields a reliable approximation to the true solution. The first step in developing this procedure, or **algorithm**, is the reduction of the relevant equation, in this case

$$\mu \ddot{r} = - \frac{\partial V_{\text{eff}}}{\partial r} \tag{1.5-1}$$

to the simplest possible form that retains all of the variability of the original problem. It is convenient to introduce a set of "natural" units such that the "reduced" variables are dimensionless. The potential energy can be expressed as

$$V(r) = \mathscr{E}\mathscr{V}(x) \qquad x = \frac{r}{\sigma} \tag{1.5-2}$$

where $\mathscr{E}$ and σ are characteristic energy and length parameters and $\mathscr{V}$ is dimensionless potential energy. Defining also the dimensionless time by $\tau = t/t_0$, we can then transform eq. (1.5-1) into

$$\frac{d^2x}{d\tau^2} = - \frac{\mathscr{E}t_0^2}{\mu\sigma^2} \frac{d}{dx} \left[\mathscr{V}(x) + \frac{1}{2} \frac{\ell^2}{\mu\sigma^2\mathscr{E}x^2} \right] \tag{1.5-3}$$

Let us arbitrarily choose the unit of time to be $t_0 = (\mu\sigma^2/\mathscr{E})^{1/2}$; then eq. (1.5-3) simplifies to

$$\frac{d^2x}{d\tau^2} = -\mathscr{V}'(x) + \lambda^2 x^{-3} \tag{1.5-4}$$

where the prime connotes the derivative with respect to x, and λ is the reduced angular momentum, $\lambda = \ell(\mu\sigma^2\mathscr{E})^{-1/2}$.

We observe that the "reduced" equation of motion (1.5-4) contains only one parameter, λ, whereas the original equation of motion (1.5-1) involves four parameters, μ, σ, $\mathscr{E}$, and ℓ. For a fixed value of λ, the "reduced" trajectory $x(t)$ corresponds to a *triple infinity* of "original" trajectories $r(t)$.

$$r(t) = \sigma x = \sigma x(t\lambda\ell^{-1})$$
$$\lambda\ell^{-1} = (\mu\sigma^2\mathscr{E})^{-1/2}$$

Each of the "original" trajectories is labeled by the triplet of numbers $(\mu, \sigma, \mathscr{E})$.

We are now in a position to develop a numerical approximation to eq. (1.5-4). A simple and generally useful technique is to approximate the second derivative by a **finite-difference formula** (Norris, 1981; Spiegel, 1971). This is accomplished by expanding x at an arbitrary time $\tau \pm \Delta\tau$ in a Taylor series about the time τ.

$$x(\tau \pm \Delta\tau) = x(\tau) \pm \dot{x}(\tau)\,\Delta\tau + \tfrac{1}{2}\ddot{x}(\tau)(\Delta\tau)^2 + \cdots \qquad (1.5\text{-}5)$$

Neglecting terms of order $(\Delta\tau)^3$ and higher, we can solve eq. (1.5-5) for $\ddot{x}(\tau)$.

$$\ddot{x}(\tau) \simeq \frac{x(\tau + \Delta\tau) - 2x(\tau) + x(\tau - \Delta\tau)}{(\Delta\tau)^2} \qquad (1.5\text{-}6)$$

Equations (1.5-4) and (1.5-6) are now combined to give

$$x(\tau + \Delta\tau) \simeq 2x(\tau) - x(\tau - \Delta\tau) - (\Delta\tau)^2\{\mathscr{V}'[x(\tau)] - \lambda^2[x(\tau)]^{-3}\} \qquad (1.5\text{-}7)$$

Equation (1.5-7) is the basis of our computer algorithm. It predicts the value of the coordinate at time $\tau + \Delta\tau$ in terms of its values at $\tau - \Delta\tau$ and τ. This suggests that we partition the entire interval of time, say $[0, \tau_{max}]$, over which we wish to determine the trajectory, into subintervals of length $\Delta\tau$. From the initial conditions $x(0)$ and $\dot{x}(0)$, we obtain the value of $x(\Delta\tau)$ using eqs. (1.5-4) and (1.5-5).

$$x(\Delta\tau) = x(0) + \dot{x}(0)\,\Delta\tau - \tfrac{1}{2}(\Delta\tau)^2\{\mathscr{V}'[x(0)] - \lambda^2[x(0)]^{-3}\} \qquad (1.5\text{-}8)$$

Using eq. (1.5-7), we next compute $x(2\,\Delta\tau)$. Again, using eq. (1.5-7) we compute $x(3\,\Delta\tau)$ from $x(\Delta\tau)$ and $x(2\,\Delta\tau)$. The steps are repeated until we reach τ_{max}.

This approach should yield a good approximation to $x(\tau)$ if $\Delta\tau$ is sufficiently small. The accuracy can be checked in several ways. First, we can simply make $\Delta\tau$ smaller. It should be noted, however, that there is a limit to $\Delta\tau$ imposed by the inherent precision of the computer. Alternative checks on accuracy are based on physical requirements. For example, since the total energy

$$H(\tau) = \frac{1}{2}[\dot{x}(\tau)]^2 + \mathscr{V}[x(\tau)] + \frac{\lambda^2}{2[x(\tau)]^2} \qquad (1.5\text{-}9)$$

must be conserved, $H(\tau)$ can be monitored to make sure that it is constant in time. We also know that the system is reversible, that is, if the velocity is reversed at any time, the trajectory should retrace itself. Reversibility, in general, is a more stringent test of accuracy than is energy conservation.

To test the algorithm, we have applied it to the linear harmonic oscillator, for which the reduced angular momentum $\lambda = 0$. The potential energy $V(r) = \frac{1}{2}\mu\omega^2 r^2$ becomes $\mathscr{V}(x) = \frac{1}{2}x^2$ in reduced units, and $\mathscr{E} = \mu\omega^2\sigma^2$. Then eqs. (1.5-7)–(1.5-9) take the specific forms

$$x(\Delta\tau) \simeq x(0) + \dot{x}(0)\,\Delta\tau - \tfrac{1}{2}x(0)(\Delta\tau)^2 \qquad (1.5\text{-}10a)$$

$$x(\tau + \Delta\tau) \simeq 2x(\tau) - x(\tau - \Delta\tau) - (\Delta\tau)^2 x(\tau) \qquad (1.5\text{-}10b)$$

$$H(\tau) \simeq \tfrac{1}{2}\{[\dot{x}(\tau)]^2 + [x(\tau)]^2\} \qquad (1.5\text{-}10c)$$

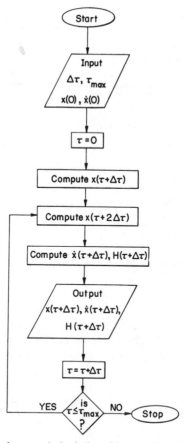

FIGURE 1.2. Algorithm for the numerical solution of Lagrange's equation for the linear harmonic oscillator.

```
50   REM   PROGRAM TO NUMERICALLY
     SOLVE LAGRANGE'S EQUATION OF
     MOTION FOR A ONE-DIMENSIONAL
     HARMONIC OSCILLATOR
100  DIM X(1100),XD(1100),H(1100)

110  PRINT "NUMERICAL SOLUTION OF
     LAGRANGE EQUATION"
115  PRINT
125  INPUT X(0),XD(0),DT,N: PRINT
     : PRINT : PRINT
150  REM  X(T) IS THE TRANSFORMED
     COORDINATE SIGMA AT TIMETAU,
      XD(T) IS TRANSFORMED VELOCI
     TY SIGMA DOT AT TAU, DT IS T
     HE TIME JUMP DELTA TAU, AND
     N IS THE NUMBER OF SUCH JUMP
     S
160  REM   N=1+(TMAX/DT)
175  X(1) = X(0) + XD(0) * DT - 0.
     5 * DT ^ 2 * X(0)

200  FOR I = 1 TO N
225  X(I + 1) = 2 * X(I) - X(I - 1
     ) - DT ^ 2 * X(I)
250  XD(I) = 0.5 * (X(I + 1) - X(I
      - 1)) / DT
275  H(I - 1) = 0.5 * (X(I - 1) ^
     2 + XD(I - 1) ^ 2)

300  NEXT I
350  PRINT  TAB( 9);"DTAU=";  TAB(
     14);DT; TAB( 23);"N=";  TAB(
     25);N: PRINT : PRINT
375  PRINT  TAB( 3);"TAU";  TAB( 1
     2);"X(TAU)";  TAB( 27);"H(TAU
     )"

400  PRINT  TAB( 3);"---";  TAB( 1
     2);"------";  TAB( 27);"-----
     -": PRINT
425  FOR I = 0 TO N - 1 STEP (N -
     1) / 10
450  PRINT  TAB( 3);I * DT; TAB(
     9);X(I); TAB( 24);H(I)
475  NEXT I
490  PRINT : PRINT
500  END
```

FIGURE 1.3. Program for the numerical solution of Lagrange's equation for the linear harmonic oscillator.

Figure 1.2 displays the flowchart for the algorithm based on eqs. (1.5-10). A BASIC program implementing the algorithm is listed in Fig. 1.3. For simplicity, the initial conditions are taken as $x(0) = 0$ and $\dot{x}(0) = 1.0$. Then the exact solution [see eq. (1.4-9)] in reduced units is $x(\tau) = \sin \tau$ and the total energy [from eq. (1.5-10c)] is *exactly* 0.5. It is easy to correlate eqs. (1.5-10) with program statements 175, 225, and 275. Note that the derivative $\dot{x}(\tau)$ required to compute the total energy is approximated in line 250 of the program by the

Table 1.1.   Numerical Solution of Lagrange's Equation for the Linear Harmonic Oscillator

τ	$\sin \tau$	$\Delta\tau = 0.1$		$\Delta\tau = 0.05$		$\Delta\tau = 0.01$	
		$x(\tau)$	$H(\tau)$	$x(\tau)$	$H(\tau)$	$x(\tau)$	$H(\tau)$
0	0	0	0.50000	0	0.50000	0	0.50000
1	0.84147	0.84275	0.50089	0.84179	0.50022	0.84148	0.50001
2	0.90930	0.91009	0.50104	0.90949	0.50026	0.90931	0.50001
3	0.14112	0.14006	0.50002	0.14085	0.50001	0.14111	0.50000
4	−0.75680	−0.75884	0.50072	−0.75731	0.50018	−0.75682	0.50001
5	−0.95892	−0.95953	0.50115	−0.95908	0.50029	−0.95893	0.50001
6	−0.27942	−0.27736	0.50010	−0.27890	0.50002	−0.27939	0.50000
7	0.65699	0.66001	0.50054	0.65774	0.50014	0.65702	0.50001
8	0.98936	0.99011	0.50123	0.98955	0.50031	0.98937	0.50001
9	0.41212	0.40921	0.50021	0.41139	0.50005	0.41209	0.50000
10	−0.54402	−0.54820	0.50038	−0.54507	0.50009	−0.54406	0.50000

central difference formula

$$\dot{x}(\tau) \simeq \frac{x(\tau + \Delta\tau) - x(\tau - \Delta\tau)}{2\,\Delta\tau}$$

Table 1.1 compares the trajectories and total energies for several different time increments $\Delta\tau$. In general, the accuracy of the computed trajectory improves with decreasing $\Delta\tau$, as expected. Also, the total energy $H(\tau)$ becomes more nearly constant and in agreement with the exact value of 0.5 as $\Delta\tau$ decreases. We may be reasonably confident of the correctness of the program at this point. As an additional verification, we could test the reversibility of the trajectory by selecting a given time point and taking as initial conditions on the reversed trajectory the position and *negative* of the velocity.

Exercises[4]

1.9.   Verify that $\mathbf{R}(t) = M^{-1}\mathbf{P}(0)t + \mathbf{R}(0)$ is the solution of Lagrange's equation for the center-of-mass coordinate of the isolated, two-particle system discussed in Section 1.3.

1.10.   Work through the steps in the derivation of eq. (1.3-4).

1.11.   Demonstrate that the constant of eq. (1.3-7) is the magnitude of the orbital angular momentum ℓ. Show that if $\ell = 0$, then the relative motion is along a straight line.

---

[4]Here and elsewhere in the book a superscript C on an exercise number indicates use of the microcomputer.

1.12. Suppose the total energy of the relative motion in Section 1.3 is equal to the minimum value attainable by the effective potential. Show that the trajectory is a circle, and that the "potential" force, $-\partial V/\partial r$, and the "centrifugal" force, $\ell^2/\mu r^3$, balance. Find the radius of the circle in the case of the Coulomb interaction.

1.13. Demonstrate that for Poisson brackets one has $\{F, GL\} = \{F, G\}L + G\{F, L\}$, where F, G, L are arbitrary dynamical variables.

1.14. If the Poisson bracket of two quantities is zero, then the two quantities are said to **commute**. Prove that for a particle any component of the angular momentum commutes with the square of the angular momentum. The quantal analog of this is very important.

1.15. Let n be an arbitrary positive integer. Show that for the harmonic oscillator

$$(i\mathscr{L})^{2n}q = (-1)^n \omega^{2n} q$$
$$(i\mathscr{L})^{2n+1}q = (-1)^n \omega^{2n} m^{-1} p$$

1.16. The classical motion of the harmonic oscillator can be described in terms of the complex variable $a = \omega m x + ip$. Find the equation of motion for $a(t)$, that is, $\dot{a}(t) = (i\mathscr{L})a(t)$, and solve this. By combining $a(t)$ and its complex conjugate $a^*(t)$ appropriately, obtain $x(t)$ and $p(t)$, and verify that they agree with the previously found solutions, eq. (1.4-9).

1.17. Demonstrate that the quantity $H = \frac{1}{2}\mu\dot{\mathbf{r}} \cdot \dot{\mathbf{r}} + V(r) + \ell^2/2\mu r^2$ is constant and, in fact, is the effective Hamiltonian for a particle in a central field. Also show that this expression becomes eq. (1.5-9) in dimensionless units.

1.18.[C] As suggested in the text, test the "reversibility" of the harmonic-oscillator trajectory using the program in Fig. 1.3. The initial velocity for the reversed trajectory can be estimated by the forward difference formula

$$\dot{x}(0) = \frac{x(\Delta\tau) - x(0)}{\Delta\tau}$$

1.19.[C] A model potential commonly used in physical chemistry is the **Morse potential**,

$$V(r) = D\{1 - \exp[-a(r - r_0)]\}^2 - D$$

Modify the program of Fig. 1.3 and apply it to the Morse potential. Compute trajectories for which the total energy is less than and greater than zero.

REFERENCES

Boas, M. L., *Mathematical Methods in the Physical Sciences*, 2nd ed., Wiley, New York, 1983, Chap. 2. Elementary treatment of the algebra of complex numbers.

Feynman, R. P., Leighton, R. B., and Sands, M., *The Feynman Lectures on Physics*, Vol. II, Addison-Wesley, Reading, 1964, Chap. 19. Brilliant lecture that explains the concept of action and Hamilton's principle.

Flanigan, F. J., *Complex Variables: Harmonic and Analytic Functions*, Dover, New York, 1983, Chap. 3. A fine review at the same level as Boas' book; there is much good material elsewhere in this book, although it is not pertinent to the present chapter. Recommended.

Marion, J. B., *Classical Dynamics of Particles and Systems*, Academic, New York, 1965, pp. 214–225. The author derives Lagrange's equations from Hamilton's principle; the derivation assumes coverage of the material on the calculus of variations from the preceding chapter.

Norris, A. C., *Computational Chemistry*, Wiley, New York, 1981, Chap. 7. This is a chapter on the numerical solution of ordinary differential equations.

Spiegel, M. R., *Theoretical Mechanics*, Schaum's Outline Series, McGraw-Hill, New York, 1967. Much more complete coverage of everything done in this chapter, with many worked examples. Highly recommended.

Spiegel, M. R., *Calculus of Finite Differences and Difference Equations*, Schaum's Outline Series, McGraw-Hill, New York, 1971, Chaps. 5 and 6. Discussion with many examples of the solution of differential equations by difference methods.

CHAPTER 2

Principles of Quantum Mechanics

The classical equations of motion outlined in Chapter 1 plus Maxwell's equations for electromagnetism lead to the prediction that atoms should collapse in a burst of radiation. The failure of classical theory to account for the stability of atoms and also for many other phenomena (e.g., blackbody radiation and atomic spectra) led in the 1920s to the development of **quantum mechanics**, the currently accepted theory of matter and radiation. In this and the subsequent three chapters we present the principles and some important applications of quantum theory in physical chemistry.

2.1 QUANTUM STATES

Since quantum mechanics is rather abstract and mathematical, it is convenient to develop the basic framework in terms of explicit assumptions, or *postulates*. These postulates are not unique and they may not be complete in the sense of formal logic. Unless otherwise stated, a postulate refers to a fixed time.

In classical mechanics the state of a system of N particles is completely determined by specifying the coordinates $\{r_i\}$ and momenta $\{p_i\}$ at an arbitrary instant. By contrast, in quantum mechanics considerably less can ever be specified for any system, even under the most ideal of circumstances. Specifically, we assume that

QM POSTULATE 1. The state of an N-particle system is described as fully as possible by a state function or wave function $\Psi(r_1, r_2, \ldots, r_N)$.

The state function has the reasonable properties of being continuous and differentiable at all points in coordinate space. In general, the state function is complex, that is,

$$\Psi(\mathbf{r}_1, \mathbf{r}_2, \ldots, \mathbf{r}_N) = \text{Re}\big[\Psi(\mathbf{r}_1, \mathbf{r}_2, \ldots, \mathbf{r}_N)\big] + i \, \text{Im}\big[\Psi(\mathbf{r}_1, \mathbf{r}_2, \ldots, \mathbf{r}_N)\big] \quad (2.1\text{-}1)$$

The designation of a state function $\Psi(\mathbf{r}_1, \mathbf{r}_2, \ldots, \mathbf{r}_N)$ is equivalent to specification of an infinite number of complex numbers, each corresponding to a possible configuration of the system (i.e., to a set of $3N$ possible values of the Cartesian coordinates). To simplify notation henceforth, we let q stand for the configuration. Thus, $\Psi(q) = \Psi(\mathbf{r}_1, \mathbf{r}_2, \ldots, \mathbf{r}_N)$.

Except for certain special cases, any system can occupy literally an infinite number of different possible states. For convenience, we often label these by a subscript. It is postulated that the state functions obey a superposition principle.

QM POSTULATE 2. If $\Psi_a(q)$ and $\Psi_b(q)$ represent any two states of the system, then

$$\Psi(q) = c_a \Psi_a(q) + c_b \Psi_b(q) \qquad (2.1\text{-}2)$$

is also a state of the system, where c_a and c_b are complex constants.

Note that if Ψ_a is proportional to Ψ_b, then Ψ as given by eq. (2.1-2) does not represent a new state. Rather Ψ_a, Ψ_b, and Ψ all describe the same state. Each possible state of a system thus has a unique label.

Since observables (i.e., properties of the system measurable in the laboratory, such as configuration q or momentum p) are real, the wave function itself can have no direct physical significance. According to the most widely held view, $\Psi(q)$ is a **probability amplitude**. The quantity

$$\Psi^*(q)\Psi(q)\,dq = |\Psi(q)|^2\,dq \qquad (2.1\text{-}3)$$

where $\Psi^*(q)$ denotes the complex conjugate of $\Psi(q)$, is the probability of finding the system in the differential element dq centered about q. That is, $|\Psi(q)|^2$ is the **probability density**. If this interpretation is to make sense, we must require

$$\int |\Psi(q)|^2\,dq = 1 \qquad (2.1\text{-}4)$$

where the integration is over the entire configuration space. Note that dq represents a $3N$-dimensional volume element, which is written in Cartesian coordinates as

$$dq = d\mathbf{r}_1\,d\mathbf{r}_2 \cdots d\mathbf{r}_N = \prod_{i=1}^{N} d\mathbf{r}_i = \prod_{i=1}^{N} dx_i\,dy_i\,dz_i$$

Equation (2.1-4) means simply that the probability of finding the particles of the system somewhere within the universe is unity. A wave function obeying eq. (2.1-4) is said to be **normalized**. Unless stated to the contrary, it is assumed henceforth that wave functions are normalized. In general, we refer to the number $\int |\Psi(q)|^2 \, dq$ as the **norm** of $\Psi(q)$.

2.2   OBSERVABLES AND OPERATORS

In Section 2.1 we indicated a tenuous connection between quantum mechanics and the "real world" of spatial location (q) or extent (dq). As will be discussed more fully in Section 2.6, we assume that a device can always be constructed to measure any chosen dynamical variable, or observable. A fundamentally novel feature of quantum mechanics that distinguishes it from classical mechanics is the manner in which observables are described.

QM POSTULATE 3.   With every observable there is associated a linear, Hermitian operator. Each such operator is constructed in the following way: Wherever classical Cartesian coordinates x_i, y_i, z_i of particle i appear, these are to be replaced by *operators* x_i, y_i, z_i, which act on a wave function by multiplying it. Wherever the components of linear momentum p_{xi}, p_{yi}, and p_{zi} of particle i appear, these are to be replaced by differential *operators* $-i\hbar \, \partial/\partial x_i$, $-i\hbar \, \partial/\partial y_i$, $-i\hbar \, \partial/\partial z_i$, where $\hbar = h/2\pi$ and h is Planck's constant.

We have already encountered the concept of operator in Section 1.4, where we introduced the Liouville operator $\mathscr{L}$. An **operator** is simply a mathematical object that acts upon one function to produce another function. If $\hat{A}$ is a quantum mechanical operator, then symbolically

$$\hat{A}\Psi(q) = \phi(q) \tag{2.2-1}$$

where $\Psi(q)$ and $\phi(q)$ are two state functions. The operator $\hat{A}$ corresponds to some dynamical variable, such as angular momentum or energy.

A **linear** operator has the property

$$\hat{A}\left[\sum_a c_a \Psi_a(q)\right] = \sum_a c_a \hat{A}\Psi_a(q) \tag{2.2-2}$$

where the c_a are generally complex constants. An operator is said to be **Hermitian** if it satisfies the following relation:

$$\int \Psi^*(q)\hat{A}\phi(q) \, dq = \int \phi(q)\hat{A}^*\Psi^*(q) \, dq \tag{2.2-3}$$

where $\Psi(q)$ and $\phi(q)$ are state functions. The motivation for postulating that quantum mechanical operators are Hermitian is not yet clear, but will become so later.

Let us now consider a specific example, however, to illustrate the mathematical content of eq. (2.2-3). For a one-dimensional, one-particle system, the linear momentum operator is $\hat{p} = -i\hbar(d/dx)$. The complex conjugate of this operator is then $\hat{p}^* = i\hbar(d/dx)$. Then for any two state functions we have

$$\int_{-\infty}^{\infty} \Psi^*(x)\left\{-i\hbar\frac{d}{dx}\right\}\phi(x)\,dx = \int_{-\infty}^{\infty} \phi(x)\left\{i\hbar\frac{d}{dx}\right\}\Psi^*(x)\,dx \quad (2.2\text{-}4)$$

We proceed to integrate the left side by parts: Let $u = \Psi^*(x)$, $dv = d\phi(x)$, then $du = d\Psi^*(x)$, $v = \phi(x)$. Hence, the left side becomes

$$-i\hbar\int_{-\infty}^{\infty} \Psi^*(x)\left[\frac{d\phi(x)}{dx}\right]dx$$

$$= -i\hbar\left\{\Psi^*(x)\phi(x)\Big|_{-\infty}^{\infty} - \int_{-\infty}^{\infty}\phi(x)\left(\frac{d\Psi^*(x)}{dx}\right)dx\right\} \quad (2.2\text{-}5)$$

The first term on the right side of eq. (2.2-5) must be zero, because if condition (2.1-4) is to be maintained for any acceptable state function, then

$$\lim_{x\to\pm\infty} \Psi(x) = \lim_{x\to\pm\infty} \phi(x) = 0$$

The right side of eq. (2.2-5) thus reduces to

$$\int_{-\infty}^{\infty} \phi(x)\left\{i\hbar\frac{d}{dx}\right\}\Psi^*(x)\,dx$$

which is identical to the right side of eq. (2.2-4). We conclude that $\hat{p}$ is indeed Hermitian, as it must be according to QM Postulate 3.

Now QM Postulate 3 tells us that every observable is to have a corresponding linear Hermitian operator. In terms of the spatial coordinates, this operator is represented as a function of $\{x_i\}$ and the partial derivatives $\{\partial/\partial x_i\}$. Alternatively, the operator can as well be represented by a matrix. To understand this, we define the **matrix element** of the operator $\hat{A}$ between two members of a *discrete set* of arbitrary state functions $\{\Psi_i(q)\}$ by

$$A_{ij} \equiv \int \Psi_i^*(q)\hat{A}\Psi_j(q)\,dq$$

that is, A_{ij} is the i, j element of a matrix **A**. Since $\hat{A}$ is Hermitian, we have from eq. (2.2-3)

$$A_{ij} = A_{ji}^* \quad (2.2\text{-}6)$$

Equation (2.2-6) constitutes the definition of a **Hermitian matrix** (Dence, 1975; Spiegel, 1971). Thus, quantum mechanical operators can be represented by Hermitian matrices.

The set of Hermitian matrices representing the various observables of interest is not unique, since the state functions $\{\Psi_i(q)\}$ are not specifically defined or in any sense well ordered. To achieve order, it is helpful to introduce sets of special state functions called eigenfunctions.

2.3   EIGENFUNCTIONS AND EIGENVALUES OF A HERMITIAN OPERATOR

For any operator $\hat{A}$, there may exist functions $\{\Psi_i(q)\}$ and numbers $\{\lambda_i\}$ such that

$$\hat{A}\Psi_i(q) = \lambda_i \Psi_i(q) \qquad (2.3\text{-}1)$$

In words, the functions $\{\Psi_i(q)\}$ are such that when acted upon by $\hat{A}$, a multiple of the original function is returned. We refer to any $\Psi_i(q)$ and the corresponding λ_i in eq. (2.3-1) as an **eigenfunction** and **eigenvalue** of $\hat{A}$. If more than one eigenfunction corresponds to the same eigenvalue, the eigenvalue is said to be **degenerate**.

In case $\hat{A}$ is Hermitian, the eigenvalues and eigenfunctions have special properties.

Property 1.   The eigenvalues of a Hermitian operator are real numbers.

Let us multiply both sides of eq. (2.3-1) on the left by the complex conjugate of $\Psi_i(q)$ and then integrate over all configuration space

$$\int \Psi_i^*(q)\hat{A}\Psi_i(q)\,dq = \int \Psi_i^*(q)\lambda_i\Psi_i(q)\,dq \qquad (2.3\text{-}2)$$
$$= \lambda_i$$

since λ_i is just a number and $\Psi_i(q)$ is assumed to be normalized. On the other hand, the complex conjugate of eq. (2.3-1) reads

$$\hat{A}^*\Psi_i^*(q) = \lambda_i^*\Psi_i^*(q)$$

and multiplication on the left of this by $\Psi_i(q)$ and integration gives

$$\int \Psi_i(q)\hat{A}^*\Psi_i^*(q)\,dq = \int \Psi_i(q)\lambda_i^*\Psi_i^*(q)\,dq \qquad (2.3\text{-}3)$$
$$= \lambda_i^*$$

But if $\hat{A}$ is Hermitian, then from eq. (2.2-3) we can rewrite eq. (2.3-3) as

$$\int \Psi_i^*(q)\hat{A}\Psi_i(q)\,dq = \lambda_i^*$$

Comparison of this with eq. (2.3-2) then shows that $\lambda_i = \lambda_i^*$. Since a number can be equal to its complex conjugate only if the number is real, we conclude that λ_i is real.

A second important property of Hermitian operators is Property 2.

Property 2.   Eigenfunctions associated with *distinct* eigenvalues of a Hermitian operator are orthogonal.

Let λ_i and λ_j denote two different eigenvalues of a Hermitian operator $\hat{A}$, and let $\Psi_i(q)$ and $\Psi_j(q)$ be their corresponding eigenfunctions. We construct the matrix elements A_{ij} and A_{ji}.

$$A_{ij} = \int \Psi_i^*(q)\hat{A}\Psi_j(q)\,dq = \int \Psi_i^*(q)\lambda_j\Psi_j(q)\,dq$$

$$= \lambda_j \int \Psi_i^*(q)\Psi_j(q)\,dq \qquad (2.3\text{-}4a)$$

$$A_{ji} = \int \Psi_j^*(q)\hat{A}\Psi_i(q)\,dq = \int \Psi_j^*(q)\lambda_i\Psi_i(q)\,dq$$

$$= \lambda_i \int \Psi_j^*(q)\Psi_i(q)\,dq \qquad (2.3\text{-}4b)$$

Since $\hat{A}$ is Hermitian, we know from eq. (2.2-6) that $A_{ij} = A_{ji}^*$. Thus, by eq. (2.3-4b) we have

$$A_{ij} = \lambda_i^* \int \Psi_j(q)\Psi_i^*(q)\,dq = \lambda_i \int \Psi_j(q)\Psi_i^*(q)\,dq \qquad (2.3\text{-}4c)$$

where the second equality follows from the fact that λ_i is real. Comparison of eqs. (2.3-4a) and (2.3-4c) then shows that

$$(\lambda_j - \lambda_i)\int \Psi_i^*(q)\Psi_j(q)\,dq = 0 \qquad (2.3\text{-}5)$$

By hypothesis $\lambda_i \neq \lambda_j$, and so eq. (2.3-5) implies that

$$\int \Psi_i^*(q)\Psi_j(q)\,dq = 0 \qquad (2.3\text{-}6)$$

Two wave functions that satisfy eq. (2.3-6) are said to be **orthogonal**. In the case of a *degenerate* eigenvalue, where several linearly independent eigenfunctions correspond to the same eigenvalue, there are no unique eigenfunctions. Any set of linearly independent linear combinations are good eigenfunctions. It can be shown, however, that the linear combinations can be so constructed that the degenerate eigenfunctions are orthogonal to one another.

A terse way of indicating the normalization and the orthogonality of eigenfunctions is by means of the **Kronecker delta**, δ_{ij}.

$$\int \Psi_i^*(q)\Psi_j(q)\, dq = \delta_{ij} = \begin{cases} 1 & i = j \\ 0 & i \neq j \end{cases} \tag{2.3-7}$$

A set of functions $\{\Psi_n(q)\}$ that satisfies eq. (2.3-7) is said to be an **orthonormal set**.

Exercises

2.1. A particle confined to a box of dimensions $a \times b \times c$ is described quantum mechanically by the ground-state eigenfunction

$$\Psi(x, y, z) = \left(\frac{8}{abc}\right)^{1/2} \sin\frac{\pi x}{a} \sin\frac{\pi y}{b} \sin\frac{\pi z}{c}$$

Show that Ψ is normalized.

2.2. For systems that possess symmetry, it may be more convenient to express the wave functions in non-Cartesian coordinates. Equation (2.1-3) then requires that the volume element dq also be expressed in the new coordinate system. The ground state of the hydrogen atom is described in Cartesian coordinates by the eigenfunction

$$\Psi(x, y, z) = \left(\pi a_0^3\right)^{-1/2} \exp\left(-\frac{\sqrt{x^2 + y^2 + z^2}}{a_0}\right)$$

where the constant a_0 is the Bohr radius. Transform this to spherical polar coordinates and show that the wave function is normalized.

2.3. Which of the following operators $\hat{A}$ are linear, and which are not:

(a) $\hat{A}f(x) \equiv i\dfrac{d}{dx}f(x)$ (b) $\hat{A}f(x) \equiv \ln f(x)$

(c) $\hat{A}f(x) \equiv \exp[if(x)]$ (d) $\hat{A}f(x) \equiv \left[x + \dfrac{d^2}{dx^2}\right]f(x)$

2.4. For a single particle moving in three dimensions, the classical kinetic energy is in Cartesian coordinates

$$T = \frac{1}{2}\sum_{i=1}^{3} m\dot{x}_i^2 = \sum_{i=1}^{3} \frac{p_i^2}{2m}$$

Use QM Postulate 3 to show that the quantum mechanical kinetic-energy operator is

$$\hat{T} = -\frac{\hbar^2}{2m}\nabla^2$$

where ∇^2, the **Laplacian**, is defined as

$$\nabla^2 = \frac{\partial^2}{\partial x^2} + \frac{\partial^2}{\partial y^2} + \frac{\partial^2}{\partial z^2}$$

2.5. Prove that the operator $\hat{T}$ for a one-particle system is Hermitian.

2.6. Which of the following is a Hermitian matrix?

$$\begin{bmatrix} 0 & 3 & i \\ 3 & 6-i & -2 \\ -i & -2 & -1 \end{bmatrix} \quad \text{or} \quad \begin{bmatrix} 4 & -2i & 7 \\ 2i & 0 & 1+i \\ 7 & 1-i & 0 \end{bmatrix}$$

2.7. For a particle in a one-dimensional box of length L, the eigenfunctions are

$$\Psi_n(x) = \left(\frac{2}{L}\right)^{1/2} \sin\frac{n\pi x}{L} \qquad n = 1,2,3,\ldots$$

The particle possesses only kinetic energy, T. From eq. (1.2-18) and the result of Exercise 2.4, deduce the form of the Hamiltonian operator $\hat{H}$ for this system. When the system is in its ground state, what is the eigenvalue of $\hat{H}$? Show explicitly that two different eigenfunctions of $\hat{H}$ are indeed orthogonal, as Property 2 states.

2.8. A one-dimensional system $(-\infty < x < \infty)$ can exist in any of only three states. The following are three normalized wave functions for the system: $\Psi_1(x) = (4/\pi)^{1/4}x\,\exp(-x^2/2)$, $\Psi_2(x) = (16/9\pi)^{1/2}x^2\exp(-x^2/2)$ and $\Psi_3(x) = (64/225\pi)^{1/4}x^3\exp(-x^2/2)$. Find the matrix representation of $\hat{p}_x$ in the basis $\{\Psi_n(x)\}$.

2.4   REPRESENTATION OF STATE FUNCTIONS AND OPERATORS

According to the superposition principle, QM Postulate 2, any linear combination of distinct wave functions yields another possible wave function. Turning this idea around, we may suppose that any *arbitrary* state function $\Psi(q)$ can be expressed as a linear combination of other state functions

$$\Psi(q) = \sum_i c_i \Psi_i(q) \tag{2.4-1}$$

where i labels the members of some suitable set of state functions and the c_i are complex constants, as usual. Since $\Psi(q)$ is an arbitrary function of q, within certain restrictions (mentioned in QM Postulate 1), we expect that the set of functions must be "complete" in some sense, in order to describe all the reasonable possible variations in Ψ as a function of q. It can be shown that if $\Psi_i(q)$ are the *eigenfunctions* of a Hermitian operator $\hat{A}$, then eq. (2.4-1) is indeed valid.

To determine the coefficients c_i in the expansion (2.4-1) we simply multiply both sides of eq. (2.4-1) on the left by any eigenfunction $\Psi_j^*(q)$ of the Hermitian operator $\hat{A}$ and integrate over all configuration space. We obtain

$$\int \Psi_j^*(q)\Psi(q)\, dq = \sum_i c_i \int \Psi_j^*(q)\Psi_i(q)\, dq$$

$$= \sum_i c_i \delta_{ij} \qquad (2.4\text{-}2)$$

where the second line follows from relation (2.3-7). The summation in eq. (2.4-2) can be explicitly carried out. Since δ_{ij} is zero unless $i = j$, the only nonzero term in the sum is that for which $i = j$. Thus we have

$$c_j = \int \Psi_j^*(q)\Psi(q)\, dq \qquad (2.4\text{-}3)$$

Upon substituting eq. (2.4-3) back into eq. (2.4-1), we have [after changing the dummy index in eq. (2.4-3) from j to i]

$$\Psi(q) = \sum_i \left[\int \Psi_i^*(q')\Psi(q')\, dq' \right] \Psi_i(q) \qquad (2.4\text{-}4)$$

which can be rearranged to

$$\Psi(q) = \int \left[\sum_i \Psi_i^*(q')\Psi_i(q) \right] \Psi(q')\, dq'$$

We have the interesting result that the "function" of q and q' defined by

$$\delta(q' - q) = \sum_i \Psi_i^*(q')\Psi_i(q) \qquad (2.4\text{-}5)$$

has the property

$$\Psi(q) = \int \delta(q' - q)\Psi(q')\, dq'$$

for any reasonable function $\Psi(q)$. This means that the only configuration at which $\delta(q' - q)$ has a nonzero value is at $q' = q$. Moreover, since the contribution to the integral at this one point is nonvanishing, the magnitude of

$\delta(q' - q)$ must be *infinite* there. This "function," called the **Dirac delta function** (after the English physicist Paul A. M. Dirac, 1901–1984), is considered in more detail in Section 2.5. In summary, $\delta(q)$ can be written

$$\delta(q) = \begin{cases} 0 & q \neq 0 \\ \infty & q = 0 \end{cases} \tag{2.4-6}$$

Equation (2.4-5), which holds for any set of eigenfunctions of a Hermitian operator, is called the **completeness** or **closure relation**.

If all of the coefficients c_i in eq. (2.4-1) are given, then the state function $\Psi(q)$ is uniquely determined. This suggests an alternative way to represent $\Psi(q)$. We have been accustomed to thinking of $\Psi(q)$ as a set of complex values of the wave function at points in coordinate space. However, we can as well represent the state by a column vector of the expansion coefficients in eq. (2.4-1).

$$\Psi = \begin{pmatrix} c_1 \\ c_2 \\ c_3 \\ \vdots \\ c_i \\ \vdots \end{pmatrix} \tag{2.4-7}$$

We say that the set of eigenfunctions $\{\Psi_i(q)\}$ is the **basis** of this particular representation. The elements of the column vector, that is, the coefficients c_i, are called the **representatives** of the state function in this basis. Clearly, we have as many bases for different, yet equivalent, representations as there are Hermitian operators. Our choice of basis is usually dictated by convenience.

While state functions can be represented by *vectors*, operators can be represented by *matrices*. This is easily seen as follows. Consider the operator equation

$$\hat{B}\Psi(q) = \phi(q) \tag{2.4-8}$$

Let us expand $\Psi(q)$ in the basis of eigenfunctions of some Hermitian operator $\hat{A}$ according to eq. (2.4-1) and then substitute this into eq. (2.4-8).

$$\hat{B}\sum_i c_i\Psi_i(q) = \phi(q)$$

We multiply both sides of this on the left by $\Psi_j^*(q)$ and integrate. The result is

$$\sum_i \left[\int \Psi_j^*(q)\hat{B}\Psi_i(q)\,dq\right]c_i = \int \Psi_j^*(q)\phi(q)\,dq$$

$$= d_j \tag{2.4-9}$$

where d_j is the coefficient of Ψ_j in the expansion (2.4-1) of $\phi(q)$. There are as many equations (2.4-9) as there are possible eigenfunctions $\Psi_j(q)$ of $\hat{A}$. These equations can, therefore, collectively be written in matrix form as

$$
\begin{bmatrix}
B_{11} & B_{12} & \cdots & B_{1n} \\
B_{21} & B_{22} & \cdots & B_{2n} \\
\vdots & & & \vdots \\
B_{n1} & \cdots & & B_{nn}
\end{bmatrix}
\begin{bmatrix}
c_1 \\
c_2 \\
\vdots \\
c_n
\end{bmatrix}
=
\begin{bmatrix}
d_1 \\
d_2 \\
\vdots \\
d_n
\end{bmatrix}
\tag{2.4-10}
$$

or more tersely, as $\mathbf{B}\Psi = \phi$.

The column vectors Ψ and ϕ represent states Ψ and ϕ; $\mathbf{B}$ is a *square* matrix, whose elements are simply the matrix elements of the operator $\hat{B}$ between the possible eigenfunctions of operator $\hat{A}$. Thus, the matrix $\mathbf{B}$ represents the operator $\hat{B}$ in the basis $\{\Psi_i(x)\}$. Note that if $\hat{B} \equiv \hat{A}$, then we have from eqs. (2.3-1) and (2.3-7)

$$
A_{ij} = \int \Psi_i^*(q) \hat{A} \Psi_j(q) \, dq = \lambda_j \delta_{ij}
\tag{2.4-11}
$$

that is, $\mathbf{A}$ is *diagonal*. The diagonal elements are the possible eigenvalues of the operator $\hat{A}$. In other words, in the basis of its own eigenfunctions, an operator is represented by a diagonal matrix.

2.5 MORE ON THE DIRAC DELTA FUNCTION

The Dirac delta function is actually not a *function*, according to the strict definition of that term, since if $\delta(q)$ vanishes everywhere *except at a single point*, the integral $\int \delta(q)\, dq$ is necessarily zero (Spiegel, 1969). Therefore, expression (2.4-6) is purely symbolic and must be interpreted with great care. Because the Dirac delta function is an essential ingredient of quantum mechanics, it is worthwhile to explore it more deeply from a strictly mathematical viewpoint. To simplify notation, we restrict the consideration to one dimension, so $q = x$ and $-\infty < x < \infty$.

Let $f(x)$ be a bounded, differentiable function with a bounded derivative and let $\{\delta_n(x)\}$ be a sequence of functions that are continuous and normalized so that

$$
\int \delta_n(x) \, dx = 1
\tag{2.5-1}
$$

We examine the limit of the sequence

$$
I_n \equiv \int_{-\infty}^{\infty} \delta_n(x) f(x) \, dx
\tag{2.5-2}
$$

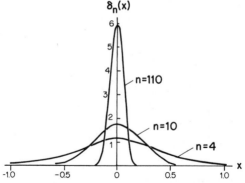

FIGURE 2.1. A delta sequence of functions.

and show that

$$\lim_{n \to \infty} I_n = f(0) \tag{2.5-3}$$

A suitable sequence of normalized functions (i.e., a **delta sequence**) is

$$\delta_n(x) = \left(\frac{n}{\pi}\right)^{1/2} \exp(-nx^2)$$

several of which are plotted in Fig. 2.1. To prove relation (2.5-3), we consider the absolute difference

$$\left| \int_{-\infty}^{\infty} [f(x)\, \delta_n(x) - f(0)]\, dx \right| = \left| \int_{-\infty}^{\infty} \left(\frac{n}{\pi}\right)^{1/2} e^{-nx^2} [f(x) - f(0)]\, dx \right|$$

$$\leq \int_{-\infty}^{\infty} \left(\frac{n}{\pi}\right)^{1/2} e^{-nx^2} |x| \left| \frac{[f(x) - f(0)]}{x} \right| dx \tag{2.5-4}$$

where the first line follows from the normalization requirement (2.5-1) and the second line from the fact that $\exp(-nx^2)$ is everywhere positive. Now the **mean value theorem for derivatives** (Spiegel, 1969) states that there must be at least one point x_0 in the interval $0 \leq x_0 \leq x$ such that the derivative $f'(x_0)$ at x_0 is equal to $[f(x) - f(0)]/x$. From Fig. 2.2, it is clear that $f'(x_0)$ may not be the maximum attainable by $f'(x)$ in the interval. Therefore,

$$|\max[f'(x)]| \geq \left| \frac{f(x) - f(0)}{x} \right| \tag{2.5-5}$$

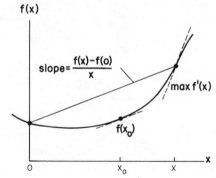

FIGURE 2.2. Illustrating that the average slope in an interval is generally less than the maximum value of the derivative there.

Combining eqs. (2.5-4) and (2.5-5), we have

$$\int_{-\infty}^{\infty} |\delta_n(x)f(x) - f(0)|\, dx \leq |\max[f'(x)]| \int_{-\infty}^{\infty} \left(\frac{n}{\pi}\right)^{1/2} e^{-nx^2} |x|\, dx$$

$$= 2|\max[f'(x)]| \int_{0}^{\infty} \left(\frac{n}{\pi}\right)^{1/2} e^{-nx^2} x\, dx$$

$$= \frac{1}{(n\pi)^{1/2}} |\max[f'(x)]| \qquad (2.5\text{-}6)$$

Since f' is bounded by hypothesis, it follows immediately from eq. (2.5-6) that

$$\lim_{n \to \infty} \int_{-\infty}^{\infty} \delta_n(x)f(x)\, dx = f(0) \qquad (2.5\text{-}7)$$

We see then that the proper way to interpret the Dirac delta function is through the relation

$$\int_{-\infty}^{\infty} \delta(x)f(x)\, dx = \lim_{n \to \infty} \int_{-\infty}^{\infty} \delta_n(x)f(x)\, dx \qquad (2.5\text{-}8)$$

The fact that

$$\lim_{n \to \infty} \left(\frac{n}{\pi}\right)^{1/2} \exp(-nx^2) = \begin{cases} 0 & x \neq 0 \\ \infty & x = 0 \end{cases} \qquad (2.5\text{-}9)$$

provides a justification for eq. (2.4-6), but we must bear in mind that the order of the limit and integration in eq. (2.5-8) cannot be interchanged.

The sequence of Gaussians employed previously is not the only suitable one. In fact, the Dirac delta function is properly defined more generally as the *class* of all equivalent sequences of functions that result in a mapping of functions $f(x)$ onto numbers $f(0)$ (Lighthill, 1970). From this more general perspective, the Dirac delta function is just one example of what are called **generalized functions** (Hoskins, 1979).

2.6 THE MEASUREMENT PROCESS

We now make the connection between quantum mechanics and laboratory measurements in the "real" world. Any realistic experiment is quite complex and we usually do not have control of all variables. Our knowledge is generally incomplete. This state of affairs is conveniently handled by introducing the notion of an **ensemble** of systems, which is a *virtual* (i.e., mental) collection of a large number of systems isolated from each other and identical in every respect to the system under investigation. The systems in the ensemble are distributed through the possible states $\Psi(q)$ with probabilities p_Ψ consistent with our knowledge of the condition, or state, of the system. That is, if the ensemble contains $\mathcal{N}$ systems and $\mathcal{N}_\Psi$ are in state $\Psi(q)$, the probability that any system drawn from the ensemble "at random" is in state $\Psi(q)$ is

$$p_\Psi = \frac{\mathcal{N}_\Psi}{\mathcal{N}} \tag{2.6-1}$$

In case we know with certainty that the system is in state Ψ then $p_\Psi = 1$, while for all other possible states ϕ, $p_\phi = 0$. The system is said to be in a **pure state** Ψ. For the remainder of the present discussion we focus on this ideal case. Later, in Chapter 6 we deal with systems in **mixed states**, that is, those for which we know only the probability $p_\Psi \neq 1$ of the system's being in a certain state Ψ.

In the ideal situation under consideration, we perform a measurement of some observable A, to which the Hermitian operator $\hat{A}$ is associated, by the following procedure. The system is first prepared in the state Ψ. It is then subjected to an "A-determining" device. We assume

QM POSTULATE 4. A measurement of observable A yields one of the eigenvalues λ_i of $\hat{A}$.

The measurement causes the state of the system to change from $\Psi(q)$ to $\Psi_i(q)$ in a way beyond the experimenter's control. If A is again measured before the system is perturbed by other external interactions, the same value λ_i is obtained. In other words, if the system is in a given eigenstate Ψ_i of the operator $\hat{A}$, a measurement of A must yield the corresponding eigenvalue. The motivation for QM Postulate 3 is now clear. The measurement must yield a real number and the eigenvalues of a Hermitian operator are real.

Repeated measurements of the observable A for a system in the pure state Ψ yields a distribution of eigenvalues of $\hat{A}$. The **average value** or **expected**

value, of A is by definition the sum of the eigenvalues obtained divided by the number of measurements. We assume that this average value can be computed from Ψ as follows.

QM POSTULATE 5.   For a system in pure state $\Psi(q)$, the expected value of an observable A is given by

$$\langle A \rangle = \int \Psi^*(q)\hat{A}\Psi(q)\,dq \qquad (2.6\text{-}2)$$

where $\Psi(q)$ is assumed to be normalized.

By expanding $\Psi(q)$ in terms of the eigenstates of $\hat{A}$ according to eq. (2.4-1), we can recast eq. (2.6-2) as

$$\langle A \rangle = \sum_i c_i^* \sum_j A_{ij} c_j \qquad (2.6\text{-}3)$$

where c_i is given by eq. (2.4-3). Using relation (2.4-11), we can simplify eq. (2.6-3) to

$$\langle A \rangle = \sum_i |c_i|^2 \lambda_i$$

Thus, we have expressed the average value as a sum over all the eigenvalues (i.e., possible values of A that can be measured), weighted by real numbers.

$$p_i = |c_i|^2 = \left| \int \Psi_i^*(q)\Psi(q)\,dq \right|^2$$

Since $\Psi(q)$ is assumed to be normalized, we have

$$1 = \int \Psi^*(q)\Psi(q)\,dq = \sum_i c_i^* \sum_j \left[\int \Psi_i^*(q)\Psi_j(q)\,dq \right] c_j$$

$$= \sum_i |c_i|^2 \qquad (2.6\text{-}4)$$

Observe that eq. (2.6-4) can be regarded as a special case of eq. (2.6-3) with $\hat{A} = 1$. Further, since $p_i \geq 0$ for all i, it follows from eq. (2.6-4) that $p_i \leq 1$ for all i. In summary, the numbers p_i satisfy the following conditions: (a) $0 \leq p_i \leq 1$, all i;   (b) $\sum_i p_i = 1$. The p_i's have the characteristics of probabilities and we give them that interpretation.

QM POSTULATE 6. If a system is prepared in a state $\Psi(q)$, then the probability that a measurement of A yields the value λ_i is given by

$$p_i = \left| \int \Psi_i^*(q) \Psi(q) \, dq \right|^2 \tag{2.6-5}$$

Note that relation (2.1-3) is a special case of eq. (2.6-5).

The average value of the observable A is then

$$\langle A \rangle = \sum_i p_i \lambda_i \tag{2.6-6}$$

that is, just the sum of the possible values λ_i of A weighted by the probabilities that the system is found in the states $\Psi_i(q)$ having those respective values. Equation (2.6-6) displays the inherently probabilistic nature of quantum mechanics. Only if we prepare the system in a particular eigenstate Ψ_i can we be certain of measuring a particular value λ_i for that observable. In this case $p_i = 1$ and $p_j = 0$ for $j \neq i$. In general, when the system is not in an eigenstate of $\hat{A}$, we cannot know in advance the particular value of A that a single measurement will yield. We can know only the *probability* of obtaining it.

The view of the measurement process that we adopt here is the more or less commonly accepted one. However, it does leave open a number of deep, unsettling questions that are beyond the scope of this text. The interpretation of quantum mechanics continues to be a topic of lively debate (Newton, 1980; Peres, 1984).

2.7 QUANTIZATION OF THE ENERGY: THE ONE-DIMENSIONAL HARMONIC OSCILLATOR

The energy-eigenvalue problem

$$\hat{H}\Psi_E(q) = E\Psi_E(q) \tag{2.7-1}$$

is paramount in physical chemistry. Most introductory texts present the solutions of eq. (2.7-1) for the various *model* Hamiltonians that have extensive application to chemical systems, for example, the particle in a box, the harmonic oscillator, and the hydrogen atom. Rather than repeat standard treatments of all these models, we focus on just one, the harmonic oscillator, and take a less common approach that reveals precisely how *quantization* arises naturally from the postulates laid down so far.

Our treatment makes use of the **commutator** of two operators $\hat{A}$ and $\hat{B}$, defined by

$$[\hat{A}, \hat{B}] \equiv \hat{A}\hat{B} - \hat{B}\hat{A} \tag{2.7-2}$$

If $[\hat{A}, \hat{B}] = 0$, $\hat{A}$ and $\hat{B}$ are said to **commute**. This is the quantum mechanical analog of the classical relation $\{A, B\} = 0$, mentioned in Exercise 1.14. However, it is not generally the case that two operators commute. Consider, for example, the action of $[\hat{x}, \hat{p}]$ on the one-dimensional state function $\Psi(x)$:

$$[\hat{x}, \hat{p}]\Psi(x) = -i\hbar x \frac{d\Psi}{dx} + i\hbar \frac{d}{dx}[x\Psi(x)] = i\hbar\Psi(x)$$

Since $\Psi(x)$ is arbitrary, we have

$$[\hat{x}, \hat{p}] = i\hbar \tag{2.7-3}$$

Clearly, $\hat{x}$ does not commute with $\hat{p}$ and herein lies the *essential* difference between quantum and classical mechanics. Indeed, eq. (2.7-3) can be taken as an alternative statement of QM Postulate 3 and is the origin of quantization.

Now, the Hamiltonian for the harmonic oscillator assumes the particular form [see eq. (1.4-8)]

$$\hat{H} = -\frac{\hbar^2}{2m}\frac{d^2}{dx^2} + \frac{1}{2}m\omega^2 x^2 \tag{2.7-4}$$

It is convenient to introduce the auxiliary operators

$$\hat{a} = (2\hbar\omega m)^{-1/2}(\omega m\hat{x} + i\hat{p}) \tag{2.7-5a}$$

$$\hat{a}^\dagger = (2\hbar\omega m)^{-1/2}(\omega m\hat{x} - i\hat{p}) \tag{2.7-5b}$$

For reasons that will become apparent, $\hat{a}$ is called the **destruction** or **lowering operator** and $\hat{a}^\dagger$ the **creation** or **raising operator**; $\hat{a}$ and $\hat{a}^\dagger$ are also referred to as **ladder operators**. Since $\hat{x}$ and $\hat{p}$ are Hermitian, $\hat{a}$ and $\hat{a}^\dagger$ can be shown to be non-Hermitian. Using eq. (2.7-3) and the definitions (2.7-5), one can verify the commutator relation

$$[\hat{a}, \hat{a}^\dagger] = 1 \tag{2.7-6}$$

From eq. (2.7-6) it follows that

$$[\hat{a}, \hat{a}^\dagger\hat{a}] = \hat{a} \tag{2.7-7a}$$

$$[\hat{a}^\dagger, \hat{a}^\dagger\hat{a}] = -\hat{a}^\dagger \tag{2.7-7b}$$

Moreover, the Hamiltonian, eq. (2.7-4), can be recast in terms of the ladder operators as

$$\hat{H} = \hbar\omega(\hat{a}^\dagger\hat{a} + \tfrac{1}{2}) \tag{2.7-8}$$

Then from eqs. (2.7-7) and (2.7-8), we have

$$[\hat{a}, \hat{H}] = \hbar\omega\hat{a} \tag{2.7-9a}$$

$$[\hat{a}^\dagger, \hat{H}] = -\hbar\omega\hat{a}^\dagger \tag{2.7-9b}$$

Now let us suppose that $\Psi_E(x)$ is an eigenfunction of $\hat{H}$, that is, $\Psi_E(x)$ satisfies eq. (2.7-1)

$$\hat{H}\Psi_E(x) = E\Psi_E(x) \tag{2.7-10}$$

where E labels the state of energy E. Expanding the commutator in eq. (2.7-9a), rearranging the equation and allowing both sides to operate on $\Psi_E(x)$, we obtain

$$\hat{H}\hat{a}\Psi_E(x) = \hat{a}(\hat{H} - \hbar\omega)\Psi_E(x) = \hat{a}\hat{H}\Psi_E(x) - \hbar\omega\hat{a}\Psi_E(x) \tag{2.7-11}$$

Using eq. (2.7-10), we can rewrite eq. (2.7-11) as

$$\hat{H}\hat{a}\Psi_E(x) = (E - \hbar\omega)\hat{a}\Psi_E(x) \tag{2.7-12}$$

Equation (2.7-12) expresses the following remarkable fact. If $\Psi_E(x)$ is an eigenfunction of $\hat{H}$ with eigenvalue E, then the operator $\hat{a}$ acts on $\Psi_E(x)$ to produce another eigenfunction of $\hat{H}$, with the eigenvalue reduced to $E - \hbar\omega$. Replacing $\Psi_E(x)$ by $\hat{a}\Psi_E(x)$ in eq. (2.7-11) and invoking eq. (2.7-12), we obtain

$$\begin{aligned}\hat{H}[\hat{a}^2\Psi_E(x)] &= \hat{a}\hat{H}[\hat{a}\Psi_E(x)] - \hbar\omega\hat{a}^2\Psi_E(x) \\ &= (E - 2\hbar\omega)[\hat{a}^2\Psi_E(x)]\end{aligned} \tag{2.7-13}$$

that is, $\hat{a}^2\Psi_E(x)$ is an eigenfunction of $\hat{H}$ with eigenvalue $E - 2\hbar\omega$. Proceeding in this fashion, we conclude that if $\Psi_E(x)$ is an eigenfunction of $\hat{H}$, then $\hat{a}^n\Psi_E(x)$ is also an eigenfunction having eigenvalue $E - n\hbar\omega$. Thus, by repeated application of the lowering operator to any given eigenfunction, we generate an apparently *unlimited* number of additional eigenstates, all lower in energy, with each eigenvalue separated from the previous one by $\hbar\omega$.

By arguments paralleling those presented previously, we deduce that $(\hat{a}^\dagger)^n\Psi_E(x)$ is also an eigenfunction of $\hat{H}$ with eigenvalue $E + n\hbar\omega$. Thus, by repeatedly applying $\hat{a}^\dagger$ to a given eigenfunction, we can produce a *limitless* number of additional eigenstates, all associated with larger eigenvalues, differing successively by $\hbar\omega$.

To this point it would seem there is no restriction on the eigenvalues, except that they be real. Such a restriction is now sought by making sure that the QM Postulates other than 3 are obeyed. In particular, consider the norm (see Section 2.1) of $\phi(x) = \hat{a}\Psi_E(x)$, which can be written

$$\int_{-\infty}^{\infty} |\phi(x)|^2 \, dx = \int_{-\infty}^{\infty} [\hat{a}^*\Psi_E^*(x)][\hat{a}\Psi_E(x)] \, dx \tag{2.7-14}$$

Using the definition (2.7-5a) of $\hat{a}$ and integrating by parts, we convert eq. (2.7-14) to

$$\int_{-\infty}^{\infty} |\phi|^2 \, dx = \int_{-\infty}^{\infty} \Psi_E^*(x) \hat{a}^\dagger \hat{a} \Psi_E(x) \, dx \qquad (2.7\text{-}15)$$

By means of eqs. (2.7-8) and (2.7-10), eq. (2.7-15) can be rewritten

$$\int_{-\infty}^{\infty} |\phi|^2 \, dx = \left(\frac{E}{\hbar\omega} - \frac{1}{2} \right) \int_{-\infty}^{\infty} |\Psi_E(x)|^2 \, dx \qquad (2.7\text{-}16)$$

Since the norms of both ϕ and Ψ_E must be nonnegative, it follows from eq. (2.7-16) that the energy eigenvalue cannot be less than $E = \hbar\omega/2$. Furthermore, $E/\hbar\omega$ must be half-integral, since otherwise the destruction operator $\hat{a}$ could "step" the state down to an eigenvalue less than $\hbar\omega/2$. We conclude that the *only* observable values of the energy for the harmonic oscillator are given by

$$E_n = \hbar\omega\left(n + \tfrac{1}{2}\right) \qquad n = 0, 1, 2, \ldots \qquad (2.7\text{-}17)$$

The corresponding eigenfunctions will be denoted by $\Psi_n(x)$ henceforth. Note that when $n = 0$, $\phi = \hat{a}\Psi_0 \equiv 0$ according to eq. (2.7-16). This shows that Ψ_0 is the lowest possible state, since $\hat{a}^n \phi \equiv 0$, that is, no new lower-energy states are produced.

Equation (2.7-10) can now be rewritten more explicitly as

$$\hbar\omega\left(\hat{a}^\dagger \hat{a} + \tfrac{1}{2}\right)\Psi_n(x) = \hbar\omega\left(n + \tfrac{1}{2}\right)\Psi_n(x)$$

From this it is clear that

$$\hat{a}^\dagger \hat{a} \Psi_n(x) = n \Psi_n(x) \qquad (2.7\text{-}18)$$

that is, $\Psi_n(x)$ is the eigenfunction of the operator

$$\hat{N} = \hat{a}^\dagger \hat{a}$$

which is referred to as the **number operator**. If $\hbar\omega$ is viewed as a *quantum* of energy, then $\hat{N}$ acts on $\Psi_n(x)$ to give the number n of quanta contained in the oscillator in that state. Figure 2.3 depicts the quantization of the harmonic oscillator.

Since $n = 0$ is the *lowest allowed* state of the harmonic oscillator, the action of $\hat{a}$ on $\Psi_0(x)$ must yield zero identically. That is,

$$\hat{a}\Psi_0(x) \equiv 0 \qquad (2.7\text{-}19)$$

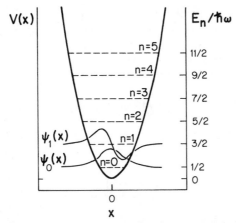

FIGURE 2.3. The quantized one-dimensional harmonic oscillator.

Using eq. (2.7-5a) and QM Postulate 3, we can recast eq. (2.7-19) as

$$\frac{d\Psi_0}{dx} = -\omega m \hbar^{-1} x \Psi_0(x) \qquad (2.7\text{-}20)$$

This equation can be integrated by standard techniques to give

$$\Psi_0(x) = c \exp\left(-\tfrac{1}{2}\omega m \hbar^{-1} x^2\right) \qquad (2.7\text{-}21)$$

where the constant c is fixed by the normalization requirement

$$\int_{-\infty}^{\infty} \Psi_0^*(x)\Psi_0(x)\, dx = 1 \qquad (2.7\text{-}22)$$

From eqs. (2.7-21) and (2.7-22) it follows that

$$|c|^2 = \left(\pi^{-1}\hbar^{-1}\omega m\right)^{1/2} \qquad (2.7\text{-}23)$$

Therefore,

$$c = \left(\pi^{-1}\hbar^{-1}\omega m\right)^{1/4} \exp(i\alpha) \qquad (2.7\text{-}24)$$

where α is a real number. The normalization constant c, and hence the eigenfunction, are determined only to within an *arbitrary phase* α. This is always the case. Since observable quantities do not depend upon the phase, we set $\alpha = 0$ for convenience and take

$$\Psi_0(x) = \left(\pi^{-1}\hbar^{-1}\omega m\right)^{1/4} \exp\left(-\tfrac{1}{2}\omega m \hbar^{-1} x^2\right) \qquad (2.7\text{-}25)$$

To obtain $\Psi_1(x)$, we can simply operate upon $\Psi_0(x)$ with $\hat{a}^\dagger$ to raise the quantum number $n = 0$ by one. However, there is a problem involving the normalization. Returning to the general case, we know only that $\hat{a}^\dagger\Psi_n(x)$ has the eigenvalue $\hbar\omega[(n + 1) + \frac{1}{2}]$, that is, $\hat{a}^\dagger\Psi_n(x)$ is *proportional* to $\Psi_{n+1}(x)$. In other words,

$$\hat{a}^\dagger\Psi_n(x) = c\Psi_{n+1}(x) \tag{2.7-26}$$

where c is an unknown constant. To find c, we form the norm of each side of eq. (2.7-26):

$$\int_{-\infty}^{\infty} \Psi_n^*(x)\hat{a}\hat{a}^\dagger\Psi_n(x)\, dx = |c|^2\int_{-\infty}^{\infty} \Psi_{n+1}^*(x)\Psi_{n+1}(x)\, dx \tag{2.7-27}$$

That the left side of eq. (2.7-27) is correct follows from reasoning similar to that used in going from eq. (2.7-14) to eq. (2.7-15). Making use of eqs. (2.7-6) and (2.7-18), we can rewrite eq. (2.7-27) as

$$(n + 1)\int_{-\infty}^{\infty}|\Psi_n|^2\, dx = |c|^2\int_{-\infty}^{\infty}|\Psi_{n+1}|^2\, dx$$

Again invoking the requirement that Ψ_n and Ψ_{n+1} be normalized, we deduce that

$$c = (n + 1)^{1/2} \tag{2.7-28}$$

where the phase of c has been chosen to be zero, as discussed following eq. (2.7-24). Combining eqs. (2.7-26) and (2.7-28), we obtain

$$\hat{a}^\dagger\Psi_n(x) = (n + 1)^{1/2}\Psi_{n+1}(x) \tag{2.7-29}$$

By parallel reasoning it may be shown that

$$\hat{a}\Psi_n(x) = (n)^{1/2}\Psi_{n-1}(x)$$

Now applying eq. (2.7-29) to the case $n = 0$, we have

$$\Psi_1(x) = \hat{a}^\dagger\Psi_0(x)$$
$$= (2\hbar\omega m)^{-1/2}\left(\omega m x - \hbar\frac{d}{dx}\right)\Psi_0(x) \tag{2.7-30}$$

where the second line follows from eq. (2.7-5b) and QM Postulate 3. Substitution of eq. (2.7-20) into eq. (2.7-30) yields

$$\Psi_1(x) = (2\omega m\hbar^{-1})^{1/2}x\Psi_0(x)$$
$$= [4\omega^3 m^3\pi^{-1}\hbar^{-3}]^{1/4}x \exp\left[-\tfrac{1}{2}\omega m\hbar^{-1}x^2\right] \tag{2.7-31}$$

For the case $n = 2$, eq. (2.7-29) reads

$$\Psi_2(x) = (2)^{-1/2}\hat{a}^\dagger\Psi_1(x)$$
$$= (2!)^{-1/2}(\hat{a}^\dagger)^2\Psi_0(x) \tag{2.7-32}$$

Equation (2.7-32) can be generalized to

$$\Psi_n(x) = (n!)^{-1/2}(\hat{a}^\dagger)^n\Psi_0(x) \tag{2.7-33}$$

Thus, all of the higher eigenfunctions can be generated from the *lowest* one by successive application of the raising operator. The eigenfunctions can be succinctly expressed in terms of the Hermite polynomials H_n (Sneddon, 1980)

$$\Psi_n(x) = \left(\beta^{-1/2}\pi^{1/2}2^n n!\right)^{-1/2} H_n\left(\beta^{1/2}x\right)\exp\left(-\tfrac{1}{2}\beta x^2\right) \tag{2.7-34}$$

where $\beta \equiv \omega m\hbar^{-1}$.

Exercises

2.9. Consider a system having only two energy states, that is, $\hat{H}\Psi_n = E_n\Psi_n$, $n = 1, 2$, and where $\int\Psi_n^*(q)\Psi_m(q)\,dq = \delta_{nm}$. Assume that the operator $\hat{A}$ is represented by the matrix

$$\begin{bmatrix} 0 & a \\ a & 0 \end{bmatrix}$$

in the basis of eigenfunctions of the Hamiltonian $\hat{H}$.
(a) What is the matrix representation of $\hat{H}$?
(b) Show that in the new basis defined by

$$\phi_+ = 2^{-1/2}[\Psi_1(q) + \Psi_2(q)]$$
$$\phi_- = 2^{-1/2}[\Psi_1(q) - \Psi_2(q)]$$

the matrix representations of $\hat{H}$ and $\hat{A}$ are

$$\mathbf{H} = \tfrac{1}{2}\begin{bmatrix} E_1 + E_2 & E_1 - E_2 \\ E_1 - E_2 & E_1 + E_2 \end{bmatrix}$$

$$\mathbf{A} = \begin{bmatrix} a & 0 \\ 0 & -a \end{bmatrix}$$

(c) If A is an observable, what can we say about a?

2.10. A set of vectors $\mathbf{a}, \mathbf{b}, \mathbf{c}, \ldots,$ is said to be **linearly dependent** if and only if there are numbers $A, B, C, \ldots,$ not all equal to zero, such that

$$A\mathbf{a} + B\mathbf{b} + C\mathbf{c} + \cdots + = 0$$

If the equation holds only when $A = B = C = \cdots = 0$, then the vectors are **linearly independent**. Prove that the members of a complete orthonormal set [see eq. (2.3-7)] are linearly independent. (*Hint:* Assume, to the contrary, that $\Psi_a = B\Psi_b + C\Psi_c$ and look for a contradiction.)

2.11. Let $\{\mathbf{a}_i\}$ be a set of n linearly independent vectors in an n-dimensional space S, and let $\mathbf{x}$ be any other vector in S. Prove that the expansion of $\mathbf{x}$ as a linear combination of the $\mathbf{a}_i$ is unique. In view of Exercises 2.10, this result may be paraphrased to read that the expansion of any state function as a linear combination of the eigenfunctions of some operator is unique.

2.12. The eigenfunctions of a particle in a one-dimensional box of length L are $(2/L)^{1/2}\sin(n\pi x/L)$, $n = 1, 2, 3, \ldots$. In view of Exercises 2.10 and 2.11, find the unique expansion of the function $\Psi(x) = 1$ in terms of the one-dimensional particle-in-a-box eigenfunctions. Find a closed expression for the expansion coefficients.

2.13.[C] Write a computer program to evaluate $\Psi(x) = 1$ at several values of x in the interval $[0, L]$, using the expansion determined in Exercise 2.12. Investigate the accuracy as a function of the number of terms taken in the expansion. Plot out a number of $\Psi(x)$ curves.

2.14. Evaluate the following integrals:

(a) $\int_{-1}^{1} \delta(x)\, dx$
(b) $\int_{2}^{\infty} e^{-x} \delta(x - 1)\, dx$
(c) $\int_{-\infty}^{\infty} e^{-x^2} \delta(2x - 1)\, dx$
(d) $\int_{-\infty}^{\infty} \delta(x - x_1)\, \delta(x_1 - x_2)\, dx_1$

2.15. Consider the sequence

$$\delta_n(t) = \begin{cases} \frac{1}{2}n & |t| < n^{-1} \\ 0 & |t| > n^{-1} \end{cases}$$

Use the **mean value theorem for integrals** (Spiegel, 1969) and show that this sequence is suitable for defining the Dirac delta function.

2.16. Examine the sequence of functions $\delta_n(x) = \frac{1}{2}n\,\mathrm{sech}^2 nx$. On the same set of axes plot the first three members of the sequence. Does $\lim_{n \to \infty} \delta_n(x)$ exist (discuss both the $x \neq 0$ and the $x = 0$ cases)? Does the sequence exhibit the first criterion defining the delta function, that is, does the definite integral

$$\int_{-\infty}^{\infty} \delta_n(x)\, dx$$

exist and equal unity? Carry out an analysis similar to that given in Section 2.5.

2.17. There is another pictorial way of thinking about the Dirac delta function. Make graphs of the following for a few integral values of n.

$$\frac{1}{2} + \frac{1}{\pi} \tan^{-1}(nx)$$

$$\pi^{-1/2} \int_{-x}^{\infty} n \exp(-n^2 u^2)\, du$$

What shape do these plots seem to approach as $n \rightarrow \infty$? Do the derivatives of these functions form a delta sequence?

2.18. Show that $\hat{a}$ and $\hat{a}^{\dagger}$ are *not* Hermitian.

2.19. Show that the harmonic-oscillator Hamiltonian eq. (2.7-4) can be recast as eq. (2.7-8).

2.20. Prove commutator relation (2.7-6).

2.21. Show that

$$\int_{-\infty}^{\infty} \phi^*(x)\phi(x)\, dx = \int_{-\infty}^{\infty} [\hat{a}^* \Psi_E^*(x)][\hat{a}\Psi_E(x)]\, dx$$

$$= \int_{-\infty}^{\infty} \Psi_E^* \hat{a}^{\dagger} \hat{a} \Psi_E(x)\, dx$$

2.22. By separating the variables Ψ_0 and x, integrate eq. (2.7-20). Show that the normalization constant is given by eq. (2.7-24).

2.23. Derive the relation $\hat{a}\Psi_n(x) = (n)^{1/2}\Psi_{n-1}(x)$.

2.24. Find the matrix representations of the operators $\hat{x}$, $\hat{p}$, $\hat{a}$, and $\hat{a}^{\dagger}$ in the basis of harmonic-oscillator energy eigenfunctions.

2.8 ENERGY STATES OF STANDARD QUANTUM MECHANICAL MODELS

From QM Postulate 3 and the classical expression (1.1-10), we can write the Hamiltonian operator for a general N-particle system as

$$\hat{H} = -\sum_{i=1}^{N} \frac{\hbar^2}{2m_i}\left(\frac{\partial^2}{\partial x_i^2} + \frac{\partial^2}{\partial y_i^2} + \frac{\partial^2}{\partial z_i^2}\right) + \sum_{i=1}^{N} V_i(\mathbf{r}_i)$$

$$+ \sum_{i=1}^{N}\sum_{j>i}^{N} V_{ij}(|\mathbf{r}_i - \mathbf{r}_j|) \tag{2.8-1}$$

where $\mathbf{r}_i = x_i\mathbf{e}_x + y_i\mathbf{e}_y + z_i\mathbf{e}_z$, as usual. This can be expressed more compactly in terms of the Laplacians as

$$\hat{H} = -\sum_{i=1}^{N} \frac{\hbar^2}{2m_i} \nabla_i^2 + \sum_{i=1}^{N} V_i(\mathbf{r}_i) + \sum_{i=1}^{N} \sum_{j>i}^{N} V_{ij}(|\mathbf{r}_i - \mathbf{r}_j|) \quad (2.8\text{-}2)$$

In general, it is impossible to find the eigenvalues and eigenfunctions of $\hat{H}$ in closed form (as was done in Section 2.7 for the harmonic oscillator). However, under special circumstances the problem can be greatly simplified. We now consider these in a systematic manner.

If the interparticle interactions V_{ij} are neglected, then $\hat{H}$ is *separable*, that is,

$$\hat{H} = \sum_{i=1}^{N} \hat{H}(i) \quad (2.8\text{-}3)$$

where

$$\hat{H}(i) \equiv -\frac{\hbar^2}{2m_i} \nabla_i^2 + V_i(\mathbf{r}_i)$$

refers only to particle i. In this case the eigenfunctions can be written as *products* of eigenfunctions of the individual $\hat{H}(i)$:

$$\Psi(\mathbf{r}_1, \mathbf{r}_2, \ldots, \mathbf{r}_N) = \Psi^{(1)}(\mathbf{r}_1)\Psi^{(2)}(\mathbf{r}_2) \cdots \Psi^{(N)}(\mathbf{r}_N)$$

$$= \prod_{i=1}^{N} \Psi^{(i)}(\mathbf{r}_i) \quad (2.8\text{-}4)$$

where

$$\hat{H}(i)\Psi^{(i)}(\mathbf{r}_i) = E(i)\Psi^{(i)}(\mathbf{r}_i)$$

Substitution of expressions (2.8-3) and (2.8-4) into the eigenvalue equation (2.7-1) yields

$$\hat{H}\Psi(\mathbf{r}_1, \mathbf{r}_2, \ldots, \mathbf{r}_N) = \sum_{i=1}^{N} \hat{H}(i)\left(\prod_{j=1}^{N} \Psi^{(j)}(\mathbf{r}_j) \right)$$

$$= \sum_{i}^{N} \left[\prod_{j \neq i}^{N} \Psi^{(j)}(\mathbf{r}_j) \right] \hat{H}(i)\Psi^{(i)}(\mathbf{r}_i)$$

$$= \sum_{i}^{N} E(i)\left[\prod_{j=1}^{N} \Psi^{(j)}(\mathbf{r}_j) \right]$$

$$= \left[\sum_{i}^{N} E(i) \right] \Psi(\mathbf{r}_1, \mathbf{r}_2, \ldots, \mathbf{r}_N) \quad (2.8\text{-}5)$$

Thus, *the possible eigenvalues are simply sums of the individual eigenvalues.*

If the potential energy $V(q)$ is itself separable, then the single-particle eigenfunction can be further factored. For example, for a particle in a three-dimensional box, $V = 0$ and the eigenfunction can be written as a product of three one-dimensional eigenfunctions (see Table 2.1). For systems in which V does not vanish but possesses certain symmetry, V is separable in terms of *generalized coordinates*. A case of frequent interest is the central field, where $V = V(r)$. Then, in spherical polar coordinates (see Section 1.3) the Hamiltonian becomes

$$\hat{H} = -\frac{\hbar^2}{2\mu} \left\{ \frac{1}{r^2} \frac{\partial}{\partial r}\left(r^2 \frac{\partial}{\partial r}\right) + \frac{1}{r^2}\left[\frac{1}{\sin\theta} \frac{\partial}{\partial\theta}\left(\sin\theta \frac{\partial}{\partial\theta}\right) \right.\right.$$
$$\left.\left. + \frac{1}{\sin^2\theta} \frac{\partial^2}{\partial\phi^2}\right]\right\} + V(r) \quad (2.8\text{-}6)$$

Upon substituting $\Psi(r, \theta, \phi) = R(r)Y(\theta, \phi)$ into eq. (2.8-6), we find that the eigenvalue equation separates into angular and radial portions:

$$-\left[\frac{1}{\sin\theta} \frac{\partial}{\partial\theta}\left(\sin\theta \frac{\partial}{\partial\theta}\right) + \frac{1}{\sin^2\theta} \frac{\partial^2}{\partial\phi^2}\right]Y(\theta, \phi) = \lambda Y(\theta, \phi) \quad (2.8\text{-}7a)$$

$$\left\{-\frac{\hbar^2}{2\mu}\left[\frac{1}{r^2} \frac{d}{dr}\left(r^2 \frac{d}{dr}\right)\right] + \frac{\hbar^2\lambda}{2\mu} \frac{1}{r^2} + V(r)\right\}R(r) = ER(r) \quad (2.8\text{-}7b)$$

Here, the separation constant λ is in fact the eigenvalue of $\hat{\ell}^2$, where $\hat{\ell}$ is the orbital angular momentum operator (see Section 1.3). The solutions for the special cases are given in Table 2.1. For the rigid rotor, r is fixed and V is a constant that we set to zero for convenience. The eigenfunctions are really just the eigenfunctions of the angular momentum operator $\hat{\ell}^2$. For the hydrogen atom, the total eigenfunction is the product of the eigenfunction of $\hat{\ell}^2$ and the solution of eq. (2.8-7b) with $V(r)$ replaced by the Coulomb potential.

A feature common to all of the models summarized in Table 2.1 is that the original three-dimensional eigenvalue equation is separable into three one-dimensional equations. Moreover, each of the one-dimensional eigenvalue problems is a particular example of the general second-order differential equation

$$\frac{d}{dx}\left[p(x)\frac{dy}{dx}\right] - q(x)y + \lambda\rho(x)y = 0$$
$$p(x) > 0; \quad \rho(x) > 0; \quad \lambda \text{ real} \quad (2.8\text{-}8)$$

subject to the boundary conditions

$$y'(a) - h_1 y(a) = 0$$
$$y'(b) - h_2 y(b) = 0 \quad (2.8\text{-}9)$$
$$h_1, h_2 > 0 \quad a < b$$

Table 2.1. Eigenfunctions and Eigenvalues of Some Standard Models

System	Eigenvalue Equation	Eigenfunctions	Eigenvalues
Particle of mass m in a one-dimensional box of length L	$-\dfrac{\hbar^2}{2m}\dfrac{d^2\psi}{dx^2} = E\psi$	$\psi_n(x) = \left(\dfrac{2}{L}\right)^{1/2}\sin\dfrac{n\pi x}{L}$	$E_n = n^2 h^2/8mL^2$ $n = 1, 2, 3, \ldots$
Particle in three-dimensional box $(L \times W \times H)$	$-\dfrac{\hbar^2}{2m}\nabla^2\psi = E\psi$	$\psi_{n,p,q}(x, y, z) = \left(\dfrac{8}{LWH}\right)^{1/2}\sin\dfrac{p\pi y}{W}\sin\dfrac{q\pi z}{H}$ $\times \sin\dfrac{p\pi y}{W}\sin\dfrac{q\pi z}{H}$	$E_{n,p,q} = \dfrac{\hbar^2\pi^2}{2m}\left\{\dfrac{n^2}{L^2} + \dfrac{p^2}{W^2} + \dfrac{q^2}{H^2}\right\}$ $n, p, q = 1, 2, 3, \ldots$
One-dimensional harmonic oscillator of reduced mass μ, and fundamental frequency ω [a]	$-\dfrac{\hbar^2}{2\mu}\dfrac{d^2\psi}{dx^2} + \dfrac{1}{2}\mu\omega^2 x^2\psi = E\psi$	$\psi_n(x) = \left(\dfrac{\sqrt{\beta}}{2^n n!\sqrt{\pi}}\right)^{1/2}\exp(-\beta x^2/2)\,H_n(\sqrt{\beta}\,x),$ $\beta = \mu\omega/\hbar$	$E_n = \left(n + \tfrac{1}{2}\right)\hbar\omega$ $n = 0, 1, 2, \ldots$
Rigid rotor of length R and reduced mass μ [b]	$-\dfrac{\hbar^2}{2I}\left\{\dfrac{1}{\sin\theta}\dfrac{\partial}{\partial\theta}\left(\sin\theta\dfrac{\partial}{\partial\theta}\right) + \dfrac{1}{\sin^2\theta}\dfrac{\partial^2}{\partial\phi^2}\right\}Y = EY$	$Y_{J,m}(\theta,\phi) = (2\pi)^{-1/2}\left\{\left(\dfrac{2J+1}{2}\right)\left(\dfrac{(J-|m|)!}{(J+|m|)!}\right)\right\}^{1/2}$ $\times P_J^{|m|}(\cos\theta)\,e^{im\phi}$	$E_J = J(J+1)\dfrac{\hbar^2}{2I},\ I = \mu R^2$ $J = 0, 1, 2, \ldots$ $m = 0, \pm 1, \pm 2, \ldots \pm J$
R-equation of the hydrogenic atom of nuclear charge Ze [c]	$-\dfrac{\hbar^2}{2\mu}\left\{\dfrac{d^2}{dr^2} + \dfrac{\ell(\ell+1)}{r^2}\right\}(rR) + \dfrac{\mu Ze^2}{2\pi\varepsilon_0\hbar^2 r}(rR) = E(rR)$	$R_{n,\ell}(r) = -\left\{\left(\dfrac{2Z}{na_0}\right)^3\dfrac{(n-\ell-1)!}{2n[(n+\ell)!]^3}\right\}^{1/2}$ $\times e^{-Zr/na_0}\left(\dfrac{2Zr}{na_0}\right)^{\ell}L_{n+\ell}^{2\ell+1}\left(\dfrac{2Zr}{na_0}\right)$	$E_n = -\mu e^4 Z^2/8n^2\varepsilon_0^2 h^2$ $n = 1, 2, 3, \ldots$ $\ell = 0, 1, 2, \ldots n-1$

[a] H_n is the Hermite polynomial of degree n.

[b] $P_J^{|m|}$ is the associated Legendre polynomial of degree J and order $|m|$, where $P_J^{|m|}(x) = (1 - x^2)^{|m|/2}(2^\ell \ell!)^{-1}\,d^{\ell+|m|}(x^2 - 1)^\ell/dx^{\ell+|m|}$.

[c] $L_{n+\ell}^{2\ell+1}$ is the associated Laguerre polynomial of degree $(n-\ell-1)$ and order $n+\ell$, where $L_{n+\ell}^{2\ell+1}(x) = d^{2\ell+1}[e^x d^{n+\ell}(x^{n+\ell}e^{-x})/dx^{n+\ell}]/dx^{2\ell+1}$.

Equations (2.8-8) and (2.8-9) constitute the so-called **Sturm–Liouville problem**. It can be demonstrated that the only nontrivial solutions occur for restricted values of λ, that is, eigenvalues, that are bounded from below and that the corresponding eigenfunctions $\{y_n\}$ satisfy an orthonormality relation

$$\int_a^b \rho(x) y_n(x) y_m(x)\, dx = \delta_{nm}$$

where $\rho(x)$ is a density or weighting function. It can also be shown that any suitably behaved function $f(x)$ defined on the interval $[a, b]$ can be expanded as a linear combination of the $\{y_n(x)\}$. These are just the properties possessed by the eigenvalues and eigenfunctions of Hermitian quantum mechanical operators (see Section 2.3).

2.9 TIME DEPENDENCE OF QUANTUM STATES

Until now all of our considerations have referred to a single arbitrary instant; time has played no essential role. However, many phenomena of chemical interest are time dependent in nature. For example, **spectroscopy** (which we do not consider in this book) is the study of the time-dependent response of a material system to electromagnetic radiation. This section and the next are concerned with the temporal development of quantum systems.

The starting point is a new postulate (Dicke and Wittke, 1960). To motivate the form of this postulate, we consider a particle moving in a region of constant potential energy. The particle has a fixed momentum $\mathbf{p}$ and fixed total energy E. On the other hand, we may describe the particle by a wave of the form

$$\Psi = A\, e^{-i(\omega t - \mathbf{k} \cdot \mathbf{r})}$$

where A is the amplitude of the wave, ω is its angular frequency, $\mathbf{k}$ is the wave vector of magnitude $k = 2\pi/\lambda$, and t is the time. The vector $\hbar \mathbf{k}$ can be interpreted as the momentum of the particle with which it is associated, since

$$\hbar k = \frac{2\pi\hbar}{\lambda} = \frac{h}{\lambda} = |\mathbf{p}|$$

by de Broglie's hypothesis. The energy of the particle and the angular frequency of the associated wave are related by the **Planck–Einstein formula**: $E = h\nu = \hbar\omega$. Hence, the wave can be reexpressed as

$$\Psi = A\, e^{-i(Et - \mathbf{p} \cdot \mathbf{r})/\hbar} \tag{2.9-1}$$

We now seek a suitable wave equation (second-order differential equation) of which eq. (2.9-1) is a solution. Application of the Laplacian to eq. (2.9-1) yields

$$\nabla^2 \Psi = -\frac{p^2}{\hbar^2} \Psi \tag{2.9-2}$$

while differentiation of eq. (2.9-1) with respect to time gives

$$\frac{\partial \Psi}{\partial t} = -\frac{iE}{\hbar} \Psi \tag{2.9-3}$$

But the total energy can be written as $E = p^2/2m + V$, and upon substituting this into eq. (2.9-2), we obtain

$$\nabla^2 \Psi = -\frac{2m(E - V)}{\hbar^2} \Psi \tag{2.9-4}$$

Finally, elimination of E between eqs. (2.9-3) and (2.9-4) yields

$$\left(-\frac{\hbar^2}{2m} \nabla^2 + V \right) \Psi = i\hbar \frac{\partial \Psi}{\partial t}$$

Schrödinger postulated that this result could be extended as follows to cover any system of particles:

QM POSTULATE 7. The wave function $\Psi(q, t)$ of a system obeys the equation

$$\hat{H}(t)\Psi(q, t) = i\hbar \frac{\partial \Psi(q, t)}{\partial t} \tag{2.9-5}$$

where $\hat{H}(t)$ is the Hamiltonian operator.

Equation (2.9-5) is known as **Schrödinger's equation**, or as the **time-dependent Schrödinger equation** (after the Austrian physicist Erwin Schrödinger, 1887–1961). The argument t of $\hat{H}(t)$ emphasizes that the Hamiltonian may depend *explicitly* on the time if the system experiences external forces. In any case, the wave function must carry the additional argument t, which stands for time. Given the wave function $\Psi(q, t_0)$ at an *arbitrary initial* time t_0, we can in principle solve eq. (2.9-5) to obtain the wave function at any other time t. For convenience, we take $t_0 = 0$ henceforth.

It is customary to define the **time-evolution** or **time-displacement operator** $\hat{U}$ by

$$\Psi(q, t) = \hat{U}(t)\Psi(q, 0) \tag{2.9-6}$$

$\hat{U}$ has the effect of "displacing," or "evolving," the wave function in time from the initial time $t_0 = 0$ to the final time t. Substituting eq. (2.9-6) into eq. (2.9-5), we obtain

$$\hat{H}(t)[\hat{U}(t)\Psi(q,0)] = i\hbar \frac{\partial \hat{U}(t)}{\partial t}\Psi(q,0)$$

Since the initial wave function $\Psi(x,0)$ can be chosen arbitrarily, the "bare" operator equation

$$\hat{H}(t)\hat{U}(t) = i\hbar \frac{\partial \hat{U}}{\partial t} \tag{2.9-7}$$

must be equivalent to Schrödinger's equation. Depending upon the nature of the system, two cases must be considered: (a) if the system is isolated, its total energy is constant and $\hat{H}$ does not depend upon t; (b) if external forces are present, the total energy is not conserved and $\hat{H}$ depends explicitly upon t. In what follows, we shall focus upon the first case.

If $\hat{H}$ does not depend upon time, then

$$\hat{U}(t) = \exp[-i\hbar^{-1}\hat{H}t] \tag{2.9-8}$$

is a solution of eq. (2.9-7). Note that the exponential in eq. (2.9-8) has the same significance as in the case of the classical Liouville operator (see Section 1.4). That expression (2.9-8) solves eq. (2.9-7) is shown simply by differentiating both sides with respect to t.

If $\Psi(q,0)$ is an eigenfunction $\Psi_E(q)$ of $\hat{H}$ [see eq. (2.7-1)], then

$$\Psi(q,t) = \exp(-it\hbar^{-1}E)\Psi_E(q)$$

This follows from eqs. (2.9-6), (2.9-8), and the relation

$$(\hat{H})^n \Psi_E(q) = (E)^n \Psi_E(q)$$

The time dependence of the wave function is completely contained within the phase factor. According to QM Postulate 5 [see eq. (2.6-2)], the expected value of an observable A at time t is given by

$$\langle A \rangle_t = \int_{-\infty}^{\infty} \Psi^*(q,t)\hat{A}\Psi(q,t)\, dq$$

$$= \int_{-\infty}^{\infty} \Psi_E^*(q)\hat{A}\Psi_E(q)\, dq \tag{2.9-9}$$

where the second line shows that $\langle A \rangle_t$ does not in fact change in time. The system is said to be in a **stationary state**.

Now consider the quantity

$$S(t) \equiv \int \chi^*(q, t) \Psi(q, t) \, dq = \int [\hat{U}(t) \chi(q, 0)]^* \Psi(q, t) \, dq \quad (2.9\text{-}10)$$

where χ and Ψ are arbitrary state functions. From eq. (2.9-8) and the definition (1.4-6) of the exponential, it follows that

$$S(t) \equiv \sum_{k=0}^{\infty} \frac{(i\hbar^{-1}t)^k}{k!} \int \left[(\hat{H}^*)^k \chi^*(q, 0) \right] \Psi(q, t) \, dq \quad (2.9\text{-}11)$$

Since $\hat{H}$ is Hermitian, and any power of $\hat{H}$ is also Hermitian, we can use relation (2.2-3) and the fact that $\hat{H}^* = \hat{H}$ to rewrite eq. (2.9-11) as

$$S(t) = \sum_{k=0}^{\infty} \frac{(i\hbar^{-1}t)^k}{k!} \int \chi^*(q, 0)(\hat{H}^*)^k \Psi(q, t) \, dq$$

$$= \int \chi^*(q, 0) \hat{U}^*(t) \Psi(q, t) \, dq \quad (2.9\text{-}12)$$

Substitution of expression (2.9-6) for $\Psi(q, t)$ into eq. (2.9-12) yields

$$S(t) = \int \chi^*(q, 0) [\hat{U}^*(t) \hat{U}(t)] \Psi(q, 0) \, dq \quad (2.9\text{-}13)$$

From eq. (2.9-8) it would appear "obvious" that

$$\hat{U}^*(t) \hat{U}(t) = \exp(i\hbar^{-1}t\hat{H}) \exp(-i\hbar^{-1}t\hat{H}) = 1 \quad (2.9\text{-}14)$$

if operators could be treated as scalars. However, in general it is *not* true that

$$\exp(\hat{A}) \exp(\hat{B}) = \exp(\hat{A} + \hat{B}) \quad (2.9\text{-}15)$$

for any two operators $\hat{A}$ and $\hat{B}$. In fact, eq. (2.9-15) holds only if $[\hat{A}, \hat{B}] = 0$. Since $\hat{A} = \hat{B} = \hat{H}$ in the present case, eq. (2.9-14) is valid. Combining eqs. (2.9-13) and (2.9-14), we have

$$S(t) = S(0) \quad (2.9\text{-}16)$$

which tells us that S is constant in time. If $\Psi(q, 0) = \chi(q, 0)$, we conclude, in particular that the norm of $\Psi(t)$ is preserved or, in other words, *the total probability of finding the system somewhere in the universe remains unity as time passes.*

Let us now consider a *conservative* system prepared in *pure* state $\Psi(q,0)$. At time t we measure A. The expected value is given by (QM Postulate 5)

$$\langle A \rangle_t = \int [\hat{U}(t)\Psi(q,0)]^* \hat{A}[\hat{U}(t)\Psi(q,0)] \, dq \qquad (2.9\text{-}17)$$

Following a line of reasoning parallel to that leading from eq. (2.9-10) to eq. (2.9-13), we can convert eq. (2.9-17) into

$$\langle A \rangle_t = \int \Psi^*(q,t)\hat{A}\Psi(q,t) \, dq$$

$$= \int \Psi^*(q,0)\hat{A}_H(t)\Psi(q,0) \, dq \qquad (2.9\text{-}18)$$

where the *time-dependent operator* is defined by

$$\hat{A}_H(t) = \hat{U}^*(t)\hat{A}\hat{U}(t) \qquad (2.9\text{-}19)$$

Equation (2.9-18) suggests two alternative views of the temporal behavior of quantum systems. On the one hand, the wave function $\Psi(q,t)$ changes in time and the operator $\hat{A}$ is constant. This is called the **Schrödinger picture**. On the other hand, the wave function $\Psi(q,0)$ remains constant and the operator $\hat{A}_H(t)$ changes in time. This is the **Heisenberg picture**. The subscript H on $\hat{A}_H(t)$ connotes that $\hat{A}$ is in the Heisenberg picture. Note that $\hat{A}$ in the Schrödinger picture may also depend upon the time explicitly.

Differentiating eq. (2.9-19) with respect to t, we obtain

$$\frac{\partial \hat{A}_H(t)}{\partial t} = \frac{\partial \hat{U}^*}{\partial t}\hat{A}\hat{U} + U^*\hat{A}\frac{\partial \hat{U}}{\partial t} + \hat{U}^*\frac{\partial \hat{A}}{\partial t}\hat{U} \qquad (2.9\text{-}20)$$

where we have allowed for possible time dependence in $\hat{A}$. Combining eqs. (2.9-8), (2.9-14), and (2.9-20) and noting that $\hat{H}$ commutes with $\hat{U}$, we have

$$\frac{\partial \hat{A}_H(t)}{\partial t} = i\hbar^{-1}[\hat{H}_H(t), \hat{A}_H(t)] + \left(\frac{\partial \hat{A}}{\partial t}\right)_H \qquad (2.9\text{-}21)$$

where

$$\left(\frac{\partial \hat{A}}{\partial t}\right)_H \equiv \hat{U}^*\frac{\partial \hat{A}}{\partial t}\hat{U}$$

Equation (2.9-21) is known as **Heisenberg's equation of motion** (after the German physicist Werner Heisenberg, 1901–1976). Since $[\hat{H}, \hat{U}] = 0$, it follows from eq. (2.9-19) that $\hat{H}_H(t) = \hat{H}$.

Let us assume that $\hat{A}$ does not depend on time. Then we observe that if $\hat{A}$ commutes with $\hat{H}$, $\partial \hat{A}_H(t)/\partial t = 0$, that is, $\hat{A}_H(t)$ is constant. As in the classical theory (see Section 1.4), we call such an observable a **constant of the motion**. In particular, as we have already seen, $\hat{H}$ is a constant of the motion.

We can express eq. (2.9-21) as

$$\frac{\partial \hat{A}_H(t)}{\partial t} = i\hat{\mathscr{L}}\hat{A}_H(t) + \left(\frac{\partial \hat{A}}{\partial t}\right)_H \qquad (2.9\text{-}22)$$

where the **Liouville operator** $\hat{\mathscr{L}}$ is defined by

$$\hat{\mathscr{L}}\hat{A} \equiv \hbar^{-1}[\hat{H}, \hat{A}] \qquad (2.9\text{-}23)$$

Equation (2.9-22) is the quantal analog of the classical equation of motion (1.4-4) and can be obtained therefrom by the replacement

$$\mathscr{L} \leftrightarrow \hat{\mathscr{L}}$$

Liouville's operator is really a **superoperator** in that it acts on *operators* and *not* state functions, as do operators corresponding to observables.

If $\hat{A}$ does not depend on t, eq. (2.9-22) can be integrated formally to yield

$$\hat{A}_H(t) = \exp[it\hat{\mathscr{L}}]\hat{A} \qquad (2.9\text{-}24)$$

in analogy to eq. (1.4-5). By expanding the exponentials and evaluating the commutators, one can convince oneself that eq. (2.9-24) is equivalent to eq. (2.9-19).

2.10 TRANSITION RATES

The aim of this section is to derive an approximate expression for the rate of change of the state of a system in terms of the microscopic properties of the system. For this purpose it is convenient to partition the Hamiltonian as

$$\hat{H} = \hat{H}^{(0)} + \hat{H}^{(1)} \qquad (2.10\text{-}1)$$

where $\hat{H}^{(0)}$ is the **unperturbed**, or **zero-order** Hamiltonian that refers to a (model) system whose eigenstates, $\Psi_n^{(0)}(q)$, are either available or can be readily determined. That is, $\Psi_n^{(0)}(q)$ are the stationary wave functions associated with $\hat{H}^{(0)}$. The rest of the Hamiltonian, $\hat{H}^{(1)}$, is a **perturbation**, generally assumed to be "small" compared to $\hat{H}^{(0)}$, that causes the states $\Psi_n^{(0)}$ to be nonstationary.[1] In other words, if the system starts out (at $t = 0$) in unperturbed state $\Psi_i^{(0)}$, it will evolve into other possible states $\Psi_f^{(0)}$. We seek a quantitative measure of the rate of this evolution.

[1] Perturbation theory is placed on a more rigorous basis in Section 3.1.

We begin by recasting Schrödinger's equation (2.9-5) in the basis of zeroth-order eigenstates

$$\left[\mathbf{H}^{(0)} + \mathbf{H}^{(1)}(t)\right]\Psi(t) = i\hbar\dot{\Psi}(t) \tag{2.10-2}$$

where Ψ is given by eq. (2.4-7). This is a special case of eq. (2.4-10). Note that, in general, $\mathbf{H}^{(1)}$ may depend explicitly on t. Moreover, the matrix elements $(\mathbf{H}^{(0)})_{nm}$ of $\mathbf{H}^{(0)}$ are given by

$$(\mathbf{H}^{(0)})_{nm} = E_n^{(0)}\delta_{nm} \tag{2.10-3}$$

since $\mathbf{H}^{(0)}$ represents $\hat{H}^{(0)}$ in the basis of eigenfunctions of $\hat{H}^{(0)}$ [see eq. (2.4-11)]. To simplify eq. (2.10-2) we introduce an auxiliary vector ϕ, defined by

$$\Psi(t) = \exp[-i\hbar^{-1}t\mathbf{H}^{(0)}]\phi(t) \tag{2.10-4}$$

Substituting eq. (2.10-4) into eq. (2.10-2), we obtain

$$\dot{\phi}(t) = -i\hbar^{-1}\mathbf{W}(t)\phi(t) \tag{2.10-5}$$

where

$$\mathbf{W}(t) \equiv \exp[i\hbar^{-1}t\mathbf{H}^{(0)}]\mathbf{H}^{(1)}(t)\exp[-i\hbar^{-1}t\mathbf{H}^{(0)}]$$

Equation (2.10-5) can be integrated formally to yield

$$\phi(t) = \phi(0) - i\hbar^{-1}\int_0^t \mathbf{W}(t')\phi(t')\,dt' \tag{2.10-6}$$

We now solve eq. (2.10-6) approximately by iteration. If ϕ does not change much over the period of time $[0, t]$ of interest, then the *first-order* approximation to $\phi(t)$ is obtained simply by replacing $\phi(t')$ in the integrand of eq. (2.10-6) by the *zero-order* $\phi(0)$

$$\phi(t) \simeq \left[1 - i\hbar^{-1}\int_0^t \mathbf{W}(t')\,dt'\right]\phi(0) \tag{2.10-7}$$

The *second-order* approximation is found by replacing $\phi(t')$ in eq. (2.10-6) by the *first-order* approximation eq. (2.10-7). This procedure can be repeated indefinitely to obtain any desired order of approximation. For the present purpose it is sufficient to stop at first order.

We now ask for the probability of finding the system in state $\Psi_f^{(0)}$ at time t, given that it is in state $\Psi_i^{(0)}$ at time $t = 0$. From QM Postulate 6, eq. (2.6-5), the answer is

$$P_{fi}(t) = \left|\int \Psi_f^{(0)*}(q)\Psi(q,t)\,dq\right|^2 = [\Psi(t)]_f^*\,[\Psi(t)]_f$$

where $[\Psi]_f$ denotes the fth element of column vector Ψ. From eqs. (2.10-3) and (2.10-4)

$$P_{fi}(t) = [\phi(t)]_f^* [\phi(t)]_f \tag{2.10-8}$$

Since the system is initially in state $\Psi_i^{(0)}$, then

$$[\phi(0)]_n = \delta_{ni}$$

It follows from eq. (2.10-7) that

$$[\phi(t)]_f \simeq \delta_{fi} - i\hbar^{-1} \int_0^t [W(t')]_{fi} \, dt' \tag{2.10-9}$$

Taking the case $f \ne i$, we have from eqs. (2.10-8) and (2.10-9)

$$P_{fi}(t) \simeq \hbar^{-2} \left| \int_0^t W_{fi}(t') \, dt' \right|^2 \tag{2.10-10}$$

The approximate equality in this expression reminds us that the probability is correct only to first order in the iteration process.

If $\hat{H}$ does not depend on the time, we can write eq. (2.10-10) explicitly as

$$P_{fi}(t) \simeq \hbar^{-2} |V_{fi}|^2 \left| \int_0^t \exp(i\omega_{fi}t') \, dt' \right|^2 \tag{2.10-11}$$

where

$$\omega_{fi} \equiv \frac{E_f^{(0)} - E_i^{(0)}}{\hbar}$$

$$V_{fi} = \int \Psi_f^{*(0)}(q) \hat{H}^{(1)} \Psi_i^{(0)}(q) \, dq$$

The integration over t' in eq. (2.10-11) can be done in closed form to give

$$P_{fi}(t) = \hbar^{-2} |V_{fi}|^2 F(t, \omega_{fi}) \tag{2.10-12}$$

where

$$F(t, \omega) = \frac{\sin^2(\omega t/2)}{(\omega/2)^2} \tag{2.10-13a}$$

The behavior of the function $F(t, \omega)$ is curious and is explored in detail in Exercise 2.31. A plot of $F(t, \omega)$ as a function of ω for fixed t is shown in Fig. 2.4. There is a central peak of height t^2 and width $\simeq 2\pi/t$, with much lower

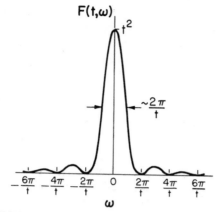

FIGURE 2.4. Plot of $F(t, \omega)$ defined by eq. (2.10-13).

peaks in the "wings." As time increases, the central peak becomes higher and narrower while the peaks in the wings collapse and shift toward $\omega = 0$. This means that the probability of finding the system in a state differing significantly in energy from the initial state is small and decreases rapidly with time [see eq. (2.10-12)]. In fact, it can be shown that as t becomes infinite

$$\lim_{t \to \infty} F(t, \omega) = 2\pi t \, \delta(\omega) \tag{2.10-13b}$$

where $\delta(\omega)$ is the Dirac delta function discussed in Section 2.5. Thus, in the limit of infinitely long time, there is nonzero probability of finding the system only in states $\Psi_f^{(0)}$ for which $\omega_{fi} = 0$. In other words, *only transitions $i \to f$ that conserve energy are permitted.*

The **transition rate** is defined as the probability per unit time of the transition and is expressed as

$$w_{fi} = \lim_{t \to \infty} \frac{P_{fi}(t)}{t} = 2\pi\hbar^{-2}|V_{fi}|^2 \, \delta(\omega_{fi})$$

$$= 2\pi\hbar^{-1}|V_{fi}|^2 \, \delta\left[E_f^{(0)} - E_i^{(0)}\right] \tag{2.10-14}$$

Note that since $\hat{H}^{(1)}$ is Hermitian, $|V_{fi}|^2 = |V_{if}|^2$. Furthermore, from eq. (2.10-13a) we see that $F(t, -\omega) = F(t, \omega)$. Since the Dirac delta is symmetric in its argument, that is, $\delta(-\omega) = \delta(\omega)$ (see Section 2.5), it follows from eq. (2.10-14) that

$$w_{if} = w_{fi} \tag{2.10-15}$$

In words, the "forward" and "backward" transition rates among states of the same energy are equal. This is known as the **principle of microscopic reversibility** or **detailed balance**.

In any real macroscopic system, the eigenstates of $\hat{H}^{(0)}$ are essentially continuously distributed in energy. For example, a gas atom in a cubic three-dimensional box that has energy E can occupy states characterized by triplets (n, p, q) (see Table 2.1) consistent with $E = \hbar^2 \pi^2 [n^2 + p^2 + q^2]/2mL^2$. For an atom of *thermal energy* k_BT (where k_B is Boltzmann's constant) in a box of macroscopic size, a large number of triplets satisfy this relation, that is, the level is highly *degenerate*. Thus, even though we can prepare the atom in a given initial state corresponding to a specific triplet, the perturbation, due perhaps to the roughness of the walls, causes the system to evolve into other states that are isoenergetic with it.

This situation can be handled in general by summing the individual transition rates w_{fi} given by eq. (2.10-14) over the final states. Actually, the sum is an integral over the energy E of the final states weighted by the density (i.e., number of states per unit energy) of final states. We have for the *total* transition rate

$$w = \sum_f w_{fi} = \int \rho \left[E_f^{(0)} \right] w_{fi} \, dE_f^{(0)} \tag{2.10-16}$$

Substituting eq. (2.10-14) for w_{fi} and invoking property (2.5-3) of the Dirac delta function, we simplify eq. (2.10-16) to

$$w = 2\pi \hbar^{-1} |V_{fi}|^2 \rho \left(E_i^{(0)} \right) \tag{2.10-17}$$

where $\rho(E_i^{(0)})$ is the density of final states having energy $E_i^{(0)}$. Expression (2.10-17) is known as **Fermi's Golden Rule** and is the most commonly employed approximation to the transition rate.

Exercises

2.25. The orbital angular momentum ℓ was defined in eq. (1.3-5). Obtain expressions for the quantum mechanical operators $\hat{\ell}_x$ and $\hat{\ell}^2$ in spherical polar coordinates. Evaluate the commutator $[\hat{\ell}_x, \hat{\ell}^2]$.

2.26. Show that if $\hat{A}$ and $\hat{B}$ commute, then eq. (2.9-15) holds.

2.27. Show explicitly that $\dot{S}(t) = 0$ where $S(t)$ is given in eq. (2.9-10).

2.28. Derive Heisenberg's equations of motion for $\hat{x}$ and $\hat{p}$ for the harmonic oscillator and compare them with the corresponding classical equations. Solve the equations in closed form.

2.29. Convince yourself of the equivalence between eqs. (2.9-19) and (2.9-24). One approach is to expand the exponentials and simplify the commutators.

2.30. Consider a diatomic gas contained between the parallel plates of a condenser. Neglect rotation and translation and assume the diatomic vibrates perpendicularly to the plates, say parallel to the x axis. Take the diatomic to be harmonic, that is,

$$\hat{H}^{(0)} = -\frac{\hbar^2}{2\mu}\frac{d^2}{dx^2} + \frac{1}{2}\mu\omega^2 x^2$$

where x is the internuclear separation, μ is the reduced mass and ω is the fundamental frequency. Suppose a voltage pulse is applied to the plates at time $t = 0$ so as to produce an electric field

$$\mathbf{E} = \begin{cases} 0 & t < 0 \\ E_0 \mathbf{e}_x \exp(-t/\tau) & t \geq 0 \end{cases}$$

where $\mathbf{e}_x$ is a unit vector in the x direction. The diatomic experiences an additional interaction

$$\hat{H}^{(1)} = -\mathbf{d} \cdot \mathbf{E}$$

where the dipole moment operator $\mathbf{d} = xq\,\mathbf{e}_x$, q being the effective charge on each atom. Use first-order perturbation theory to derive an expression for the probability that a diatomic, initially in state $n = 0$, is found in state $n = 1$ at time $t = \infty$. Estimate the magnitude of this probability for HCl (dipole moment $= 1.08$ D and bond length $= 1.28$ Å) and $E_0 = 10^5$ V cm^{-1}, $\tau = 1$ μs.

2.31.[C] Write a computer program to plot $F(t, \omega)$ defined by eq. (2.10-13) as a function of ω for two different values of t. Determine the extrema. In particular, find the minima analytically. Show that the absolute maximum is at $\omega = 0$, where $F(t, 0) = t^2$. Find the first two maxima in the wings approximately numerically and show that these are $< 5\%$ of the height of the central peak.

REFERENCES

Dence, J. B., *Mathematical Techniques in Chemistry*, Wiley, New York, 1975, pp. 279–304. Review of basic material on properties of matrices.

Dicke, R. H. and Wittke, J. P., *Introduction to Quantum Mechanics*, Addison-Wesley, Reading, 1960, p. 36. Our argument motivating the form of QM Postulate 7 is taken from here. It is, in fact, the one advanced by Schrödinger in the fourth of a series of his papers that appeared in 1926 [*Ann. Physik.* **81**, 109 (1926)].

Hoskins, R. F., *Generalized Functions*, Ellis Horwood, Chichester, 1979. A slightly more complete treatment of generalized functions than that in Lighthill's slim book (see the following reference), but still at a very accessible level.

Lighthill, M. J., *An Introduction to Fourier Analysis and Generalised Functions*, Cambridge University Press, Cambridge, 1970. Read pages 1–29 for a short introduction to some basic theory of the Dirac delta function and other generalized functions; Lighthill's definition of the Dirac delta function is contained in his Definition 5 and Example 6 on p. 17.

Newton, R. G., "Probability Interpretation of Quantum Mechanics," *Am. J. Phys.*, **48**, 1029 (1980). The nature of measurement and the question of what one can know in quantum mechanics are perhaps the most vexatious interpretative problems in quantum mechanics today. Scan this article for one man's opinion.

Peres, A., "What is a State Vector"? *Am. J. Phys.*, **52**, 699 (1984). A rather abstract article, but at least one sees the point that a measurement according to Peres is not a simple, idealized affair but rather a complex macroscopic event.

Sneddon, I. N., *Special Functions of Mathematical Physics and Chemistry*, 3rd ed., Longman, London, 1980, Chap. 5. This chapter of 25 pages deals with the functions of Hermite and Laguerre; very useful. Chapter 5 is a short appendix on the Dirac delta function.

Spiegel, M. R., *Real Variables*, Schaum's Outline Series, McGraw-Hill, New York, 1969, pp. 73, 156, 166. The indicated pages give the theorem on integration of a function over a set of measure zero, and the mean value theorem for integrals and derivatives, respectively.

Spiegel, M. R., *Advanced Mathematics for Engineers and Scientists*, Schaum's Outline Series, McGraw-Hill, New York, 1971, Chap. 15. Discussion of properties and theorems on matrices at the same level as Dence.

CHAPTER 3

Approximation Methods in Quantum Chemistry

The principles given in Chapter 2 are nearly sufficient for estimating the properties of atoms and molecules. Exact results were summarized in Table 2.1 for a number of simple and idealized systems. No guidance was given, however, on how to deal with complex systems that cannot be solved exactly. In this chapter we consider methods for obtaining approximate eigenvalues and eigenfunctions of the Hamiltonian operator for such systems.

3.1 TIME-INDEPENDENT PERTURBATION THEORY

We assume that the Hamiltonian can be written in the form

$$\hat{H} = \hat{H}^{(0)} + \hat{H}^{(1)} + \hat{H}^{(2)} + \cdots + \tag{3.1-1}$$

where the zero-order term $\hat{H}^{(0)}$ is in some sense "large" compared with the higher-order terms $\hat{H}^{(1)}$, $\hat{H}^{(2)}$, and so on, regarded as **perturbations**. We further assume that the operator $\hat{H}^{(0)}$ has a set of *known*, *nondegenerate* eigenvalues $\{E_i^{(0)}\}$ (the restriction of nondegeneracy can be removed when necessary). Since the terms $\hat{H}^{(1)}$, $\hat{H}^{(2)}$, and so on, are small, we expect the eigenvalue E_i of $\hat{H}$ to lie in the neighborhood of $E_i^{(0)}$ and the eigenfunction $\Psi_i(q)$ of $\hat{H}$ to be similar to the eigenfunction $\Psi_i^{(0)}(q)$ of $\hat{H}^{(0)}$.

To keep track of the different orders of approximation, we introduce a dimensionless parameter λ, whose value is restricted to the continuous interval $0 \leq \lambda \leq 1$.

$$\hat{H}(\lambda) = \hat{H}^{(0)} + \lambda\hat{H}^{(1)} + \lambda^2\hat{H}^{(2)} + \cdots + \tag{3.1-2}$$

Eigenvalues and eigenfunctions of the Hamiltonian in eq. (3.1-2) are then expressed as power series in λ, thus giving successively higher-order approximations to the true eigenvalues and eigenfunctions of $\hat{H}(\lambda)$. Later, upon setting $\lambda = 1$, we obtain approximate solutions to the original Hamiltonian in eq. (3.1-1) (Epstein, 1954).

Let $\Psi_i(q, \lambda)$ be an eigenfunction of $\hat{H}(\lambda)$ in eq. (3.1-2) that corresponds to an arbitrary choice of λ, and let $E_i(\lambda)$ be the eigenvalue. The Maclaurin series expansions (presumed convergent) of $\Psi_i(q, \lambda)$ and $E_i(\lambda)$ in powers of λ are

$$\Psi_i(q, \lambda) = \Psi_i^{(0)}(q) + \left(\frac{\partial \Psi_i(q, \lambda)}{\partial \lambda}\right)_{\lambda=0} \lambda + \sum_{k=2}^{\infty} \frac{1}{k!} \left(\frac{\partial^k \Psi_i(q, \lambda)}{\partial \lambda^k}\right)_{\lambda=0} \lambda^k$$

$$\text{(3.1-3a)}$$

$$E_i(\lambda) = E_i^{(0)} + \left(\frac{\partial E_i(\lambda)}{\partial \lambda}\right)_{\lambda=0} \lambda + \sum_{k=2}^{\infty} \frac{1}{k!} \left(\frac{\partial^k E_i(\lambda)}{\partial \lambda^k}\right)_{\lambda=0} \lambda^k \quad \text{(3.1-3b)}$$

These equations indicate that we need the various derivatives of $\Psi_i(q, \lambda)$ and $E_i(\lambda)$.

Now consider the identity

$$E_k(\lambda) \delta_{ki} = \int \Psi_k^*(q, \lambda) \hat{H}(\lambda) \Psi_i(q, \lambda) \, dq \quad \text{(3.1-4)}$$

which follows from the fact that the $\Psi_i(q, \lambda)$ are members of a complete orthonormal set. This yields upon differentiation

$$\frac{\partial E_k}{\partial \lambda} \delta_{ki} = \int \frac{\partial \Psi_k^*(q, \lambda)}{\partial \lambda} \hat{H}(\lambda) \Psi_i(q, \lambda) \, dq + \int \Psi_k^*(q, \lambda) \hat{H}(\lambda) \frac{\partial \Psi_i(q, \lambda)}{\partial \lambda} \, dq$$
$$+ \int \Psi_k^*(q, \lambda) \frac{\partial \hat{H}(\lambda)}{\partial \lambda} \Psi_i(q, \lambda) \, dq$$

In the first integral, we simply replace $\hat{H}(\lambda)\Psi_i(q, \lambda)$ by $E_i(\lambda)\Psi_i(q, \lambda)$. In the second integral, we make use of the Hermiticity of $\hat{H}(\lambda)$ and write

$$\int \Psi_k^*(q, \lambda) \hat{H}(\lambda) \frac{\partial \Psi_i(q, \lambda)}{\partial \lambda} \, dq = \int \frac{\partial \Psi_i(q, \lambda)}{\partial \lambda} \hat{H}^*(\lambda) \Psi_k^*(q, \lambda) \, dq$$

Consequently, we have from eq. (3.1-4)

$$\frac{\partial E_k(\lambda)}{\partial \lambda} \delta_{ki} = E_i(\lambda) \int \frac{\partial \Psi_k^*(q, \lambda)}{\partial \lambda} \Psi_i(q, \lambda) \, dq$$
$$+ E_k(\lambda) \int \Psi_k^*(q, \lambda) \frac{\partial \Psi_i(q, \lambda)}{\partial \lambda} \, dq$$
$$+ \int \Psi_k^*(q, \lambda) \frac{\partial \hat{H}(\lambda)}{\partial \lambda} \Psi_i(q, \lambda) \, dq \quad \text{(3.1-5)}$$

However, since $\Psi_i(q, \lambda)$ and $\Psi_k(q, \lambda)$ are orthonormal, then

$$\int \frac{\partial \Psi_k^*(q, \lambda)}{\partial \lambda} \Psi_i(q, \lambda)\, dq = \frac{\partial}{\partial \lambda} \int \Psi_k^*(q, \lambda) \Psi_i(q, \lambda)\, dq$$

$$- \int \Psi_k^*(q, \lambda) \frac{\partial \Psi_i(q, \lambda)}{\partial \lambda}\, dq$$

$$= \frac{\partial}{\partial \lambda} \delta_{ki} - \int \Psi_k^*(q, \lambda) \frac{\partial \Psi_i(q, \lambda)}{\partial \lambda}\, dq$$

$$= - \int \Psi_k^*(q, \lambda) \frac{\partial \Psi_i(q, \lambda)}{\partial \lambda}\, dq$$

Making this replacement in eq. (3.1-5), we arrive at

$$\frac{\partial E_k(\lambda)}{\partial \lambda} \delta_{ki} = \{ E_k(\lambda) - E_i(\lambda) \} \int \Psi_k^*(q, \lambda) \frac{\partial \Psi_i(q, \lambda)}{\partial \lambda}\, dq$$

$$+ \int \Psi_k^*(q, \lambda) \frac{\partial \hat{H}(\lambda)}{\partial \lambda} \Psi_i(q, \lambda)\, dq \qquad (3.1\text{-}6)$$

In eq. (3.1-6), consider first the case $k = i$. The result,

$$\frac{\partial E_i(\lambda)}{\partial \lambda} = \int \Psi_i^*(q, \lambda) \frac{\partial \hat{H}(\lambda)}{\partial \lambda} \Psi_i(q, \lambda)\, dq \qquad (3.1\text{-}7)$$

is known as the **Hellmann–Feynman theorem**, or sometimes as the **Feynman theorem** (Feynman, 1939; Musher, 1966), and it is true for arbitrary λ. In particular, when $\lambda = 0$ one has

$$\left(\frac{\partial E_i(\lambda)}{\partial \lambda} \right)_{\lambda = 0} = \int \Psi_i^{*(0)}(q) \left(\frac{\partial H(\lambda)}{\partial \lambda} \right)_{\lambda = 0} \Psi_i^{(0)}(q)\, dq$$

$$= \int \Psi_i^{*(0)}(q) \hat{H}^{(1)} \Psi_i^{(0)}(q)\, dq$$

from eq. (3.1-2). Substitution of this into eq. (3.1-3b) then gives to first order (after setting $\lambda = 1$)

$$E_i \simeq E_i^{(0)} + \int \Psi_i^{*(0)}(q) \hat{H}^{(1)} \Psi_i^{(0)}(q)\, dq \qquad (3.1\text{-}8)$$

This important result says that *to find the first-order correction to the energy, one needs only the zero-order wave function* (Fig. 3.1).

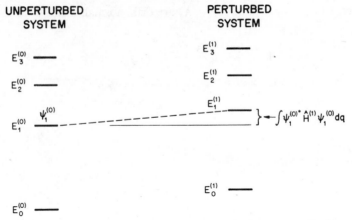

FIGURE 3.1. Perturbation to first order of a nondegenerate energy level.

To find the first-order correction to the wave function, we may proceed as follows. If $k \neq i$ in eq. (3.1-6), then

$$\int \Psi_k^*(q, \lambda) \frac{\partial \Psi_i(q, \lambda)}{\partial \lambda} \, dq = \frac{\int \Psi_k^*(q, \lambda)(\partial \hat{H}/\partial \lambda)\Psi_i(q, \lambda) \, dq}{E_i(\lambda) - E_k(\lambda)} \quad (3.1\text{-}9)$$

We now replace the derivative on the left-hand side by an expansion in the eigenfunctions $\{\Psi_j(q, \lambda)\}$:

$$\frac{\partial \Psi_i(q, \lambda)}{\partial \lambda} = \sum_j a_j(\lambda)\Psi_j(q, \lambda) \quad (3.1\text{-}10)$$

As usual, to find the $a_j(\lambda)$, we multiply both sides of eq. (3.1-10) by an arbitrary eigenfunction $\Psi_k^*(q, \lambda)$ and integrate over q. Only one term survives on the right, that where $j = k$.

$$a_k(\lambda) = \int \Psi_k^*(q, \lambda) \frac{\partial \Psi_i(q, \lambda)}{\partial \lambda} \, dq \quad (3.1\text{-}11)$$

From the right side of eq. (3.1-9), it would appear that there is trouble with the coefficient $a_i(\lambda)$. However, the following argument shows that we can choose a_i to be zero. Since $\Psi_i(q, \lambda)$ is normalized,

$$\int \Psi_i^*(q, \lambda)\Psi_i(q, \lambda) \, dq = 1$$

independent of λ. Differentiating this expression with respect to λ and rearranging the result yields

$$a_i(\lambda) = -a_i^*(\lambda) = \int \frac{\partial \Psi_i^*}{\partial \lambda} \Psi_i \, dq$$

or a_i is pure imaginary. Simply by choosing the phase of $\Psi_i(q, \lambda)$ to make Ψ_i *real*, we can force $a_i(\lambda)$ to vanish.

Now substitute eq. (3.1-11) into eq. (3.1-10) to get

$$\frac{\partial \Psi_i(q, \lambda)}{\partial \lambda} = \sum_k \Psi_k(q, \lambda) \int \Psi_k^*(q, \lambda) \frac{\partial \Psi_i(q, \lambda)}{\partial \lambda} \, dq$$

$$= \sum_k{}' \frac{\Psi_k(q, \lambda) \int \Psi_k^*(q, \lambda) \dfrac{\partial \hat{H}(\lambda)}{\partial \lambda} \Psi_i(q, \lambda) \, dq}{E_i(\lambda) - E_k(\lambda)} \qquad (3.1\text{-}12)$$

from eq. (3.1-9). The prime on the summation sign indicates omission of the term $k = i$.

Like eq. (3.1-7), eq. (3.1-12) is valid for any λ. In particular, when $\lambda = 0$ we have

$$\left[\frac{\partial \Psi_i(q, \lambda)}{\partial \lambda} \right]_{\lambda=0} = \sum_k{}' \frac{\Psi_k^{(0)}(q) \int \Psi_k^{*(0)}(q) \hat{H}^{(1)} \Psi_i^{(0)}(q) \, dq}{E_i^{(0)} - E_k^{(0)}} \qquad (3.1\text{-}13)$$

Finally, substitution of eq. (3.1-13) into eq. (3.1-3a) yields the wave function to first order (after now setting $\lambda = 1$)

$$\Psi_i(q) \simeq \Psi_i^{(0)}(q) + \sum_k{}' \frac{\Psi_k^{(0)}(q) \int \Psi_k^{*(0)}(q) \hat{H}^{(1)} \Psi_i^{(0)}(q) \, dq}{E_i^{(0)} - E_k^{(0)}} \qquad (3.1\text{-}14)$$

Note that in contrast to E_i [eq. (3.1-8)], which depends only upon a single zero-order wave function, Ψ_i depends on *all* the zero-order wave functions.

As an application of perturbation theory, let us work out the total energy to first order of a one-dimensional **anharmonic oscillator** having the potential energy $V(x) = \frac{1}{2}\mu\omega^2 x^2 - Cx^3 + Dx^4$. This $V(x)$ should be a better approximation to the true potential energy of a diatomic molecule than a simple parabolic potential (Fig. 3.2), at least in the immediate vicinity of the minimum. The energy eigenvalue equation is

$$\left[\underbrace{\left(\frac{-\hbar^2}{2\mu} \frac{d^2}{dx^2} + \frac{1}{2}\mu\omega^2 x^2 \right)}_{\hat{H}^{(0)}} + \underbrace{(-Cx^3 + Dx^4)}_{\hat{H}^{(1)}} \right] \Psi(x) = E \, \Psi(x)$$

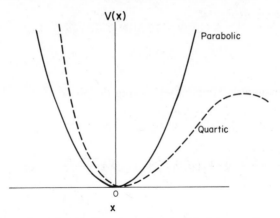

FIGURE 3.2. Comparison of parabolic and quartic potentials for a one-dimensional oscillator.

and the eigenfunctions $\{\Psi_n^{(0)}(x)\}$ of $\hat{H}^{(0)}$ are given in Table 2.1. Then to first order we have from eq. (3.1-8)

$$
E_n = E_n^{(0)} - \frac{C\beta^{1/2}}{2^n n!\sqrt{\pi}} \int_{-\infty}^{\infty} x^3 e^{-\beta x^2} H_n^2(\beta^{1/2}x) \, dx
$$

$$
+ \frac{D\beta^{1/2}}{2^n n!\sqrt{\pi}} \int_{-\infty}^{\infty} x^4 e^{-\beta x^2} H_n^2(\beta^{1/2}x) \, dx
$$

The first integral is zero because $e^{-\beta x^2} H_n^2(\beta^{1/2}x)$ is an even function of x, whereas x^3 is an odd function. Hence, after substituting ξ for $\beta^{1/2}x$ in the second integral, we have

$$
E_n = E_n^{(0)} + \frac{D}{2^n n!\sqrt{\pi}} \left(\frac{\hbar^2}{\mu^2 \omega^2} \right) \int_{-\infty}^{\infty} \xi^4 e^{-\xi^2} H_n^2(\xi) \, d\xi \qquad (3.1\text{-}15)
$$

To evaluate the integral, we make use of the following recursion relation among the Hermite polynomials (Sneddon, 1980):

$$
\xi H_n(\xi) = n H_{n-1}(\xi) + \tfrac{1}{2} H_{n+1}(\xi) \qquad (3.1\text{-}16)
$$

We multiply eq. (3.1-16) by ξ and square both sides.

$$
\xi^4 H_n^2(\xi) = \xi^2 n^2 H_{n-1}^2(\xi) + \tfrac{1}{4}\xi^2 H_{n+1}^2(\xi) + \xi^2 n H_{n-1}(\xi) H_{n+1}(\xi) \qquad (3.1\text{-}17)
$$

The left-hand side of eq. (3.1-17) is the factor needed in the integrand of eq. (3.1-15). Each term on the right-hand side of eq. (3.1-17) can be handled by further use of relation (3.1-16). For example, the first term in eq. (3.1-17) leads to

$$\int_{-\infty}^{\infty} \xi^2 n^2 H_{n-1}^2(\xi) \, e^{-\xi^2} \, d\xi = n^2 \int_{-\infty}^{\infty} e^{-\xi^2} \left[(n-1) H_{n-2}(\xi) + \tfrac{1}{2} H_n(\xi) \right]^2 d\xi$$

$$= n^2 (n-1)^2 \int_{-\infty}^{\infty} e^{-\xi^2} H_{n-2}^2(\xi) \, d\xi$$

$$+ \tfrac{1}{4} n^2 \int_{-\infty}^{\infty} e^{-\xi^2} H_n^2(\xi) \, d\xi$$

$$+ n^2 (n-1) \int_{-\infty}^{\infty} e^{-\xi^2} H_{n-2}(\xi) H_n(\xi) \, d\xi \quad (3.1\text{-}18)$$

Values of the three integrals in eq. (3.1-18) can be written down immediately because the Hermite polynomials form an orthonormal set with respect to the weighting function $e^{-\xi^2}$, that is,

$$\int_{-\infty}^{\infty} H_m(\xi) H_n(\xi) \, e^{-\xi^2} \, d\xi = \left\{ \pi m! n! 2^{n+m} \right\}^{1/2} \delta_{mn}$$

The final result is

$$E_n = E_n^{(0)} + \frac{3(2n^2 + 2n + 1)}{4} \left(\frac{\hbar^2}{\mu^2 \omega^2} \right) D \qquad (3.1\text{-}19)$$

For the ground state ($n = 0$) eq. (3.1-19) becomes

$$E_0 = \frac{1}{2} \hbar \omega + \frac{3 \hbar^2 D}{4 \mu^2 \omega^2}$$

Observation of the fundamental vibration and of the first overtone in the infrared spectrum of a diatomic molecule would permit the evaluation of ω and D.

3.2 SPIN ANGULAR MOMENTUM

In Chapter 1 it was pointed out that a particle moving in a central field, such as the electron of a one-electron atom, possesses an orbital angular momentum ℓ that is a constant of the motion. We now ask: Is orbital angular momentum the only kind of angular momentum possible for a bound electron? The Dutch physicists G. E. Uhlenbeck (b. 1900) and S. Goudsmit (1902–1978) pointed out that certain unexplained features of atomic spectra could be rationalized if an electron were postulated to possess an inherent **spin angular momentum** (Uhlenbeck and Goudsmit, 1926). As a property of the particle, this angular momentum would be independent of the choice of origin of a coordinate

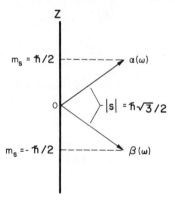

FIGURE 3.3. Electron spin angular momentum has two possible independent projections along a preferred direction in space.

system, and originally was pictured in classical terms as due to the electron's spinning about a hypothetical axis through its "center." This quickly led to difficulties that forced the abandonment of any classical picture.

Accordingly, in the nonrelativistic mechanics that we are presenting here, spin of the electron appears strictly as a *postulate*. By analogy with orbital angular momentum, electron spin angular momentum is represented by a Hermitian operator $\hat{s}$ that obeys the eigenvalue equations

$$\hat{s}^2\phi = s(s+1)\hbar^2\phi \tag{3.2-1a}$$

$$\hat{s}_z\phi = m_s\hbar\phi \tag{3.2-1b}$$

where s is the **spin quantum number** of the electron, ϕ is a spin eigenfunction, and m_s is one of the permitted values (in units of $\hbar$) of the projection of the electron spin angular momentum vector onto the z axis (Fig. 3.3). To obtain agreement with atomic spectra, one must assign a value of $\frac{1}{2}$ to the electron spin quantum number s, and then m_s takes on the possible values of $\pm\frac{1}{2}$.

It is conventional to designate $\alpha(\omega)$ and $\beta(\omega)$ as the spin eigenfunctions [ϕ in eq. (3.2-1)] corresponding to $m_s = +\frac{1}{2}, -\frac{1}{2}$, respectively. The variable ω stands for the "spin configuration," analogous to the spatial configuration q. Since $\alpha(\omega)$ and $\beta(\omega)$ are eigenfunctions of a Hermitian operator, they have the usual property of orthonormality.[1]

$$\int\alpha^*(\omega)\alpha(\omega)\,d\omega = \int\beta^*(\omega)\beta(\omega)\,d\omega = 1 \qquad \int\alpha^*(\omega)\beta(\omega)\,d\omega = 0 \tag{3.2-2}$$

Dirac showed in 1928 that a relativistic theory of one-electron systems could be constructed in which the spin of the electron arises in a natural way

---

[1]The integrations in eq. (3.2-2) are best interpreted symbolically, since the spin configuration ω has no clear physical significance, as does q.

Table 3.1. The Spin Quantum Number of Selected Particles

Class 1[a]		s	Class 2[a]		s
Pi meson	π^0	0^b	Neutrino	ν	$\frac{1}{2}$[b]
Photon	γ	1^b	Neutron	n	$\frac{1}{2}$
Deuteron	H^2	1	Electron	e^-	$\frac{1}{2}$
Nitrogen	N^{14}	1	Proton	p	$\frac{1}{2}$
Chlorine	Cl^{36}	2	Carbon	C^{13}	$\frac{1}{2}$
Boron	B^{10}	3	Boron	B^{11}	$\frac{3}{2}$
Sodium	Na^{24}	4	Oxygen	O^{17}	$\frac{5}{2}$

[a] R. C. Weast and M. J. Astle, Eds., *CRC Handbook of Chemistry and Physics*, 61st ed., CRC Press, Boca Raton, 1980, p. *E-71*.
[b] See footnote (*a*), pp. *F-269* to *F-271*.

mathematically. This theory has led to the common belief that all fundamental particles possess a characteristic spin quantum number. Table 3.1 gives the spin quantum number s of several particles; these values have been deduced experimentally in various ways, including atomic beam resonance measurements, nuclear magnetic resonance, and microwave spectroscopy. All known particles so far have been found to possess spin quantum numbers that are either even multiples of $\frac{1}{2}$, including 0 (Class 1 of Table 3.1), or odd multiples of $\frac{1}{2}$ (Class 2). No particle has a spin quantum number of $\frac{1}{3}$, for example. Particles of Class 1 are called **bosons** (after the Indian physicist S. N. Bose, 1894–1974), and particles of Class 2 are called **fermions** (after the Italian physicist Enrico Fermi, 1901–1954).

3.3 THE FUNDAMENTAL SYMMETRY PRINCIPLE

When two or more identical particles are in close proximity, it becomes impossible to distinguish them because of interference effects arising from the wavelike nature of the particles. It follows that physical observables cannot depend upon the labeling of the particles. This has a far-reaching implication for the wave functions of any system of identical particles.

The wave function for a many-body system is a function of the spatial and spin coordinates of all the particles. We can write such a wave function for N particles symbolically as

$$\Psi(q, \omega) \equiv \Psi(\mathbf{r}_1, \mathbf{r}_2, \ldots, \mathbf{r}_N; \omega_1, \omega_2, \ldots, \omega_N) \equiv \Psi(1, 2, \ldots, N) \quad (3.3\text{-}1)$$

We now interchange the coordinates of, say, particles 1 and 2 in $\Psi(q, \omega)$. Since the particles cannot be distinguished experimentally, the energy must be invariant. This implies that the Hamiltonian must be symmetric with respect to the exchange of the coordinates of the two particles, for otherwise the expectation value of $\hat{H}$ would change.

Now let $\hat{P}_{ij}$ stand for the **permutation operator** that exchanges the sets of coordinates for particles i and j, for example,

$$\hat{P}_{12}\Psi(1,2,3,\ldots,N) = \Psi(2,1,3,\ldots,N) \qquad (3.3\text{-}2)$$

It then follows that

$$\hat{H}\{\hat{P}_{12}\Psi(1,2,3,\ldots,N)\} = \hat{H}\{\Psi(2,1,3,\ldots,N\} = E\Psi(2,1,3,\ldots,N)$$

On the other hand, the quantity $\hat{P}_{12}\{\hat{H}\Psi(1,2,3,\ldots,N)\}$ becomes

$$\hat{P}_{12}\{E\Psi(1,2,3,\ldots,N)\} = E\hat{P}_{12}\Psi(1,2,3,\ldots,N) = E\Psi(2,1,3,\ldots,N)$$

since the energy is invariant under exchange of the particle labels. Hence, we see that $\hat{P}_{ij}$ and $\hat{H}$ commute. The set of eigenfunctions of $\hat{P}_{ij}$ may thus be chosen identical to that of $\hat{H}$,[2] and one can write

$$\hat{P}_{ij}\Psi(1,2,3,\ldots,N) = \lambda\Psi(1,2,3,\ldots,N) \qquad (3.3\text{-}3)$$

Since $\hat{P}_{ij} \cdot \hat{P}_{ij}$ must leave the wave function unchanged (a double exchange returns the system to its original configuration), we have $\lambda^2 = 1$, or $\lambda = \pm 1$.

The operator $\hat{P}_{ij}$ does not depend explicitly upon the time. Equation (2.9-21) can, therefore, be written

$$\frac{\partial\left(\hat{P}_{ij}\right)_H}{\partial t} = i\hbar^{-1}\left[\hat{H},\left(\hat{P}_{ij}\right)_H\right]$$

Further, $\hat{H}$ does not depend explicitly on the time; hence, since it commutes with $\hat{P}_{ij}$, it commutes with $(\hat{P}_{ij})_H$ also, and we infer from this equation that $\hat{P}_{ij}$ must be constant in time. The property of evenness ($\lambda = +1$) or oddness ($\lambda = -1$) under the permutation operator is thus a permanent, inherent property of the system.

With only two possibilities for the eigenvalue λ, a connection with Table 3.1 is strongly suspected. In all cases studied so far, it is found that any system of identical *bosons* has $\lambda = +1$ and any system of identical *fermions* has $\lambda = -1$. These correlations are summarized in the **Fundamental Symmetry Principle**:

Wave functions for systems of identical fermions must be antisymmetric upon exchange of the labels of any two particles, whereas wave functions for systems

---

[2]We are invoking here the theorem that if any two operators $\hat{A}$ and $\hat{B}$ commute, then the eigenfunctions of $\hat{A}$ may be chosen to be identical to those of $\hat{B}$. The proof may be found in many books on quantum mechanics (e.g., see Dicke and Wittke, 1960).

of identical bosons must be symmetric upon exchange of the labels of any two of them.

Since electrons are fermions, electronic wave functions for atoms and molecules must be antisymmetric. This is often called the **Pauli Antisymmetry Principle**, because it was first given in somewhat more restricted form (as the celebrated Exclusion Principle) by the Austrian physicist Wolfgang Pauli (1900–1958). The Fundamental Symmetry Principle is not a postulate, but is regarded as having been proved (by Pauli, himself); the proof is necessarily relativistic in nature and is outside the scope of the book.

Now consider the wave function of eq. (3.3-1) again. In the **orbital approximation** each electron is treated separately in the following sense. For a given electron i, one defines a function of the form

$$\phi(i) = \chi(i)\begin{Bmatrix} \alpha(i) \\ \beta(i) \end{Bmatrix} \tag{3.3-4}$$

where $\chi(i)$ is a function of the spatial configuration q of electron i, and $\alpha(i)$ or $\beta(i)$ is an appropriate spin eigenfunction for that electron. The function $\phi(i)$ can be thought of as an eigenfunction of a one-electron Hamiltonian, (see Section 2.6) and in the case of atoms, $\phi(i)$ has been termed by the American physical chemist Robert S. Mulliken (1896–1986) an **atomic spin-orbital** (Mulliken, 1932). John C. Slater (American physicist, 1900–1976) showed that for an atom a wave function obeying the Pauli Antisymmetry Principle can be constructed by forming an appropriate determinant of atomic spin-orbitals (Slater, 1929). If the atom contains $2N$ electrons, then the determinantal function is

$$\Psi(1, 2, \ldots, 2N) = \frac{1}{\sqrt{(2N)!}}$$

$$\times \begin{bmatrix} \chi_a(1)\alpha(1) & \chi_a(1)\beta(1) & \chi_b(1)\alpha(1) & \chi_b(1)\beta(1) \cdots \chi_N(1)\beta(1) \\ \chi_a(2)\alpha(2) & \chi_a(2)\beta(2) & \chi_b(2)\alpha(2) & \chi_b(2)\beta(2) \cdots \chi_N(2)\beta(2) \\ \chi_a(3)\alpha(3) & \chi_a(3)\beta(3) & \chi_b(3)\alpha(3) & \chi_b(3)\beta(3) \cdots \\ \vdots & \vdots & \vdots & \vdots & \vdots \\ \chi_a(2N)\alpha(2N) & \chi_a(2N)\beta(2N) & \chi_b(2N)\alpha(2N) & \chi_b(2N)\beta(2N)\chi_N(2N)\beta(2N) \end{bmatrix}$$

$$\tag{3.3-5}$$

and is known as a **Slater determinant**. It is a normalized wave function provided the spin-orbitals $\phi(i)$ are orthonormal. An abbreviated way of writing a Slater determinant is

$$\Psi(1, 2, \ldots, 2N) = |\chi_a \bar{\chi}_a \chi_b \bar{\chi}_b \cdots \chi_N \bar{\chi}_N|$$

where the horizontal bars indicate β spin functions and the vertical lines mean formation of the determinant and multiplication by $[(2N)!]^{-1/2}$.

Note the pattern of a Slater determinant. In a given row there are all possible spin-orbitals for a given electron, while in a given column there are all possible electrons for a given spin-orbital. Note also that a function such as eq. (3.5-5) is *not* an exact, or even necessarily a very good, wave function; only the correctness of its symmetry is guaranteed. That eq. (3.3-5) is antisymmetric follows because exchange of two electrons amounts to interchanging the corresponding rows.[3]

The most spectacular consequence of the Pauli Antisymmetry Principle is its effect on the energies (and hence, the chemistry) of atoms and molecules. Consider the ground state of the lithium atom. If the Pauli Antisymmetry Principle were not operative, the electronic configuration would be $(1s)^3$. We may estimate the energy of such an hypothetical atom in the following way. Exclusive of the interelectronic repulsions, the Hamiltonian of the atom is

$$\hat{H}^{(0)} = -\sum_{i=1}^{3}\left(\frac{\hbar^2}{2m}\nabla_i^2 + \frac{3e^2}{4\pi\varepsilon_0 r_i}\right) \tag{3.3-6}$$

and a ground-state eigenfunction ($\Psi_0^{(0)}(q)$) and eigenvalue ($E_0^{(0)}$) are

$$\Psi_0^{(0)}(1,2,3) = \chi_{1s}(1)\chi_{1s}(2)\chi_{1s}(3)\alpha(1)\beta(2)\alpha(3)$$

$$E_0^{(0)} = -\frac{3Z^2 e^2}{8n^2\pi\varepsilon_0 a_0}$$

$$= -\frac{27}{8\pi}\frac{e^2}{\varepsilon_0 a_0} \quad (-35{,}444 \text{ kJ mol}^{-1})$$

We now regard the interelectronic repulsions, which are necessarily present, as a perturbation on this hypothetical system.

$$\hat{H}^{(1)} = \frac{e^2}{4\pi\varepsilon_0}\sum_{i=1}^{2}\sum_{j>i}^{3}\frac{1}{r_{ij}} \tag{3.3-7}$$

The first-order approximation to the energy is then calculated from eq. (3.1-8). We omit the details; the final result is (Levine, 1983)

$$E_0 = E_0^{(0)} + \frac{45}{32\pi}\frac{e^2}{\varepsilon_0 a_0}$$

$$= -\frac{63}{32\pi}\frac{e^2}{\varepsilon_0 a_0} \quad (-20{,}674 \text{ kJ mol}^{-1})$$

[3]A basic theorem in the theory of determinants states that the interchange of any two rows or of any two columns in a determinant multiplies the value of the original determinant by -1 (Lipschutz, 1968).

We now repeat these steps, taking into account the Pauli Antisymmetry Principle. For the hypothetical three-electron atom in its ground level and with no interelectronic repulsions, an eigenfunction of the Hamiltonian in eq. (3.3-6) is given by the Slater determinant

$$\Psi_0^{(0)}(1, 2, 3) = |\chi_{1s}(1)\bar{\chi}_{1s}(2)\chi_{2s}(3)| \qquad (3.3\text{-}8)$$

The associated eigenvalue is

$$E_0^{(0)} = 2 \times \frac{-3^2 e^2}{8\pi\varepsilon_0 a_0} + \frac{-3^2 e^2}{8(2^2)\pi\varepsilon_0 a_0}$$

$$= -\frac{81}{32\pi} \frac{e^2}{\varepsilon_0 a_0} \qquad (-26{,}581 \text{ kJ mol}^{-1})$$

The first-order correction to $E_0^{(0)}$ is again given by eq. (3.3-7), but evaluation is more involved since the unperturbed wave function $\Psi_0^{(0)}(1, 2, 3)$ now consists of six terms. The result is (Levine, 1983)

$$E_0 = E_0^{(0)} + \frac{5965}{7776\pi} \frac{e^2}{\varepsilon_0 a_0}$$

$$= -\frac{6859}{3888\pi} \frac{e^2}{\varepsilon_0 a_0} \qquad (-18{,}525 \text{ kJ mol}^{-1})$$

Thus, the effect of the Pauli Antisymmetry Principle has been to raise the energy of an otherwise hypothetical atom by about 2150 kJ mol^{-1} (roughly 500 kcal mol^{-1})! The results are shown in Fig. 3.4.

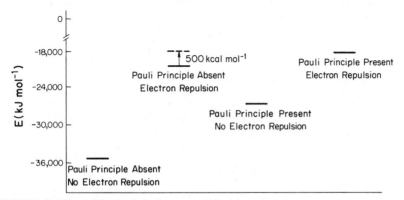

FIGURE 3.4. Effect of the Pauli Antisymmetry Principle and of electron repulsion on the energy of a lithium atom in its ground state.

Exercises[4]

3.1. Let the Hermitian operator $\hat{A}$ and the Hamiltonian $\hat{H}$ both depend on the parameter λ and let $\Psi_i(q, \lambda)$ be a normalized eigenfunction of $\hat{H}$. By considering $\partial\langle\hat{A}\hat{H}\rangle/\partial\lambda$, prove that

$$\left\langle \hat{A}\frac{\partial\hat{H}}{\partial\lambda} \right\rangle = \langle A\rangle\left(\frac{\partial E_i}{\partial\lambda} \right) - \int \Psi_i^*(q, \lambda)[\hat{A}, \hat{H}]\frac{\partial\Psi_i(q, \lambda)}{\partial\lambda}\, dq$$

How is this related to the Feynman theorem, eq. (3.1-7)? See Balasubramanian (1984).

3.2. Verify the result in eq. (3.1-19). For HCl the fundamental vibration and the first overtone occur at 2885.9 and 5668.0 cm^{-1}, respectively. Evaluate ω and D for HCl, deduce the location of the second overtone, and compare with the experimental value 8347 cm^{-1}.

3.3. A particle in a one-dimensional box of length L experiences a "top-heavy" field that adds to the Hamiltonian a perturbation

$$\hat{H}^{(1)} = \begin{cases} \mathscr{E} & \frac{1}{2}L < x \leq L \\ -\mathscr{E} & 0 \leq x < \frac{1}{2}L \end{cases}$$

Using perturbation theory, show that the ground-state wave function to first order is given by (refer to Table 2.1 if necessary)

$$\Psi_1(x) = \sqrt{\frac{2}{L}}\sin\frac{\pi x}{L} - \frac{32m\mathscr{E}L\sqrt{2L}}{\pi h^2}\sum_{k=2}^{\infty}\frac{k}{(1 - k^2)^2}\cos\frac{k\pi}{2}\sin\frac{k\pi x}{L}$$

Sketch curves for $\Psi_1^{(0)}(x)$ and $\Psi_1(x)$ if $\mathscr{E} = h^2/32mL^2$. In which half of the box is the particle more likely to be found?

3.4.* To derive the second-order correction to the energy, proceed as follows. Start with eq. (3.1-7) and differentiate with respect to λ. This yields three terms. Into two of these terms substitute eq. (3.1-12). Now let $\lambda = 0$ in order to obtain $[\partial^2 E_i(\lambda)/\partial^2\lambda]_{\lambda=0}$ and then insert this into eq. (3.1-3b), set $\lambda = 1$, and obtain finally

$$E_i \simeq E_i^{(0)} + \int \Psi_i^{*(0)}(q)\hat{H}^{(1)}\Psi_i^{(0)}(q)\, dq$$

$$+ \sum_k{}' \frac{\int \Psi_i^{*(0)}(q)\hat{H}^{(1)}\Psi_k^{(0)}(q)\, dq \int \Psi_k^{*(0)}(q)\hat{H}^{(1)}\Psi_i^{(0)}(q)\, dq}{E_i^{(0)} - E_k^{(0)}}$$

$$+ \int \Psi_i^{*(0)}(q)\hat{H}^{(2)}\Psi_i^{(0)}(q)\, dq$$

---

[4]An asterisk after an exercise number indicates a moderately difficult exercise.

3.5.   A diatomic molecule is placed in a constant electric field of strength E and direction parallel to the internuclear axis. We consider only vibration, and let the molecule have a dipole moment that is given by $\mu = \mu_{eq} + q\mathbf{x}$, where q is the magnitude of the effective charge on each atom and $\mathbf{x}$ is the displacement of the internuclear separation from equilibrium. The molecule experiences a perturbation

$$\hat{H}^{(1)} = \mu \cdot \mathbf{E}$$

Using first-order perturbation theory, find the shift in the $0 \rightarrow 1$ vibrational transition frequency. Consider the specific case of HCl, and use $|\mathbf{E}| = 10^7$ V m^{-1}.

3.6.   The formal theory of the spin angular momentum operators was developed by Pauli in 1925. The theory follows from the presumed commutation relations

$$[\hat{s}_x, \hat{s}_y] = i\hbar\hat{s}_z$$
$$[\hat{s}_y, \hat{s}_z] = i\hbar\hat{s}_x$$
$$[\hat{s}_z, \hat{s}_x] = i\hbar\hat{s}_y$$

Prove, using these relations alone, that the operator for the square of the spin angular momentum commutes with $\hat{s}_x$, $\hat{s}_y$, and $\hat{s}_z$. This result says (in view of footnote 2 in Section 3.3) that it is possible to choose eigenfunctions that belong simultaneously to $\hat{s}^2$ and $\hat{s}_z$, that is, $\hat{s}^2\phi = a\hbar^2\phi$ and $\hat{s}_z\phi = m_s\hbar\phi$. It follows that $a \geq m_s^2$. Why?

3.7.   Define the **creation operator** $\hat{s}_+ = \hat{s}_x + i\hat{s}_y$. First show that $[\hat{s}_z, \hat{s}_+] = \hbar\hat{s}_+$. Now multiply both sides of $\hat{s}_z\phi = m_s\hbar\phi$ by $\hat{s}_+$ and make use of the first result to obtain $\hat{s}_z\hat{s}_+ = \hbar(m_s + 1)\hat{s}_+$. For some spin function ϕ_{max} the eigenvalue m_s has a maximum value m_s^{max}, otherwise the inequality in Exercise 3.6 would be violated. It follows that $\hat{s}_+\phi_{max} = 0$. Why?

3.8.   Define the **annihilation operator** $\hat{s}_- = \hat{s}_x - i\hat{s}_y$. Prove that

$$\hat{s}_-\hat{s}_+ = \hat{s}^2 - \hat{s}_z^2 - \hbar\hat{s}_z$$

Using this result, operate on both sides of $\hat{s}_+\phi_{max} = 0$ with $\hat{s}_-$ to obtain $\hat{s}_-\hat{s}_+\phi_{max} = 0$. It follows that $a = m_s^{max}(m_s^{max} + 1)$. Why? For the electron, experiment indicates that $m_s^{max} = \frac{1}{2}$; hence, the eigenvalue of $\hat{s}^2$ is $\frac{3}{4}$, and we recover eq. (3.2-1a).

3.9.   If $\alpha(\omega)$ and $\beta(\omega)$ are the one-electron spin eigenfunctions, deduce (a) expected values of $\hat{s}_x$ and $\hat{s}_y$ for both α and β states, (b) expected values of $\hat{s}_x^2$ and $\hat{s}_y^2$ for both α and β states, (c) $\hat{s}_x\alpha(\omega)$, $\hat{s}_x\beta(\omega)$ (*Hint:* $\hat{s}_x$ is a Hermitian operator), (d) $\hat{s}_y\alpha(\omega)$, $\hat{s}_y\beta(\omega)$, (e) $\int\beta(\omega)\hat{s}\alpha(\omega)\,d\omega$, where $\hat{s} = \hat{s}_x e_x + \hat{s}_y e_y + \hat{s}_z e_z$, and (f) $[\int \beta(\omega)\hat{s}\alpha(\omega)\,d\omega] \cdot [\int \alpha(\omega)\hat{s}\beta(\omega)\,d\omega]$.

3.10. Verify the form and numerical magnitude of the eigenvalue for the eigenfunction $\Psi_0^{(0)}(1, 2, 3)$ indicated following eq. (3.3-6).

3.11. Account for the factor $[(2N)!]^{-1/2}$ that occurs in the Slater determinant of eq. (3.3-5). Can a Slater determinant for a $2N$-electron atom generally be expressed as the product of a function only of spatial coordinates times a function only of spin coordinates? Can one do this in the particular case of helium?

3.12. Consider the 4×4 determinant

$$\begin{bmatrix} 2 & 1 & 0 & 4 \\ 3 & 5 & -1 & -1 \\ 0 & -4 & 2 & -1 \\ 7 & 6 & -3 & 1 \end{bmatrix}$$

Demonstrate, as implied in the text, that interchange of two rows multiplies the value of the original determinant by -1.

3.13. Write out the forms of the integrals that would have to be evaluated in order to obtain E_0 to first order using the unperturbed wave function of eq. (3.3-8).

3.4 THE BORN–OPPENHEIMER APPROXIMATION

For a system of electrons and nuclei the energy eigenvalue equation takes the form [see eq. (2.8-1)]

$$\left(-\sum_\alpha \frac{\hbar^2}{2M_\alpha}\nabla_\alpha^2 - \sum_i \frac{\hbar^2}{2m}\nabla_i^2 + \sum_\alpha \sum_{\beta>\alpha} \frac{Z_\alpha Z_\beta e^2}{4\pi\varepsilon_0 R_{\alpha\beta}} \right.$$

$$\left. -\sum_\alpha \sum_i \frac{Z_\alpha e^2}{4\pi\varepsilon_0 r_{i\alpha}} + \sum_i \sum_{j>i} \frac{e^2}{4\pi\varepsilon_0 r_{ij}} \right) \Psi(Q, q) = E_T \Psi(Q, q) \quad (3.4\text{-}1)$$

Here, α and β label the nuclei and i and j label the electrons; Q stands for the configuration of the nuclei and q for the configuration of the electrons. We emphasize that the wave functions depend on both sets of coordinates. It would be most convenient if eq. (3.4-1) could be reorganized as a problem involving the two sets separately. The German physicist Max Born (1882–1970) and his American student J. Robert Oppenheimer (1904–1967) showed that to a high degree of approximation eq. (3.4-1) could be so treated (Slater, 1963).

Suppose one first solves an equation identical to eq. (3.4-1) but with the nuclear kinetic energy terms omitted.

$$\left(-\sum_i \frac{\hbar^2}{2m}\nabla_i^2 + \sum_\alpha \sum_{\beta > \alpha} \frac{Z_\alpha Z_\beta e^2}{4\pi\varepsilon_0 R_{\alpha\beta}} - \sum_\alpha \sum_i \frac{Z_\alpha e^2}{4\pi\varepsilon_0 r_{i\alpha}}\right.$$

$$\left. + \sum_i \sum_{j>i} \frac{e^2}{4\pi\varepsilon_0 r_{ij}}\right)\chi(Q, q) = \mathscr{E}(Q)\chi(Q, q) \quad (3.4\text{-}2)$$

This amounts to describing the behavior of the electrons at a fixed nuclear configuration Q, and is justified physically by the fact that the much heavier nuclei move more slowly than the electrons. Equation (3.4-2) yields a set of energies $\mathscr{E}(Q)$ that depend parametrically on Q, plus a set of eigenfunctions $\chi(Q, q)$ that depend both on Q (parametrically) and on the continuously varying electronic coordinates q.

Next, for a given $\chi(Q, q)$ we use the corresponding $\mathscr{E}(Q)$ as the potential energy function in an energy eigenvalue equation for the nuclei

$$\left[-\sum_\alpha \frac{\hbar^2}{2M_\alpha}\nabla_\alpha^2 + \mathscr{E}(Q)\right]\phi(Q) = W\phi(Q) \quad (3.4\text{-}3)$$

Now, the eigenvalues W contain no parameters at all and the eigenfunctions $\phi(Q)$ are functions only of the nuclear coordinates. The **adiabatic Born–Oppenheimer approximation** consists in supposing that the E_T of eq. (3.4-1) is given approximately by the W of eq. (3.4-3), and that $\Psi(Q, q) \approx \chi(Q, q)\phi(Q)$.

We now substitute the expression $\Psi(Q, q) = \chi(Q, q)\phi(Q)$ back into eq. (3.4-1). The first term in eq. (3.4-1) requires us to operate on $\chi(Q, q)\phi(Q)$ with the nuclear Laplacian ∇_α^2. From vector calculus there is the very useful identity (Spiegel, 1959)[5]

$$\nabla^2(fg) = f\nabla^2 g + g\nabla^2 f + 2(\nabla f \cdot \nabla g) \quad (3.4\text{-}4)$$

Making the identifications $f \equiv \chi(Q, q)$ and $g \equiv \phi(Q)$, we obtain after division by $\chi(Q, q)\phi(Q)$ and transposition of terms

$$-\sum_\alpha \frac{\hbar^2}{2M_\alpha} \frac{1}{\phi}\nabla_\alpha^2\phi - \underbrace{\sum_\alpha \frac{\hbar^2}{M_\alpha} \frac{1}{\chi\phi}(\nabla_\alpha\chi \cdot \nabla_\alpha\phi) - \sum_\alpha \frac{\hbar^2}{2M_\alpha} \frac{1}{\chi}\nabla_\alpha^2\chi - E_T}_{A}$$

$$= \sum_j \frac{\hbar^2}{2m} \frac{1}{\chi}\nabla_j^2\chi - \{PE\} \quad (3.4\text{-}5)$$

---

[5] The relation follows from the fact that if f and g are two scalar functions, then grad $fg = f$ grad $g + g$ grad f, and if **B** is a vector function, then div $f\mathbf{B} = \mathbf{B} \cdot$ grad $f + f$ div **B**. Finally, $\nabla^2 = $ div grad.

where {PE} stands for the three potential energy terms in the braces of eq. (3.4-2). Comparison of eq. (3.4-5) with eq. (3.4-2) shows that the right-hand side of eq. (3.4-5) is just $-\mathscr{E}(Q)$. Now eq. (3.4-5) does not have the dependent variables χ and ϕ strictly separated, but if the quantity A were negligible, then eq. (3.4-5) would reduce to

$$-\sum_\alpha \frac{\hbar^2}{2M_\alpha} \frac{1}{\phi} \nabla_\alpha^2 \phi + \mathscr{E}(Q) = E_T \qquad (3.4\text{-}6)$$

which becomes identical to eq. (3.4-3) if one approximates E_T by W.

That A is small may be seen as follows. The quantity $\nabla_\alpha^2 \chi(Q, q)$ is of the same order of magnitude as $\nabla_j^2 \chi(Q, q)$ because $\chi(Q, q)$ depends mainly on $(q_j - Q_\alpha)$, the coordinates of the electrons with respect to the nuclei. The quantity $(-\hbar^2/2m)\nabla_j^2\chi(Q, q)$ would then be roughly equal to the kinetic energy of an electron, and so $(-\hbar^2/2M_\alpha)\nabla_\alpha^2(Q, q)$ is smaller than this by the factor m/M_α, a number of the order of magnitude 10^{-3}. A similar argument applies to the $\nabla_\alpha\chi \cdot \nabla_\alpha\phi$ term. We therefore expect A to be negligible compared to E_T. Equation (3.4-2), which includes the nuclear repulsions, is the Born–Oppenheimer energy-eigenvalue equation for the electronic motion and the $\mathscr{E}(Q)$ are the electronic energy eigenvalues. This equation is the usual starting point for molecular electronic structure calculations.

A plot of $\mathscr{E}(Q)$ versus Q for a diatomic molecule (or an analogous plot in higher dimensions for a polyatomic molecule) is referred to as the **potential-energy curve** (or **surface**) for the molecule. This is misleading in one sense because $\mathscr{E}(Q)$ contains the kinetic energy of the electrons. However, the use of the term is justified because *within the context of the Born–Oppenheimer approximation* the $\mathscr{E}(Q)$ plays the role of a potential energy in the nuclear energy–eigenvalue equation (3.4-3). Note that it is the W of eq. (3.4-3) that one should identify as the approximate *total energy* of the molecule. In the usual applications at a fixed nuclear configuration, the reported molecular energy, $\mathscr{E}(Q)$, represents this total energy minus the nuclear kinetic energy. Equation (3.4-3) is the usual starting point for the discussion of the vibrations and rotations of molecules (see Chapter 7).

3.5   THE VARIATION THEOREM

Let a system described by a Hamiltonian $\hat{H}$ possess a complete orthonormal set of eigenfunctions $\Psi_0(q), \Psi_1(q), \Psi_2(q), \ldots$, and a corresponding set of eigenvalues $E_0 \le E_1 \le E_2 \le \ldots$. In general, we may not know any of the $\{\Psi_n(q)\}$ or any of the $\{E_n\}$. However, if $\Psi_{tr}(q)$ is a *normalized trial* wave function, it can in principle (see Section 2.4) be expanded in terms of the $\{\Psi_n(q)\}$.

$$\Psi_{tr}(q) = \sum_{k=0}^{\infty} c_k \Psi_k(q) \qquad (3.5\text{-}1)$$

Using the trial wave function, we calculate the expected value $\langle H \rangle$ of the Hamiltonian (QM Postulate 5, Section 2.6).

$$\langle H \rangle = \int \left[\sum_{k=0}^{\infty} c_k^* \Psi_k^*(q) \right] \hat{H} \sum_{k=0}^{\infty} c_k \Psi_k(q) \, dq$$

$$= \sum_{k=0}^{\infty} \sum_{j=0}^{\infty} c_k^* c_j \int \Psi_k^*(q) \hat{H} \Psi_j(q) \, dq$$

$$= \sum_{k=0}^{\infty} \sum_{j=0}^{\infty} c_k^* c_j E_j \, \delta_{kj}$$

$$= \sum_{j=0}^{\infty} |c_j|^2 E_j \tag{3.5-2}$$

Since $|c_j|^2 \geq 0$ for all j and the $\{E_n\}$ are ordered, the following inequality holds,

$$\langle H \rangle = \sum_{j=0}^{\infty} |c_j|^2 E_j \geq E_0 \sum_{j=0}^{\infty} |c_j|^2$$

$$= E_0$$

where the second line follows because $\Psi_{tr}(q)$ is presumed normalized. Thus, we have the **variation theorem**

$$\langle H \rangle = \int \Psi_{tr}^*(q) \hat{H} \Psi_{tr}(q) \, dq \geq E_0 \tag{3.5-3}$$

In words, the variation theorem says that *no trial wave function can yield an expected value for the energy that is less than the true ground-state energy.* This theorem is the basis of an alternative approximation technique to perturbation theory. We emphasize that as stated the theorem pertains only to the exact *ground-state* energy E_0.

The choice of trial wave function is arbitrary, provided it is finite, continuous, single valued, and normalizable. A trial wave function that is too rigid in form may give a decent value for the ground-state energy but poor values for other calculated ground-state properties. In practice, flexibility is introduced into $\Psi_{tr}(q)$ by expressing it as a function of one or more parameters $\lambda_1, \lambda_2, \ldots, \lambda_n$. The optimum choice of these parameters must obey the condition for minimization of the expected value of the Hamiltonian.

$$\frac{\partial}{\partial \lambda_i} \int \Psi_{tr}^*(q) \hat{H} \Psi_{tr}(q) \, dq = 0 \qquad (i = 1, 2, \ldots, n) \tag{3.5-4}$$

From the equations so obtained, one can in principle solve for the

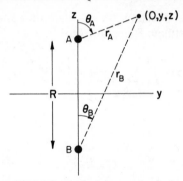

FIGURE 3.5. The hydrogen molecule–ion.

$\lambda_1, \lambda_2, \ldots, \lambda_n$, and thereby find the "best" $\Psi_{tr}(q)$ and the lowest calculated energy.

An instructive illustration of the variation theorem that permits a "double barreled" use of it is provided by the hydrogen molecule–ion H_2^+ (Fig. 3.5). We invoke the Born–Oppenheimer approximation so that the electron is considered to move in the field of two fixed nuclei. A simple trial wave function for H_2^+ in its ground state is the *linear* combination

$$\Psi_{tr}(q) = c_A 1s_A(q_A) + c_B 1s_B(q_B)$$

where $1s_A$ and $1s_B$ are identical, normalized, real hydrogenic atomic orbitals centered on nuclei A and B. Note that the argument $q_A(q_B)$ indicates the electron's position is referred to nucleus A (B) as an origin. The function $\Psi_{tr}(q)$ is not immediately normalized, but becomes so if we divide it by $(c_A^2 + 2c_A c_B S_{AB} + c_B^2)^{1/2}$, where the **overlap integral** S_{AB} is

$$S_{AB} = \int 1s_A(q_A) 1s_B(q_B)\, dq$$

The expected value of the energy is given by

$$\mathscr{E} = \frac{\int [c_A 1s_A(q_A) + c_B 1s_B(q_B)] \hat{H} [c_A 1s_A(q_A) + c_B 1s_B(q_B)]\, dq}{c_A^2 + 2c_A c_B S_{AB} + c_B^2} \qquad (3.5\text{-}5)$$

Defining the matrix elements

$$H_{AA} = H_{BB} = \int 1s_A(q_A) \hat{H} 1s_A(q_A)\, dq$$

$$H_{AB} = H_{BA} = \int 1s_A(q_A) \hat{H} 1s_B(q_B)\, dq$$

we can rewrite eq. (3.5-5) as

$$\mathscr{E} = \frac{\left(c_A^2 + c_B^2\right)H_{AA} + 2c_A c_B H_{AB}}{c_A^2 + 2c_A c_B S_{AB} + c_B^2}$$

Application of eq. (3.5-4) to this expression ($\lambda_1 = c_A$, $\lambda_2 = c_B$) leads to the simultaneous equations

$$(H_{AA} - \mathscr{E})c_A + (H_{AB} - \mathscr{E}S_{AB})c_B = 0$$
$$(H_{AB} - \mathscr{E}S_{AB})c_A + (H_{AA} - \mathscr{E})c_B = 0 \qquad (3.5\text{-}6)$$

In order for eq. (3.5-6) to have a nontrivial solution, the determinant of the coefficients of the unknowns c_A and c_B must vanish,[6] that is,

$$\begin{bmatrix} H_{AA} - \mathscr{E} & H_{AB} - \mathscr{E}S_{AB} \\ H_{AB} - \mathscr{E}S_{AB} & H_{AA} - \mathscr{E} \end{bmatrix} = 0 \qquad (3.5\text{-}7)$$

The left side of eq. (3.5-7) is referred to as the **secular determinant**. Equation (3.5-7) is satisfied by two values of $\mathscr{E}$, namely,

$$\mathscr{E} = \frac{H_{AA} \pm H_{AB}}{1 \pm S_{AB}} \qquad (3.5\text{-}8)$$

Upon evaluating the matrix elements H_{AA}, H_{AB}, and S_{AB} (which is done shortly in this chapter), it is found that the smaller of the values of $\mathscr{E}$ corresponds to the $+$ sign in eq. (3.5-8). This is the best approximation to the true ground-state energy of H_2^+ attainable with a trial function of the assumed form. The other root [corresponding to the minus signs in eq. (3.5-8)] is an approximation to the energy of the first excited state.

A second use of the variation theorem can be made by now giving explicit form to the atomic orbitals. We write each atomic orbital in terms of an "effective" nuclear charge Z, for example,

$$1s_A(q_A) = \left(\frac{Z^3}{\pi a_0^3}\right)^{1/2} e^{-Zr_A/a_0} \qquad (3.5\text{-}9)$$

where q_A now stands for the spherical polar coordinates relative to nucleus A. This atomic orbital is normalized. The Hamiltonian for the problem, however,

---

[6]A theorem of linear algebra states that a system of linear, homogeneous equations has a nontrivial, that is, nonzero, solution provided that the determinant of the coefficients of the unknowns is zero (Lipschutz, 1968).

must be the correct one and, therefore, will not contain "effective" values for the nuclear charges.

$$\hat{H} = -\frac{\hbar^2}{2m}\nabla^2 - \frac{e^2}{4\pi\varepsilon_0}\left(r_A^{-1} + r_B^{-1} - R^{-1}\right) \tag{3.5-10}$$

Combination of eqs. (3.5-9) and (3.5-10) requires the evaluation of three integrals, which we deal with separately.

First, consider the overlap integral S_{AB}, which takes the form

$$S_{AB} = \frac{Z^3}{\pi a_0^3}\int e^{-Z(r_A+r_B)/a_0}\,dq$$

where the integration variable q is not explicit. Note that as q ranges over all points of three-dimensional space, r_B varies in a way depending on r_A and vice versa. In view of the symmetry inherent in H_2^+, it is natural to use **confocal elliptical coordinates**. We now introduce these.

As shown in Fig. 3.5, the molecule is positioned along the z axis with the nuclei at $(0, 0, \pm\frac{1}{2}R)$. From elementary geometry we have

$$r_A^2 - \left[z - \frac{R}{2}\right]^2 = r_B^2 - \left[z + \frac{R}{2}\right]^2$$

or

$$z = \frac{r_B^2 - r_A^2}{2R}$$

$$= \frac{R}{2}\frac{(r_B + r_A)}{R}\frac{(r_B - r_A)}{R} \tag{3.5-11}$$

We next define two new variables ξ and η,

$$\xi = \frac{r_B + r_A}{R} \qquad 1 \leq \xi < \infty \tag{3.5-12a}$$

$$\eta = \frac{r_B - r_A}{R} \qquad -1 \leq \eta \leq 1 \tag{3.5-12b}$$

From Fig. 3.5 the z coordinate of an arbitrary point is given by

$$z = \frac{R}{2} + r_A\cos\theta_A = \frac{R}{2}\left(1 + \frac{2r_A}{R}\cos\theta_A\right) \tag{3.5-13}$$

Subtracting eq. (3.5-12b) from (3.5-12a) gives $2r_A/R = \xi - \eta$. Then using this relation in eq. (3.5-13) and equating the two expressions, eqs. (3.5-11) and

(3.5-13), for z, we obtain

$$\tfrac{1}{2}R\xi\eta = \tfrac{1}{2}R\left[1 + (\xi - \eta)\cos\theta_A\right]$$

or

$$\cos\theta_A = \frac{\xi\eta - 1}{\xi - \eta} \tag{3.5-14}$$

It follows that

$$
\begin{aligned}
\sin\theta_A &= \left(1 - \cos^2\theta_A\right)^{1/2} \\
&= \left[1 - \left(\frac{\xi\eta - 1}{\xi - \eta}\right)^2\right]^{1/2} \\
&= \frac{\sqrt{\xi^2 - 1}\,\sqrt{1 - \eta^2}}{\xi - \eta}
\end{aligned}
$$

We have finally for the transformation equations between Cartesian and confocal elliptical coordinates

$$
\begin{aligned}
x &= r_A\sin\theta_A\cos\phi = \tfrac{1}{2}R\sqrt{\xi^2 - 1}\,\sqrt{1 - \eta^2}\,\cos\phi \\
y &= r_A\sin\theta_A\sin\phi = \tfrac{1}{2}R\sqrt{\xi^2 - 1}\,\sqrt{1 - \eta^2}\,\sin\phi \\
z &= \frac{R}{2} + r_A\cos\theta_A = \tfrac{1}{2}R\xi\eta
\end{aligned}
\tag{3.5-15}
$$

The volume element in confocal elliptical coordinates may be shown by the use of Jacobians to be $dq = \tfrac{1}{8}R^3(\xi^2 - \eta^2)\,d\xi\,d\eta\,d\phi$ (Dence, 1975).

The overlap integral then becomes

$$
\begin{aligned}
S_{AB} &= \frac{Z^3}{\pi a_0^3}\int_0^{2\pi}d\phi\int_1^\infty\frac{R^3}{8}e^{-R\xi Z/a_0}\,d\xi\int_{-1}^1(\xi^2 - \eta^2)\,d\eta \\
&= \frac{R^3 Z^3}{2a_0^3}\int_1^\infty\left(\xi^2 - \frac{1}{3}\right)e^{-R\xi Z/a_0}\,d\xi \\
&= e^{-RZ/a_0}\left[1 + \left(\frac{RZ}{a_0}\right) + \frac{1}{3}\left(\frac{RZ}{a_0}\right)^2\right]
\end{aligned}
\tag{3.5-16}
$$

The matrix element H_{AA} is given by

$$H_{AA} = \frac{Z^3}{\pi a_0^3} \int e^{-Zr_A/a_0} \left[\frac{-\hbar^2}{2m} \nabla^2 - \frac{e^2}{4\pi e_0} \left(r_A^{-1} + r_B^{-1} - R^{-1} \right) \right] e^{-Zr_A/a_0} \, dq$$

$$= \frac{Z^3}{\pi a_0^3} \left[\int e^{-Zr_A/a_0} \left(\frac{-\hbar^2}{2m} \nabla^2 - \frac{e^2}{4\pi \varepsilon_0 r_A} \right) e^{-Zr_A/a_0} \, dq \right.$$

$$\left. - \frac{e^2}{4\pi \varepsilon_0} \int r_B^{-1} e^{-2Zr_A/a_0} \, dq \right] + \frac{e^2}{4\pi \varepsilon_0 R}$$

The first integral on the right may be handled by means of spherical polar coordinates because only r_A appears in the integrand. The second integral requires the confocal elliptical coordinate system again because both r_A and r_B appear in the integrand. The final result is

$$H_{AA} = -\frac{\hbar^2}{ma_0^2} \left[-\frac{Z^2}{2} + Z - \frac{a_0}{R} \left(1 + \frac{RZ}{a_0} \right) e^{-2RZ/a_0} \right] \quad (3.5\text{-}17)$$

Finally, H_{AB} is given by

$$H_{AB} = \frac{Z^3}{\pi a_0^3} \left[\int e^{-Zr_A/a_0} \left(\frac{-\hbar^2}{2m} \nabla^2 - \frac{e^2}{4\pi \varepsilon_0 r_B} \right) e^{-Zr_B/a_0} \, dq \right.$$

$$\left. - \frac{e^2}{4\pi \varepsilon_0} \int r_A^{-1} e^{-Z(r_A + r_B)/a_0} \, dq \right] + \frac{e^2 S_{AB}}{4\pi \varepsilon_0 R}$$

Here confocal elliptical coordinates are needed for each term. The final result is

$$H_{AB} = -\frac{\hbar^2}{ma_0^2} \left[Z(2 - Z) e^{-RZ/a_0} \left(\frac{RZ}{a_0} + 1 \right) + S_{AB} \left(\frac{Z^2}{2} - \frac{a_0}{R} \right) \right] \quad (3.5\text{-}18)$$

Upon inserting eqs. (3.5-16)–(3.5-18) into eq. (3.5-8), one expresses the energy as a function of the parameters R and Z. Optimization of these parameters should yield a minimum value for $\mathscr{E}$. In an early paper based on this method, values of $R = 1.06$ Å, $Z = 1.228$, and $\mathscr{E} = -15.78$ eV were obtained (Finkelstein and Horowitz, 1928). Currently accepted best values for R and $\mathscr{E}$ are 1.0569 Å and -16.398 eV (Schaad and Hicks, 1970). The crude treatment using the variation theorem and a very simple wave function yields values that agree quite well with these. A plot of total energy $\mathscr{E}(R)$ as a function of internuclear distance R is shown in Fig. 3.6.

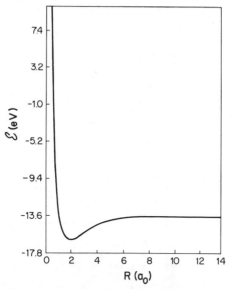

FIGURE 3.6. The total energy curve for the ground state of the hydrogen molecule–ion [calculated from results in H. Wind, *J. Chem. Phys.*, **42**, 2371 (1965)].

3.6   COMPUTER HIGHLIGHT: OPTIMIZATION OF Z FOR H_2^+

Optimization of Z and R for the hydrogen molecule–ion could be done directly by means of eq. (3.5-4). The procedure is a bit messy, however, and the problem is more conveniently handled numerically on a computer. To simplify the problem somewhat, let us take the value of R to be the previously quoted precise value of 1.0569 Å; the optimization then involves a search within a one-variable, rather than two-variable, space.

Figure 3.7 displays a computer program to determine the optimum value of Z and the value of $\mathscr{E}$ corresponding to it. The procedure employed is to search for a minimum in $\mathscr{E}$ over a coarse grid of Z values, and then successively to refine this grid by subdividing the interval within which each succeeding minimum occurs. Thus, to begin, an interval (0.6, 1.6) with Z increasing in jumps of 0.1 is examined. This yields a minimum in $\mathscr{E}$ at $Z = 1.2$. Next, an interval 0.1 on either side of this (1.1–1.3) is subdivided 10-fold and examined. This leads to an energy minimum at $Z = 1.24$. Two further refinements are carried out. Statement Number 375 in the program allows one to look at the results after each set of calculations and to decide whether or not to continue. If further calculation is deemed appropriate, then CONT (for "continue") is typed in and an $N \neq 1$ is entered for the input of line Number 400. The computer is thus instructed to return to line Number 125 and new limits on Z are entered.

```
50   REM   PROGRAM TO OPTIMIZE EF-
           FECTIVE NUCLEAR CHARGE AND
           TO CALCULATE GROUND-STATE
           ENERGY OF H2-MOLECULE-ION
75   INPUT R,AO
80   PRINT : PRINT "MOLECULAR ENER
           GY AS A FUNCTION OF Z": PRINT
           : PRINT
85   PRINT   TAB( 6);"Z"; TAB( 20);
           "E(EV)"
90   PRINT   TAB( 5);"---"; TAB( 20
           );"-----": PRINT
100  REM   R  IS BOND DISTANCE, AO
           IS BOHR RADIUS, EACH IN ANG-
           STROMS
125  INPUT L1,L2: PRINT
150  REM   L1,L2 ARE THE LOWER AND
           UPPER LIMITS ON THE RANGE
           OF Z
160  REM   E IS COMPUTED IN UNITS
           OF H^2/(4*M*AO^2*PI^2), IN
           WHICH M IS MASS OF ELECTRON
           AND H IS PLANCK'S CONSTANT
170  REM   H1=H(AA) AND H2=H(AB)
175  FOR Z = L1 TO L2 STEP (0.1 *
           (L2 - L1))
200  H1 =  - 0.5 * Z * Z + Z - AO *
           R ^  - 1 * (1 + R * Z / AO) *
           EXP ( - 2 * R * Z / AO)
225  S = (1 + R * Z / AO + (R * Z /
           AO) ^ 2 / 3) * EXP ( - R *
           Z / AO)
250  H2 = Z * (2 - Z) * (1 + R * Z
           / AO) *  EXP ( - R * Z / AO
           ) + S * (0.5 * Z * Z - AO /
           R)
275  E =  - 27.211 * (H1 + H2) / (
           1 + S)
300  PRINT   TAB( 6);Z; TAB( 16);E

310  REM   E IS PRINTED OUT IN EV
325  NEXT Z
350  PRINT
375  STOP : REM   HALTS EXECUTION
400  INPUT N
425  IF N = 1 THEN 475
450  GOTO 125
475  END
```

FIGURE 3.7. Computer program for the optimization of the effective nuclear charge in a variational treatment of the hydrogen molecule–ion.

After four divisions, the final result obtained is $Z = 1.239$ and $\mathscr{E} = -15.959$ eV. The curve in Fig. 3.6 is asymptotic to a horizontal line at $\mathscr{E} = -13.598$ eV, which corresponds to the energy of a ground-state hydrogen atom. Thus, H_2^+ in its ground electronic state dissociates to a proton plus a ground-state hydrogen atom. The true dissociation energy of H_2^+ is $16.398 - 13.598 = 2.800$ eV. Our variational calculation has accounted for $((15.959 - 13.598)/2.800) \times 100\%$ or 84.3% of this dissociation energy.

Exercises[7]

3.14. Work through the steps leading to eq. (3.4-5). Along the way, prove the relation in eq. (3.4-4).

3.15.* Consider a particle in a one-dimensional box of length L subjected to the perturbation given in Exercise 3.3. Show that first-order perturbation theory provides no correction to the ground-state energy, but that second-order perturbation theory does (see Exercise 3.4). For comparison, now estimate the ground-state energy using the variation theorem. Take as the trial wave function $\Psi_{tr}(x) = \sqrt{2/L}\,[\sin(\pi x/L) + \lambda \sin(2\pi x/L)]$, where the parameter λ is to be varied.

3.16. Verify the form of the integrals in eqs. (3.5-16) and (3.5-17).

3.17.[C] Execute the program listed in Fig. 3.7 in order to confirm the value of Z given in the text. With this value of Z calculate the energy of the first-excited state of H_2^+ at the same internuclear distance R. Is this state predicted to be stable?

3.18.[C] Select 10 values of R from 0.5 to 3 Å and for each calculate the ground-state energy. In general, a different value of Z may be needed for each R. Plot $\mathscr{E}(R)$ versus R; what is the optimum value of R?

3.19.[C]** Investigate the effect on the ground-state energy of H_2^+ of taking as the trial wave function

$$\Psi_{tr}(q) = c_1\big[1s_A(q_A) + 1s_B(q_B)\big] + c_2\big[2s_A(q_A) + 2s_B(q_B)\big]$$

The $2s$ orbitals have the form

$$2s = \left(\frac{Z^5}{96\pi a_0^5}\right)^{1/2} r\, e^{-Zr/2a_0}$$

For the $1s$ orbitals retain the value $Z = 1.239$, but for the $2s$ orbitals let the common value of their Z be a parameter to be optimized.

REFERENCES

Balasubramanian, S., "Notes on Feynman's Theorem," *Am. J. Phys.*, **52**, 1143 (1984). There appears to be a sign error in eq. (4).

Dence, J. B., *Mathematical Techniques in Chemistry*, Wiley, New York, 1975, pp. 109, 153, and 342–343. See these pages for discussion of the confocal elliptical coordinate system.

[7]A double asterisk after an exercise number indicates either a challenging exercise or an exercise requiring extensive work.

Dicke, R. H. and Wittke, J. P., *Introduction to Quantum Mechanics*, Addison-Wesley, Reading, 1960, p. 97. A proof is given here of the theorem on eigenfunctions of commuting operators.

Epstein, S. T., "Note on Perturbation Theory," *Am. J. Phys.*, **22**, 613 (1954). The author derives first- and second-order perturbation theory in a manner different from that presented in standard texts. Our Section 3.1 is based on this approach.

Feynman, R. P., "Forces in Molecules," *Phys. Rev.*, **56**, 340 (1939). This very accessible paper was written by the now eminent Nobelist in physics when he was an undergraduate at MIT.

Finkelstein, B. N. and Horowitz, G. E., "Über die Energie des He-Atoms und des positiven H_2-Ions im Normalzustande," *Z. Phys.*, **48**, 118 (1928). Minimization of the energy of H_2^+ by variation of the bond distance R and the effective nuclear charge Z.

Levine, I. N., *Quantum Chemistry*, 3rd ed., Allyn and Bacon, Boston, 1983, pp. 245 and 251–252. These pages provide more detail on the lithium atom calculation mentioned in our Section 3.3.

Lipschutz, S., *Linear Algebra*, Schaum's Outline Series, McGraw-Hill, New York, 1968, pp. 174 and 177. The first page states the theorem regarding the value of a determinant when rows or columns are interchanged, and the second page discusses the application of determinants to systems of linear equations. A useful volume.

Mulliken, R. S., "Electronic Structures of Polyatomic Molecules and Valence. II. General Considerations," *Phys. Rev.*, **41**, 49 (1932). A classic and very readable paper on the description of electronic structures of polyatomic molecules in terms of atomic and molecular orbitals. Most of this article forms the basis of qualitative lectures on covalent bonding in physical chemistry courses today.

Musher, J. I., "Comment on Some Theorems of Quantum Chemistry," *Am. J. Phys.*, **34**, 267 (1966). An interesting thumbnail history of Feynman's theorem, including some remarks on the German scientist Hellmann.

Schaad, L. J. and Hicks, W. V., "Equilibrium Bond Length in H_2^+," *J. Chem. Phys.*, **53**, 851 (1970). Accurate calculation of $\mathscr{E}$ and R for H_2^+.

Slater, J. C., "Theory of Complex Spectra," *Phys. Rev.*, **34**, 1293 (1929). Introduction of determinantal wave functions into quantum theory.

Slater, J. C., *Quantum Theory of Molecules and Solids*, Vol. 1, McGraw-Hill, Inc., New York, 1963, pp. 9–13. Our discussion of the Born–Oppenheimer approximation is based on these pages.

Sneddon, I. N., *Special Functions of Mathematical Physics and Chemistry*, 3rd ed., Longman, London, 1980, Chap. 5. The recursion relation needed is derived on p. 149.

Spiegel, M., *Vector Analysis*, Schaum, New York, 1959, Chap. 4. See the first several pages in this chapter on the gradient, divergence, and curl.

Uhlenbeck, G. E. and Goudsmit, S., "Spinning Electrons and the Structure of Spectra," *Nature* (*London*) **117**, 264 (1926). Authors point out that the experimentally observed Zeeman effect of atoms can be accounted for if the ratio of the magnetic moment to the spin angular momentum of an electron is twice what one would calculate for orbital motion.

CHAPTER 4

Energy States of Molecules

Chapter 3 presented the basic idea behind the variational method of finding approximate eigenvalues and eigenfunctions. A specialized, systematic technique for atoms and molecules is the **Hartree–Fock theory**, which is based on some very early intuitive work by D. R. Hartree in England in 1928, later improved by V. Fock in Russia in 1930, and by others. In this chapter we outline the procedure for closed-shell molecules of $2N$ electrons; more detailed discussions can be found elsewhere (Blinder, 1965).

4.1 HARTREE–FOCK THEORY

We invoke the Born–Oppenheimer approximation and consider only the total *electronic energy* E of a molecule with some fixed nuclear configuration. This energy is the total energy E_T of the molecule [see eq. (3.4-1)] minus the nuclear kinetic and potential energies

$$E = E_T - E_{\text{nucl}}$$

$$= \mathscr{E}(Q) - \sum_\alpha \sum_{\beta > \alpha} \frac{Z_\alpha Z_\beta e^2}{4\pi\varepsilon_0 R_{\alpha\beta}}$$

where α and β label the nuclei. The Hamiltonian can be partitioned into one-electron ($\hat{H}_1$) and two-electron ($\hat{H}_2$) contributions as

$$\hat{H} = \hat{H}_1 + \hat{H}_2$$

$$= \sum_{i=1}^{2N} \left(-\frac{\hbar^2}{2m}\nabla_i^2 - \sum_\alpha \frac{Z_\alpha e^2}{4\pi\varepsilon_0 r_{i\alpha}} \right) + \sum_{i=1}^{2N-1} \sum_{j>i} \frac{e^2}{4\pi\varepsilon_0 r_{ij}} \qquad (4.1\text{-}1)$$

where i and j label electrons.

Let us now take as a trial ground-state wave function, $\Psi(1, 2, \ldots, 2N)$, for the molecule a single $2N \times 2N$ Slater determinant, as in eq. (3.3-5). Then the contribution to E from $\hat{H}_1$ may be reasoned as follows. The operator $\hat{H}_1$ consists of $2N$ identical one-electron operators; each such operator represents the kinetic energy plus the potential energy of an electron in the field of the stationary nuclei. Since the $2N$ electrons are indistinguishable, the energy may be computed as a sum of terms, one for each electron, in which the electrons have been distributed among N spatial **molecular orbitals** χ_n with two electrons per spatial orbital.

$$\int \Psi^*(1, 2, \ldots, 2N) \hat{H}_1 \Psi(1, 2, \ldots, 2N) \, dq$$

$$= 2 \sum_{n=1}^{N} \int \chi_n^*(i) \left(\frac{-\hbar^2}{2m} \nabla_i^2 - \sum_{\alpha} \frac{Z_\alpha e^2}{4\pi\varepsilon_0 r_{i\alpha}} \right) \chi_n(i) \, dq_i$$

$$= 2 \sum_{n=1}^{N} H_{nn} \tag{4.1-2}$$

Here, $dq_i = d\mathbf{r}_i$ represents the volume element for a typical one (i) of the $2N$ electrons.

The contribution to E from $\hat{H}_2$ is due to repulsions between electrons. In physical terms, we may think of each such repulsion as arising from the interaction between an electron i in charge cloud $[\chi_n(i)]^2$ and an electron j in charge cloud $[\chi_m(j)]^2$ (see Fig. 4.1). For a given pair of spatial orbitals (χ_n, χ_m), four contributions are possible since in each orbital the electron may be either of α or β spin. The factor of 4 is taken into account if we write the total repulsion as the double sum

$$\sum_{n=1}^{N} \sum_{m=1}^{N} 2 \int\int \chi_n^*(i) \chi_n(i) \frac{e^2}{4\pi\varepsilon_0 r_{ij}} \chi_m^*(j) \chi_m(j) \, dq_i \, dq_j$$

$$= \sum_{n=1}^{N} \sum_{m=1}^{N} 2 J_{nm} \tag{4.1-3}$$

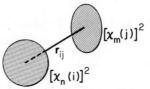

FIGURE 4.1. The interaction of two charge clouds leads to a term J_{nm} in the expectation value of $\hat{H}_2$.

One factor of 2 is accounted for because both J_{nm} and J_{mn}, which are equal by symmetry, are tallied; an additional factor of 2 appears explicitly in eq. (4.1-3). The integral J_{nm} is referred to as a **Coulomb integral**.

However, there is a correlation between electrons of like spin because the Pauli Antisymmetry Principle prevents two such electrons from occupying the same spatial orbital (see Section 3.3). The repulsion energy calculated intuitively above has implicitly ignored the Pauli Principle. Qualitatively, the extent of the correlation between two electrons in spatial orbitals χ_n and χ_m would be expected to be proportional to the overlap of the orbitals. Hence, the correction for spin correlation that must be applied to eq. (4.1-3) takes the form

$$- \sum_{n=1}^{N} \sum_{m=1}^{N} \int \int \chi_n^*(i) \chi_m^*(j) \frac{e^2}{4\pi\varepsilon_0 r_{ij}} \chi_m(i) \chi_n(j) \, dq_i \, dq_j$$

$$= - \sum_{n=1}^{N} \sum_{m=1}^{N} K_{nm} \tag{4.1-4}$$

No factor of 2 is needed here because we are considering the spins to be $\alpha\alpha$ or $\beta\beta$, but not $\alpha\beta$ or $\beta\alpha$. An integral of the type K_{nm} is called an **exchange integral**.

Equations (4.1-3) and (4.1-4) together constitute the expected value of the operator $\hat{H}_2$. The Hartree–Fock electronic energy for a closed-shell molecule of $2N$ electrons in its ground state, that is, the expected value of $\hat{H}$, may then be written as

$$E = \sum_{n=1}^{N} \left(2H_{nn} + \sum_{m=1}^{N} (2J_{nm} - K_{nm}) \right) \tag{4.1-5}$$

From the variation theorem, eqs. (3.5-3) and (3.5-4), the minimum in E is found by varying the N spatial molecular orbitals in eq. (3.3-5). Because no explicit forms have yet been given to the molecular orbitals, variation here means *variation of the functional form of the orbitals* and not merely differentiation with respect to simple independent variables. This kind of extremum problem belongs to a branch of mathematics called the **calculus of variations**, a thorough discussion of which is outside the scope of this book (Boas, 1983; Weinstock, 1974). The problem is compounded in that the $\{\chi_n\}$ cannot be varied independently because in order for eq. (4.1-5) to be correct, one must have started with a Slater determinant of the form given in eq. (3.3-5). That form presumes that the spatial molecular orbitals are orthonormal. Thus, one has **equations of constraint**

$$\int \chi_n^*(i) \chi_m(i) \, dq_i = \delta_{nm} \tag{4.1-6}$$

The equations of constraint are taken into account by introducing a set of **Lagrangian multipliers** $\{\varepsilon_{nm}\}$ and then considering the variation of the new function

$$G = 2 \sum_{n=1}^{N} H_{nn} + \sum_{n=1}^{N} \sum_{m=1}^{N} (2J_{nm} - K_{nm})$$

$$- 2 \sum_{n=1}^{N} \sum_{m=1}^{N} \varepsilon_{nm} \int \chi_n^*(i)\chi_m(i)\,dq_i$$

We omit the details (Pople and Beveridge, 1970). The minimization of G leads to the following set of *coupled integrodifferential equations*:

$$\left(-\frac{\hbar^2}{2m}\nabla_i^2 - \sum_{\alpha}\frac{Z_\alpha e^2}{4\pi\varepsilon_0 r_{\alpha i}} + \sum_{n=1}^{N}\left[2\hat{J}_n(i) - \hat{K}_n(i) \right] \right)\chi_m(i) = \varepsilon_m \chi_m(i)$$

$$(4.1\text{-}7a)$$

Equations (4.1-7a) are known as the **Hartree–Fock equations**, and the operator in parentheses is called the **Fock operator**, $\hat{F}$. Each equation in eq. (4.1-7a) is of the form of an eigenvalue equation.

$$\hat{F}\chi_m(i) = \varepsilon_m \chi_m(i) \qquad (4.1\text{-}7b)$$

However, it is a special kind of eigenvalue equation because the operator $\hat{F}$ itself *depends upon* the eigenfunctions, which we do not know in advance. This dependency arises through the **Coulomb** $[\hat{J}_n(i)]$ and **exchange** $[\hat{K}_n(i)]$ **operators**, which are given by

$$\hat{J}_n(i)\chi_m(i) = \int \chi_n^*(j)\frac{e^2}{4\pi\varepsilon_0 r_{ij}}\chi_n(j)\chi_m(i)\,dq_j \qquad (4.1\text{-}8a)$$

$$\hat{K}_n(i)\chi_m(i) = \left(\int \chi_n^*(j)\frac{e^2}{4\pi\varepsilon_0 r_{ij}}\chi_m(j)\,dq_j \right)\chi_n(i) \qquad (4.1\text{-}8b)$$

The equations in (4.1-7) are *integrodifferential* equations because the unknown functions $\{\chi_n(i)\}$ appear both as arguments of the differential operator ∇_i^2 and as integrands in the operators $\hat{J}_n(i)$ and $\hat{K}_n(i)$. The equations are *coupled* because any one such equation as unknowns all N of the eigenfunctions $\chi_n(i)$.

The eigenvalue ε_p, which has dimensions of energy, can be interpreted as an ionization energy or, more precisely, as the (approximate) energy required to remove an electron from the molecular orbital χ_p. To see this, consider the removal of an electron and suppose that in the resulting ion all spatial molecular orbitals remain unchanged. Expression (4.1-5) for the ground-state energy of the ion is modified as follows. From the first summation one of the two terms $n = p$ is absent, since only one electron (of spin α or β) remains in χ_p. From the second summation all terms in which both n and m equal p are absent; likewise all terms in which just one of n or m equals p are missing. Thus, the energy expression becomes

$$E^+ = 2\sum_n{}' H_{nn} + H_{pp} + \sum_n{}'\sum_m{}'(2J_{nm} - K_{nm})$$

$$+ \sum_m{}'(2J_{pm} - K_{pm}) \tag{4.1-9}$$

where the prime on a summation sign $\sum_n'$ indicates the omission of the term $n = p$ or $m = p$. We can get rid of the primes by rewriting eq. (4.1-9) as

$$E^+ = 2\sum_n H_{nn} - H_{pp} + \sum_n\sum_m(2J_{nm} - K_{nm}) - \sum_m(2J_{pm} - K_{pm})$$

$$= E - H_{pp} - \sum_m(2J_{pm} - K_{pm}) \tag{4.1-10}$$

where the second line follows from a comparison with eq. (4.1-5). Then the approximate ionization energy is given by

$$E^+ - E = -\left[H_{pp} + \sum_m(2J_{pm} - K_{pm}) \right] \tag{4.1-11}$$

Now consider eq. (4.1-7a) for the case $m = p$. Multiplying both sides by $\chi_p^*(i)$ and integrating, we have

$$\int \chi_p^*(i) \left(-\frac{\hbar^2}{2m}\nabla_i^2 - \sum_\alpha \frac{Z_\alpha e^2}{4\pi\varepsilon_0 r_{i\alpha}} \right) \chi_p(i)\, dq_i$$

$$+ \sum_n \left(2\int \chi_p^*(i)\hat{J}_n(i)\chi_p(i)\, dq_i - \int \chi_p^*(i)\hat{K}_n(i)\chi_p(i)\, dq_i \right)$$

$$= \varepsilon_p \int \chi_p^*(i)\chi_p(i)\, dq_i$$

or

$$H_{pp} + \sum_n(2J_{pn} - K_{pn}) = \varepsilon_p \tag{4.1-12}$$

The second line of eq. (4.1-12) follows from the first by eqs. (4.1-2) to (4.1-4), (4.1-6), and (4.1-8). Comparing relations (4.1-11) and (4.1-12), we have

$$E^+ - E = -\varepsilon_p \tag{4.1-13}$$

Table 4.1.   Application of Koopmans' Theorem to the Estimation of Ionization
Energies of Atoms and Molecules

| Substance | Ionization Energy (eV) | | |
	Orbital	Calculated	Observed[d]
Ne^a	$2s$	52.5	48.5
Ar^a	$3p$	16.1	15.8
HF^b	3σ	20.9	20.00
$O_2{}^b$	$2p\pi_g$	14.47	12.54
trans-1,3-Buta-diene^c	$2p\pi_g$	9.02	9.18
Benzene^c	$2p\pi_g$	9.35	9.38

[a] D. A. McQuarrie, *Quantum Chemistry*, University Science Books, Mill Valley, CA, 1983, p. 310.
[b] W. G. Richards and D. L. Cooper, *Ab initio Molecular Orbital Calculations for Chemists*, 2nd ed., Clarendon, Oxford, 1983, p. 53.
[c] M. J. S. Dewar, *The Molecular Orbital Theory of Organic Chemistry*, McGraw-Hill, New York, 1969, p. 274.
[d] Vertical ionization potentials.

Thus, $-\varepsilon_p$ is an approximation to the ionization energy of an electron from the pth molecular orbital. This result is known as **Koopmans' theorem** (Koopmans, 1934). More commonly, one refers to the $\{\varepsilon_m\}$ as **molecular orbital energies**.

Calculations show Koopmans' theorem to be a fair approximation to actual experimental ionization energies (Table 4.1).[1] The examples shown in the table may not be strictly comparable because the results have been taken from different sources, but they are nevertheless strongly suggestive.

Let us now return to eq. (4.1-7); further progress may seem hopeless! Nevertheless, the eigenfunctions $\{\chi_m\}$ constitute the best set of spatial molecular orbitals, subject to the approximation of representing the ground-state wave function as a single Slater determinant of spin-orbitals. In order to proceed, one *assumes* a starting set of orbitals $\{\chi_1^{(0)}, \chi_2^{(0)}, \ldots, \chi_N^{(0)}\}$. From these the operators $\hat{J}_n$ and $\hat{K}_n$ are constructed and eq. (4.1-7) now reduces to a set of *uncoupled* differential equations. These are then solved to generate a new set of orbitals, $\{\chi_1^{(1)}, \chi_2^{(1)}, \ldots, \chi_N^{(1)}\}$, and an initial set of energies, $\{\varepsilon_1^{(0)}, \varepsilon_2^{(0)}, \ldots, \varepsilon_N^{(0)}\}$. The cycle of operations is repeated until the $\{\chi_n\}$ and $\{\varepsilon_n\}$ remain constant to within a prescribed tolerance. The operations have the effect of leading to a set of molecular orbitals that are internally consistent, so that Hartree–Fock solutions are commonly called **self-consistent field** or **SCF orbitals**. Clearly, an SCF calculation is a nontrivial undertaking, but is manageable on a large computer.

[1] For molecules, one must distinguish between **vertical** and **adiabatic ionization energies**; the former refer to a process at a *fixed* nuclear configuration. Vertical ionization energies are typically greater than adiabatic ionization energies (process in which ion is produced at its most stable nuclear configuration) by about 0.2 eV.

4.2 THE ROOTHAAN EQUATIONS

The procedure just outlined has generally been employed only for atoms. With molecules (and atoms, as well) it has been found more practical to use a procedure that was pioneered by C. C. J. Roothaan in the 1950s. In this approach, each molecular orbital is expressed as a **linear combination of atomic orbitals (LCAO)**,

$$\chi_n(i) = \sum_b c_{bn}\phi_b(i) \tag{4.2-1}$$

where $\phi_b(i)$ is an atomic orbital (for electron i) centered on a particular nucleus. If the **basis set** $\{\phi_b(i)\}$ is complete, then the true molecular orbital $\chi_n(i)$ can be exactly represented by a unique linear combination of the form (4.2-1). In practice, of course, it is not possible to employ an infinite basis set. A **minimum basis set** comprises all of those atomic orbitals up to and including the valence orbitals of every atom in the molecule. In what follows we deal with a fixed basis set of B functions.

To determine the coefficients $\{c_{bn}\}$ we substitute expansion (4.2-1) into eq. (4.1-7b), multiply both sides of the resulting equation by $\phi_a^*(i)$ and integrate over q_i. This gives

$$\sum_{b=1}^{B}(F_{ab} - \varepsilon_n S_{ab})c_{bn} = 0 \qquad a = 1, 2, \ldots, B \tag{4.2-2}$$

where the overlap integral S_{ab} is given by

$$S_{ab} = \int \phi_a^*(i)\phi_b(i)\,dq_i \tag{4.2-3}$$

and the **Fock matrix element** by

$$F_{ab} = \int \phi_a^*(i)\left(-\frac{\hbar^2}{2m}\nabla_i^2 - \sum_\alpha \frac{Z_\alpha e^2}{4\pi\varepsilon_0 r_{i\alpha}}\right.$$
$$\left. + \sum_{n=1}^{N}[2\hat{J}_n(i) - \hat{K}_n(i)]\right)\phi_b(i)\,dq_i \tag{4.2-4}$$

Equations (4.2-2) are commonly known as the **Roothaan equations** (Roothaan, 1951a); they were developed independently by G. G. Hall (Hall, 1951). The Roothaan equations are a system of B^2 *homogeneous* equations in B^2 unknowns, for from B linearly independent basis functions one can construct no more than B independent linear combinations and, if each such combination involves all B basis functions, then there are $B \times B$ coefficients to be determined. In general, there are more equations than there are doubly occupied molecular orbitals. For example, in the simple case of lithium hydride (LiH), suppose a basis set consisting of $1s_H$, $1s_L$, $2s_L$, $2p_{zL}$ is employed. There are then 16 coefficients to be determined (from 16 equations), leading to 4 molecular orbitals, but only 2 of these are occupied in the ground state. The remaining molecular orbitals are called **virtual orbitals** and can be used in a description of the excited states of LiH.

The Roothaan equations can be written compactly in matrix form. Let the Fock matrix elements be arranged in a $B \times B$ matrix **F** and let the molecular orbital energies $\{\varepsilon_n\}$ be the elements of a diagonal matrix **E**. If the coefficients of each molecular orbital are arranged to form a column vector and the column vectors are combined to form a matrix **C**, eqs. (4.2-2) can be expressed as

$$\mathbf{FC} = \mathbf{SCE} \qquad (4.2\text{-}5)$$

Solution of eq. (4.2-5) proceeds along lines similar to that employed for H_2^+ (see Section 3.5). A nonzero solution exists if the **secular determinant** is zero, that is,

$$\begin{vmatrix} F_{11} - \varepsilon_n & F_{12} - \varepsilon_n S_{12} & F_{13} - \varepsilon_n S_{13} & \cdots & F_{1B} - \varepsilon_n S_{1B} \\ F_{21} - \varepsilon_n S_{21} & F_{22} - \varepsilon_n & F_{23} - \varepsilon_n S_{23} & \cdots & F_{2B} - \varepsilon_n S_{2B} \\ F_{31} - \varepsilon_n S_{31} & F_{32} - \varepsilon_n S_{32} & F_{33} - \varepsilon_n & \cdots & \\ \vdots & \vdots & \vdots & & \vdots \\ F_{B1} - \varepsilon_n S_{B1} & F_{B2} - \varepsilon_n S_{B2} & F_{B3} - \varepsilon_n S_{B3} & & F_{BB} - \varepsilon_n \end{vmatrix} = 0$$

Expansion of the determinant yields a B-order polynomial equation in ε_n with B roots. To each value of ε_n there corresponds a unique set of coefficients, $\{c_{1n}, c_{2n}, c_{3n}, \ldots, c_{Bn}\}$.

Combining eqs. (4.2-4) and (4.1-8), we see that the Fock matrix element can be written explicitly as

$$F_{ab} = H_{ab} + \sum_{n=1}^{N} \sum_{a'=1}^{B} \sum_{b'=1}^{B} c_{a'n}^* c_{b'n} [2(ab|a'b') - (aa'|bb')] \qquad (4.2\text{-}6)$$

where

$$H_{ab} = \int \phi_a^*(i) \left(-\frac{\hbar^2}{2m} \nabla_i^2 - \sum_\alpha \frac{Z_\alpha e^2}{4\pi\varepsilon_0 r_{i\alpha}} \right) \phi_b(i) \, dq_i \qquad (4.2\text{-}7)$$

and the **electron-repulsion integral** $(ab|cd)$ is

$$(ab|cd) = \int\int \phi_a^*(i) \phi_b(i) \left(\frac{e^2}{4\pi\varepsilon_0 r_{ij}} \right) \phi_c^*(j) \phi_d(j) \, dq_i \, dq_j \qquad (4.2\text{-}8)$$

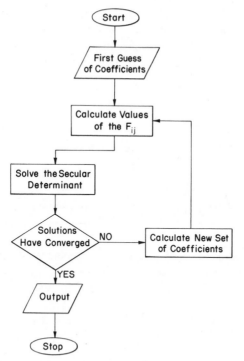

FIGURE 4.2. Algorithm for the solution of Roothaan's equations for a molecule.

Thus, F_{ab} depends on the unknown coefficients and the Roothaan equations must be solved in an iterative manner (see Fig. 4.2), as is the case with the original Hartree–Fock equations (4.1-7). It is clear that once a self-consistent solution has been obtained, this can be of no higher quality than that of the basis set. One says that the **Hartree–Fock limit** has been reached when addition of one or more basis functions has no sensible effect on the molecular orbitals and their energies. Hartree–Fock SCF calculations are not normally taken to the Hartree–Fock limit.

It is convenient to define an **electron density matrix**.

$$P_{ab} = 2 \sum_{n=1}^{N} c_{an}^{*} c_{bn} \qquad (4.2\text{-}9)$$

The elements can be given a semiquantitative interpretation. The operator for the electron density at a point $\mathbf{r}$ in space is expressed as

$$\hat{\rho}(\mathbf{r}) = \sum_{i=1}^{2N} \delta(\mathbf{r} - \mathbf{r}_i) \qquad (4.2\text{-}10)$$

where the Dirac delta function is a function of the *vector* difference between the point $\mathbf{r}$ and the position $\mathbf{r}_i$ of the ith electron. The discussion in Section 2.5 can be extended to several dimensions. In particular, in three dimensions the integral property of the Dirac delta function assumes the form

$$\int \delta(\mathbf{r} - \mathbf{r}')f(\mathbf{r}') \, d\mathbf{r}' = f(\mathbf{r})$$

where $d\mathbf{r}' \equiv dx' \, dy' \, dz'$. The expected value of $\hat{\rho}(\mathbf{r})$ is thus computed by analogy to that of $\hat{H}_1$ [see eq. (4.1-2)].

$$\int \Psi^*(q)\hat{\rho}(\mathbf{r})\Psi(q) \, dq = 2 \sum_{n=1}^{N} \chi_n^*(\mathbf{r})\chi_n(\mathbf{r}) \qquad (4.2\text{-}11a)$$

Each spatial molecular orbital $\chi_n(\mathbf{r})$ is now replaced by the LCAO (4.2-1). Equation (4.2-11a) thus becomes

$$\rho(\mathbf{r}) = 2 \sum_{n=1}^{N} \sum_{a=1}^{B} c_{an}^* \phi_a^*(\mathbf{r}) \sum_{b=1}^{B} c_{bn}\phi_b(\mathbf{r})$$

$$= \sum_{a,b=1}^{B} P_{ab}\phi_a^*(\mathbf{r})\phi_b(\mathbf{r}) \qquad (4.2\text{-}11b)$$

in view of the definition (4.2-9) of P_{ab}.

Clearly, if $\rho(\mathbf{r})$, the electron density at $\mathbf{r}$, is integrated over all possible $\mathbf{r}$, one should have the total number of electrons in the system.

$$\int \rho(\mathbf{r}) \, d\mathbf{r} = \sum_{a,b} P_{ab} \int \phi_a^*(\mathbf{r})\phi_b(\mathbf{r}) \, d\mathbf{r}$$

$$= \sum_{a,b=1}^{B} P_{ab}S_{ab}$$

$$= 2N \qquad (4.2\text{-}12)$$

We may therefore think of a term P_{aa} [corresponding to $a = b$ and $S_{aa} = 1$ in eq. (4.2-12)] as giving roughly the electron population in atomic orbital a. An off-diagonal term, $P_{ab}S_{ab}$, can then be considered the electronic population of the atomic overlap distribution $\phi_a^*\phi_b$. Qualitatively, the more positive is $P_{ab}S_{ab}$, the greater the overlap distribution contributes to chemical bonding (provided that a and b are centered on adjacent nuclei). These ideas and related ones were discussed extensively by Mulliken (Mulliken, 1955).

4.3 SLATER ORBITALS

Since the size of the basis B is severely limited in practice, one must choose the basis functions wisely (usually on intuitive grounds) to get optimal results. One popular type of basis consists of **Slater atomic orbitals**, first introduced by Slater in 1930. These have the form

$$\Phi_{n^*\ell m}(r, \theta, \phi) = N_{n^*\ell} r^{n^*-1} e^{-\xi r} Y_\ell^m(\theta, \phi) \tag{4.3-1}$$

where r is the distance of the electron from a specific nucleus, upon which the orbital is *centered*; $N_{n^*\ell}$ is a normalization constant; and ξ, the **Slater screening factor**, is treated as a parameter. The functions $Y_\ell^m(\theta, \phi)$ are the normalized spherical harmonics, identical to the wave functions of the rigid rotor (see Table 2.1).

The parameter ξ, which reflects the partial shielding of the electron from the nucleus (on which the orbital is centered) by the other electrons, is given by **Slater's rules**; other sets of rules have been proposed more recently (Silver and Nieuwpoort, 1978). The screening factor ξ is expressed according to Slater as

$$\xi = \frac{Z - \sum s_i}{n^* a_0} \tag{4.3-2}$$

where Z is the atomic number of the nucleus of interest and n^*, a modified principal quantum number, has the following relation to the hydrogenic principal quantum number n.

n	1	2	3	4	5	6
n^*	1	2	3	3.7	4.0	4.2

The quantity $\sum s_i$ for a given electron is determined as a sum of contributions from all of the other electrons as follows:

1. Electrons are divided into the following groups:

$$1s|2s, 2p|3s, 3p|3d|4s, 4p|4d, 4f|5s, 5p$$

No contribution to $\sum s_i$ is made by any electron in a group to the right of that to which the given electron belongs. For example, no contribution to $\sum s_i$ for any of the $2p$ electrons in sodium is made by the lone $3s$ electron.

2. An amount of 0.35 is contributed to $\sum s_i$ by each other electron in the group of the given electron; if this group is the $1s$ group, however, the amount is 0.30.

3. If the given electron is an s or p electron, an amount 0.85 is contributed to Σs_i by each electron in a group of principal quantum number one less, and an amount 1.00 is contributed by each electron of principal quantum number less by 2 or more. For example, for the lone $3s$ electron in sodium, an amount $8 \times 0.85 + 2 \times 1.00 = 8.8$ is contributed to Σs_i due to the eight $2s, 2p$ electrons and the two $1s$ electrons.

4. If the given electron is a d or f electron, then Rule (3) should be replaced by: An amount 1.00 is contributed to Σs_i by *every* electron in groups to the left. For example, for the $3d$ electron in scandium, an amount $18 \times 1.00 = 18.0$ is contributed to Σs_i for all of the 18 $1s, 2s, 2p, 3s, 3p$ inner electrons.

Slater orbitals permit a saving in computational time relative to hydrogenic orbitals for the many integrals [e.g., $(ab|cd)$] because the radial parts of the hydrogenic orbitals consist of a polynomial times an exponential. In refined calculations, the screening factor ξ is treated as a parameter that is to be optimized by the variation theorem. In some calculations, basis sets referred to as **double-zeta** sets are employed. In them a different screening factor is assigned to each of the two electrons described by a given type of atomic orbital. Table 4.2 shows some typical orbital screening factors that were used in a double-zeta calculation on carbon monoxide.

Although we shall use single-ξ Slater orbitals in the remainder of this chapter, it should be mentioned that many calculations reported today use basis sets comprising **Gaussian orbitals** (Boys, 1950). These have the form

$$\chi_{\ell m}(r, \theta, \phi) = Nr^{\ell}e^{-\eta r^2}Y_{\ell}^m(\theta, \phi) \qquad (4.3\text{-}3)$$

where ℓ is the orbital quantum number of the analogous hydrogenic orbital

Table 4.2   Slater Screening Factors in a Calculation on CO with a Double-Zeta Basis Set[a]

Basis Function Type	C	O
$1s$	$\begin{cases} 5.3036 \\ 8.3830 \end{cases}$	7.1651 10.6143
$2s$	$\begin{cases} 1.2696 \\ 1.8562 \end{cases}$	1.6011 2.5888
$2p\sigma$	$\begin{cases} 1.2871 \\ 2.8537 \end{cases}$	1.6515 3.6754
$2p\pi$	$\begin{cases} 1.2871 \\ 2.8537 \end{cases}$	1.6515 3.6754
$3d\sigma$	1.895	2.103
$3d\pi$	1.175	3.019

[a] From R. K. Nesbet, *J. Chem. Phys.*, **40**, 3619 (1964).

and η is a variational parameter. The mathematical behavior of a Gaussian orbital is somewhat different from that of a Slater or hydrogenic orbital; generally, two or three Gaussian functions are needed in order to imitate a single Slater function. Offsetting this deficiency of the Gaussian orbitals is the saving in computational time relative to Slater orbitals needed to evaluate any of the integrals (Poirier, Kari and Csizmadia, 1985).

Exercises

4.1. Derive eq. (4.1-2) by considering the explicit form of the wave function $\Psi(1, 2, \ldots, 2N)$.

4.2. A closed-shell diatomic molecule contains six electrons.
   (a) What is the dimensionality of the Slater determinant?
   (b) How many of the products in the expansion of the Slater determinant contain the molecular spin-orbital $\chi_1(1)\alpha(1)$? Write a few of them.
   (c) Show some examples of terms in the expansions of Ψ and Ψ^* that make nonzero contributions to $\int \Psi^*(q)\hat{H}_1\Psi(q)\,dq$.
   (d) If electrons 1 and 2 are in the same spatial orbital, how many terms in the expansion of the Slater determinant contain the factor $\chi_1(1)\alpha(1)\chi_1(2)\beta(2)$?
   (e) If electrons 1 and 2 are in different spatial orbitals, how many terms in the expansion of the Slater determinant contain the factor $\chi_1(1)\alpha(1)\chi_3(3)\alpha(3)$? Write some of them.
   (f) Show some examples of terms from the expansions of Ψ and Ψ^* that contain the factor $\chi_1(1)\alpha(1)\chi_3(3)\alpha(3)$ and that make nonzero contributions to $\int \Psi^*(q)\hat{H}_2\Psi(q)\,dq$.

4.3. Ransil (1960) calculated the total *molecular* energy $\mathscr{E}$ of lithium hydride with a minimal basis set of Slater orbitals: $\mathscr{E} = -216.78$ eV.
   (a) How much does the total *electronic* energy E amount to if $R = 1.5957$ Å (Huber and Herzberg, 1979)?
   (b) If Ransil's calculated value of $\mathscr{E}$ were accurate, what would be the electronic dissociation energy D_e (refer to Section 5.3) of LiH, given that the experimental ionization potentials of H and Li are: H(13.598 eV), Li(5.392, 75.638, 122.451 eV)? What does your result imply about the quality of Ransil's calculation? (*Hint:* Set up an appropriate thermodynamic cycle.)
   (c) The following data are tabulated in reference sources: $\Delta\tilde{H}_f^0[\text{LiH}_{(s)}] = -21.66$ kcal, $\Delta\tilde{H}_{subl}^0[\text{LiH}_{(s)}] = 55.27$ kcal, fundamental vibration frequency $\tilde{\nu}_0 = 1406$ cm^{-1}, $\Delta\tilde{H}_f^0[\text{Li}_{(g)}] = +38.40$ kcal, $\Delta\tilde{H}_f^0[\text{H}_{(g)}] = +52.10$ kcal. Estimate the electronic dissociation energy D_e in electron volts for gaseous LiH. What would have to be the value of

Ransil's calculated $\mathcal{E}$ in order to give agreement between theory and experiment for the electronic dissociation energy?

(d) Ransil obtained the following values for the orbital energies: $\varepsilon_{1_\sigma} = -66.58$ eV, $\varepsilon_{2_\sigma} = -8.26$ eV, $\varepsilon_{3_\sigma} = 0.45$ eV, $\varepsilon_{4_\sigma} = 9.50$ eV. What is the value of the total electronic repulsion energy, $\sum_{n=1}^{N}\sum_{m=1}^{N}(2J_{nm} - K_{nm})$, for ground-state LiH, according to Ransil?

(e) How does the calculated first ionization potential of LiH compare with the "experimental" value (Rosmus and Meyer, 1977) of 7.7 eV?

4.4. Consult Ransil's paper and after looking up any necessary experimental data, answer all parts of Exercise 4.3 for the N_2 molecule.

4.5. From eq. (4.2-7) what is the relationship between H_{ab} and H_{ba}, and why? How many different electron-repulsion integrals [see eq. (4.2-8)] involving a, b, c, and d are necessarily equal to $(ab|cd)$?

4.6. Consider the hydrogen molecule and let the molecular orbitals have the form $\chi_n(i) = c_{An}1s_A(i) + c_{Bn}1s_B(i)$. Write eq. (4.2-5) in explicit matrix form in as much detail as possible.

4.7. A Roothaan–Hartree–Fock calculation on HF using a minimal basis set has yielded the following set of occupied molecular orbitals.

AO/MO	1	2	3	4	5
$1s_H$	-0.0046	-0.1606	-0.5761	0	0
$1s_F$	0.9963	0.2435	-0.0839	0	0
$2s_F$	0.0163	-0.9322	0.4715	0	0
$2pz_F$	0.0024	0.0907	-0.6870	0	0
$2px_F$	0	0	0	1	0
$2py_F$	0	0	0	0	1

What is the electron population in the $1s_H$ orbital? Which makes the greater contribution to the chemical bonding: the $1s_H - 1s_F$ or the $1s_H - 2pz_F$ overlap population? Is this reasonable?

4.8. A problem with Slater orbitals in calculations *on atoms* is that the orbitals are not all orthogonal. Show that an orthogonalized Slater $2s$ orbital can be constructed from the customary Slater $1s$ and $2s$ orbitals in the following way:

$$\chi_{2s'} = (1 - S^2)^{-1/2}[\chi_{2s} - \chi_{1s}S]$$

where $S = \sqrt{192}\,\xi^{3/2}\xi'^{5/2}(\xi + \xi')^{-4}$ and ξ, ξ' are the Slater screening factors for the $1s, 2s$ orbitals, respectively.

4.9. Show that an orthogonalized Slater $3s$ orbital can be constructed in the following form:

$$\chi_{3s'} = \left(1 - S_{2'3}^2 - S_{13}^2\right)^{-1/2}[\chi_{3s} - \chi_{2s'}S_{2'3} - \chi_{1s}S_{13}]$$

where the quantities S_{13}, $S_{2'3}$ mean

$$S_{13} = \int \chi_{1s}\chi_{3s}\,dq \qquad S_{2'3} = \int \chi_{2s'}\chi_{3s}\,dq$$

4.10. An interesting, elementary paper in which Slater screening factors ξ are deduced entirely from simple mathematics without any curve fitting of atomic spectroscopic data (as was originally done by Slater) is Bessis and Bessis (1981). Consult this paper and work through the steps leading to the authors' eq. (10). [*Note:* In the equation immediately preceding the authors' eq. (4), there is a typographical error; the factor $(x_i)_0^{-3}$ should be $(x_i)_0^{-2}$.] Equation (10) calculates the parameter Σs_i in our eq. (4.3-2). Calculate ξ for $1s, 2s$ for lithium from the paper, and compare the values with those given by Ransil (see Exercise 4.3).

4.11. Sketch qualitatively the hydrogenic, Slater, and Gaussian $2s$ orbital functions, and discuss the differences between them. Certain molecular properties depend upon the electron density at the surface of a nucleus (e.g., the nuclear spin–spin coupling constant). Which orbital function would be the preferred one to use in the calculation? Ignore any consideration of computer time.

4.4 A HARTREE–FOCK TREATMENT OF LITHIUM HYDRIDE

Lithium hydride is the simplest, neutral, heteronuclear diatomic molecule to which the Hartree–Fock method in the Roothaan form can be applied. A minimal basis set employs a $1s$ orbital on hydrogen and the $1s$, $2s$, and three $2p$ orbitals on lithium. We shall do a subminimal basis set calculation here in which the $2p_x$ and $2p_y$ orbitals are deleted. The LiH molecule is oriented as

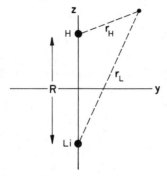

FIGURE 4.3. Coordinate system used for the LiH molecule.

Table 4.3.   Input Data for a Calculation on Lithium Hydride

Orbital	Type	ξ (in a_0^{-1})	Function
1	$1s_H$	1	$\left(\dfrac{\xi^3}{\pi}\right)^{1/2} e^{-\xi r_H}$
2	$1s_L$	2.70	$\left(\dfrac{\xi^3}{\pi}\right)^{1/2} e^{-\xi r_L}$
3	$2s_L$	0.65	$\left(\dfrac{\xi^5}{3\pi}\right)^{1/2} r_L\, e^{-\xi r_L}$
4	$2pz_L$	0.65	$\left(\dfrac{\xi^5}{\pi}\right)^{1/2} r_L\, e^{-\xi r_L}\cos\theta_L$

$$R = 1.59569 \text{ Å}$$

shown in Fig. 4.3. Each molecular orbital has the form

$$\chi_n(i) = c_{1n}\phi_1(i) + c_{2n}\phi_2(i) + c_{3n}\phi_3(i) + c_{4n}\phi_4(i) \qquad (4.4\text{-}1)$$

where the basis functions $\{\phi_m(i)\}$ are listed in Table 4.3. Four independent molecular orbitals of the form (4.4-1) can be constructed. In what follows, we work step-by-step from eq. (4.4-1) to the final listing of the eigenvalues, eigenvectors, and total molecular energy $\mathscr{E}$ of the LiH molecule at its equilibrium internuclear separation.

STEP 1.   EVALUATION OF OVERLAP AND CORE INTEGRALS

Figure 4.2 shows that solving Roothaan's equations and obtaining the eigenvectors are circularly related. In order to get started, we break the circle by omitting the electron-repulsion terms in eq. (4.2-2) and solving the simpler set of equations

$$\sum_{b=1}^{4} (H_{ab} - \varepsilon_n S_{ab})c_{bn} = 0 \qquad (4.4\text{-}2)$$

To this end, the integrals S_{ab} and H_{ab} need to be evaluated.

Henceforth, we work in **atomic units** (au). In this system of units, the quantities length, charge, mass, energy, and angular momentum are given in multiples of values appropriate to the atomic scale. The system is summarized in Table 4.4. The bond length of LiH, for example, as quoted in Table 4.3, is now 3.01542 au. The system of units amounts formally to setting $m_e = |e| = \hbar = 4\pi\varepsilon_0 = 1$ in any equation expressed in SI units.

The overlap integrals $S_{12}, S_{13}, S_{14}, S_{23}$ are the easiest to evaluate. They are best worked out in confocal elliptical coordinates in a manner similar to that

Table 4.4.   The System of Atomic Units (au)

Quantity	1 au Equals[a]	Value (in SI Units)
Mass	m_e	9.109534×10^{-31} kg
Charge	$\lvert e \rvert$	1.602189×10^{-19} C
Angular momentum	$\hbar$	1.054589×10^{-34} J s
Length	$\dfrac{4\pi\varepsilon_0 \hbar^2}{m_e e^2} (= a_0)$	5.29177×10^{-11} m
Energy	$\dfrac{e^2}{4\pi e_0 a_0}$	4.359814×10^{-18} J

[a] ε_0 is the permittivity of free space and has the value 8.85419×10^{-12} C^2 N^{-1} m^{-2}; see E. R. Cohen and B. N. Taylor, *J. Phys. Chem. Ref. Data*, **2**, 663 (1973).

used in Section 3.5, and some general formulas are available in the literature (Mulliken, et al., 1949). For example, let the 1s orbital of H with screening factor ξ_H be centered at $(0, 0, \frac{1}{2}R)$, and the 1s orbital of Li with screening factor ξ_L be centered at $(0, 0, -\frac{1}{2}R)$. Then the overlap integral S_{12} can be shown to be given by

$$S_{12} = \tfrac{1}{4}R^3(\xi_H \xi_L)^{3/2}[A_2(\alpha)B_0(\beta) - A_0(\alpha)B_2(\beta)] \qquad (4.4\text{-}3)$$

where the arguments are $\alpha = \frac{1}{2}R(\xi_L + \xi_H)$, $\beta = \frac{1}{2}R(\xi_L - \xi_H)$, and the ancillary functions $A_n(x)$ and $B_n(x)$ are defined as

$$A_n(x) = \frac{n!\, e^{-x}}{x^{n+1}} \sum_{k=0}^{n} \frac{x^k}{k!} \qquad (4.4\text{-}4a)$$

$$B_n(x) = \frac{n!}{x^{n+1}} \left(e^x \sum_{k=0}^{n} \frac{(-x)^k}{k!} - e^{-x} \sum_{k=0}^{n} \frac{x^k}{k!} \right) \qquad x \neq 0 \quad (4.4\text{-}4b)$$

Limited tabulations of $A_n(x)$ and $B_n(x)$ are available in the literature (Miller, Gerhauser and Matsen, 1959).

The functions $A_n(x)$ and $B_n(x)$ satisfy the following recurrence relations:

$$A_n(x) = \frac{n}{x}A_{n-1}(x) + A_0(x) \qquad (4.4\text{-}5a)$$

$$B_n(x) = \frac{n}{x}B_{n-1}(x) - A_0(x) + (-1)^{n+1}A_0(-x) \qquad (4.4\text{-}5b)$$

These relations enable one to obtain the higher $A_n(x)$ and $B_n(x)$ functions from knowledge of $A_0(x)$, $A_0(-x)$, and $B_0(x)$ and are convenient to pro-

Table 4.5. The Overlap Matrix for Lithium Hydride at $R = 3.01542$ au[a]

	1	2	3	4
1	1	0.094113013	0.475448356	0.554423608
2	0.094113013	1	0.166264842	0
3	0.475448356	0.166264842	1	0
4	0.554423608	0	0	1

[a] The numbers are those of the members of the basis set in Table 4.3.

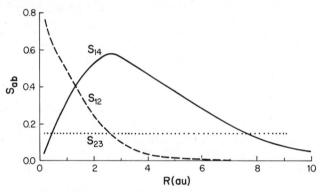

FIGURE 4.4. Plot of overlap integrals for LiH as a function of internuclear distance.

gram on a computer. With the aid of eqs. (4.4-5), the remaining three nonzero overlap integrals can be done in the same way as S_{12}. The results are gathered in the overlap matrix of Table 4.5. Figure 4.4 shows a plot of three of the overlap integrals as a function of distance R between the nuclei.

We turn next to the evaluation of the matrix elements H_{ab} [see eq. (4.2-7)]. When expressed in atomic units, the core Hamiltonian $\hat{H}_1$ for LiH assumes the explicit form [see eq. (4.1-1)]

$$\hat{H}_1(i) = -\frac{1}{2}\nabla_i^2 - \frac{3}{r_{Li}} - \frac{1}{r_{Hi}}$$

where r_{Li}, for example, is the distance (in atomic units) of electron i from the Li nucleus. In confocal elliptical coordinates, the Laplacian becomes (Dence, 1975)

$$\nabla^2 = \frac{4}{R^2(\xi^2 - \eta^2)}\left[\frac{\partial}{\partial\xi}\left((\xi^2 - 1)\frac{\partial}{\partial\xi}\right) + \frac{\partial}{\partial\eta}\left((1 - \eta^2)\frac{\partial}{\partial\eta}\right)\right.$$
$$\left. + \frac{\xi^2 - \eta^2}{(\xi^2 - 1)(1 - \eta^2)}\frac{\partial^2}{\partial\phi^2}\right]$$

Table 4.6.   The Core Hamiltonian Matrix for Lithium Hydride at $R = 3.01542$ au[a]

	1	2	3	4
1	−1.48528468	−0.418511587	−0.741763927	−0.822283124
2	−0.418511587	−4.7866285	−0.716816327	−0.012506751
3	−0.741763927	−0.716816327	−1.16849736	−0.11076307
4	−0.822283124	−0.012506751	−0.11076307	−1.08671176

[a] The numbers are those of the members of the basis set in Table 4.3.

and r_L and r_H are $r_L = \frac{1}{2}R(\xi + \eta)$, $r_H = \frac{1}{2}R(\xi - \eta)$ [see eq. (3.5-12)]. The matrix element H_{11}, for example, is then given by

$$
\begin{aligned}
H_{11} &= \frac{R^3}{8\pi} \int_0^{2\pi} d\phi \int_{-1}^{1} d\eta \int_1^{\infty} e^{-R(\xi-\eta)/2}\left(\xi^2 - \eta^2\right) \\
&\quad \times \left(\frac{-\nabla^2}{2} - \frac{6}{R(\xi+\eta)} - \frac{2}{R(\xi-\eta)}\right) e^{-R(\xi-\eta)/2}\, d\xi \\
&= \frac{R^2}{2}\left(\frac{R}{4}\left[A_0(R)B_2(-R) - A_2(R)B_0(-R)\right]\right. \\
&\qquad \left. + 3\left[A_0(R)B_1(-R) - A_1(R)B_0(-R)\right]\right)
\end{aligned}
\tag{4.4-6}
$$

where eq. (4.4-4) has been used. All of the remaining 15 core integrals can be evaluated in a similar manner. A useful check is that for the off-diagonal matrix elements one must have $H_{ab} = H_{ba}$ since the Hamiltonian is Hermitian and the orbitals are real. Another check is provided by recalculating some of the matrix elements (especially the diagonal ones) with the aid of formulas given in a very useful paper by Roothaan (Roothaan, 1951b).

Table 4.6 presents the complete matrix of the core Hamiltonian. The graph of three of the matrix elements as a function of R is displayed in Fig. 4.5. Note that the matrix elements do not all approach the same limit as $R \to \infty$.

STEP 2.   TRANSFORMATION OF THE FOCK MATRIX EQUATION INTO STANDARD EIGENVALUE FORM

We know that the Fock matrix equation (4.2-5) has a solution if the associated secular determinant is zero. However, eq. (4.2-5) is not in the most convenient form because the unknown ε appears in every term of the secular determinant. The following transformation of the matrices remedies this problem. We define matrices $\mathbf{F}'$ and $\mathbf{C}'$ by

$$
\mathbf{F}' = \mathbf{S}^{-1/2}\mathbf{F}\mathbf{S}^{-1/2} \qquad \mathbf{C}' = \mathbf{S}^{1/2}\mathbf{C}
\tag{4.4-7a}
$$

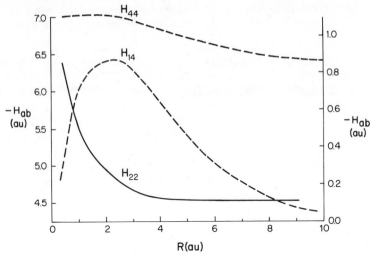

FIGURE 4.5. Plot of core Hamiltonian matrix elements for LiH as a function of internuclear distance. Solid line corresponds to left and dashed line to right scale.

whereupon, solving for **F** and **C**, we have

$$\mathbf{F} = \mathbf{S}^{1/2}\mathbf{F}'\mathbf{S}^{1/2} \qquad \mathbf{C} = \mathbf{S}^{-1/2}\mathbf{C}' \qquad (4.4\text{-}7b)$$

In these equations, $\mathbf{S}^{1/2}$ is a matrix with the property $\mathbf{S}^{1/2}\mathbf{S}^{1/2} = \mathbf{S}$, and $\mathbf{S}^{-1/2}$ is the inverse of $\mathbf{S}^{1/2}$, with the property $\mathbf{S}^{1/2}\mathbf{S}^{-1/2} = \mathbf{1}$, where $\mathbf{1}$ is the unit matrix. Substitution of eq. (4.4-7b) into eq. (4.2-5) gives

$$(\mathbf{S}^{1/2}\mathbf{F}'\mathbf{S}^{1/2})(\mathbf{S}^{-1/2}\mathbf{C}') = \mathbf{S}(\mathbf{S}^{-1/2}\mathbf{C}')\mathbf{E}$$

or

$$\mathbf{F}'\mathbf{C}' = \mathbf{C}'\mathbf{E} \qquad (4.4\text{-}8)$$

Equation (4.4-8) is now in standard eigenvalue form, and for the case of LiH we have

$$\begin{bmatrix} F'_{11} - \varepsilon & F'_{12} & F'_{13} & F'_{14} \\ F'_{21} & F'_{22} - \varepsilon & F'_{23} & F'_{24} \\ F'_{31} & F'_{32} & F'_{33} - \varepsilon & F'_{34} \\ F'_{41} & F'_{42} & F'_{43} & F'_{44} - \varepsilon \end{bmatrix} = 0 \qquad (4.4\text{-}9)$$

From the roots of the characteristic equation we obtain **E**, and then from eq. (4.4-8) the matrix **C**' is obtained. Finally, using eqs. (4.4-7), we can calculate **C**.

STEP 3.   DETERMINATION OF EIGENVALUES AND EIGENVECTORS

In order to set up the determinant in eq. (4.4-9), we must obtain the matrix $S^{1/2}$. The general problem of finding the roots of matrices is interesting. We do this by using a theorem due to the 19th century British mathematician J. J. Sylvester (1814–1897). Let M be an $N \times N$ real, *symmetric* matrix; such a matrix necessarily has real eigenvalues (Deif, 1982). Suppose further that the eigenvalues $\{\lambda_k\}$ of M are all distinct. Then the mth root of M is given by (Frazer et al., 1963)

$$M^{1/m} = \sum_{k=1}^{N} \lambda_k^{1/m} Z(\lambda_k)$$

where

$$Z(\lambda_k) \equiv \frac{\prod_{j \neq k} (\lambda_j 1 - M)}{\prod_{j \neq k} (\lambda_j - \lambda_k)} \tag{4.4-10}$$

In this way, the matrix $S^{1/2}$ can be obtained once the eigenvalues of S are calculated. Elements of the inverse matrix $S^{-1/2}$ are then calculated by a standard procedure, for example, by means of the formula (Dence, 1975)

$$S^{-1/2} = \frac{\mathrm{adj}\, S^{1/2}}{\det |S^{1/2}|} \tag{4.4-11}$$

where the **classical adjoint** (adj $S^{1/2}$) is the transpose of the matrix of cofactors of the elements of $S^{1/2}$. The final matrix $S^{-1/2}$ is given in Table 4.7.

With the matrix $S^{-1/2}$ at hand, the transformed Fock matrix F' can be set up and its eigenvalues and eigenvectors determined. The four eigenvalues are determined by analytically solving the fourth-degree equation (the **characteristic equation**) resulting from the expansion of the determinant in eq. (4.4-9). The eigenvector c_i' corresponding to each eigenvalue ε_i is obtained by solving

Table 4.7.   Elements of the Matrix $S^{-1/2}$ for Lithium Hydride at $R \equiv 3.01542$ au[a]

	1	2	3	4
1	1.34377353	−0.024647679	−0.376880406	−0.442716737
2	−0.024647679	1.01111391	−0.076730548	0.009247521
3	−0.376880406	−0.076730548	1.15396222	0.168406874
4	−0.442716737	0.009247521	0.168406874	1.19778566

[a] The numbers are those of the members of the basis set in Table 4.3.

the equation

$$\mathbf{F'c_i'} = \varepsilon_i c_i'$$

All of the resulting (column) eigenvectors are grouped together to form the matrix $\mathbf{C'}$.

$$\mathbf{C'} = \begin{bmatrix} c_{11}' & c_{12}' & c_{13}' & c_{14}' \\ c_{21}' & c_{22}' & c_{23}' & c_{24}' \\ c_{31}' & c_{32}' & c_{33}' & c_{34}' \\ c_{41}' & c_{42}' & c_{43}' & c_{44}' \end{bmatrix} \qquad \mathbf{E} = \begin{bmatrix} \varepsilon_1 & 0 & 0 & 0 \\ 0 & \varepsilon_2 & 0 & 0 \\ 0 & 0 & \varepsilon_3 & 0 \\ 0 & 0 & 0 & \varepsilon_4 \end{bmatrix}$$

$\uparrow$ $\uparrow$

Eigenvector 2   Corresponding eigenvalue 2

Our desired eigenvectors are, of course, the columns of $\mathbf{C}$. These vectors, as obtained from eq. (4.4-7b), are not necessarily normalized, although they are orthogonal. They can be normalized in each case by finding that number N_n such that

$$N_n^2 \sum_{a=1}^{4} \sum_{b=1}^{4} c_{an} c_{bn} S_{ab} = 1 \qquad (4.4\text{-}12)$$

STEP 4.   ITERATION

Recall that the matrix $\mathbf{C}$ just computed is actually the set of eigenvectors of the core Hamiltonian matrix [see eq. (4.4-2)]. Next, one constructs the full Fock matrix, each element of which has the form

$$F_{ab} = H_{ab} + \sum_{a',b'=1}^{B} P_{a'b'} \{ (ab|a'b') - \tfrac{1}{2}(aa'|bb') \} \qquad (4.4\text{-}13)$$

according to eqs. (4.2-6) and (4.2-9). This brings us to the evaluation of the electron-repulsion integrals [eq. (4.2-8)], the most difficult and time-consuming stage of any Hartree–Fock calculation. Such integrals are difficult to evaluate because of the r_{ij}^{-1} factor in the integrand, which mixes the coordinates of two different electrons. Roothaan developed a procedure for the calculation of these integrals for diatomic molecules (Roothaan et al., 1956).

For lithium hydride with a four-member basis set, there is a total of $4 \times 4 \times 4 \times 4 = 256$ electron-repulsion integrals. Many of these are equal to one another, as was suggested in Exercise 4.5. For example, $(12|23) = (21|23) = (21|32)$. Many others are equal to 0 on grounds of symmetry. In the present case, there are 47 distinct nonzero electron-repulsion integrals, whose values are given in Table 4.8. We evaluate most of these using the procedures

Table 4.8.   Nonvanishing Electron-Repulsion Integrals for Lithium Hydride
at $R = 3.01542$ au

$a\ b\ c\ d$	$(ab\|cd)$ (au)	$a\ b\ c\ d$	$(ab\|cd)$ (au)
1 1 1 1	0.625000000	1 3 2 4	0.004135098
1 1 1 2	0.035942069	1 3 3 3	0.119273783
1 1 1 3	0.219012525	1 3 3 4	0.031242251
1 1 1 4	0.296307629	1 3 4 4	0.12520^a
1 1 2 2	0.327670092	1 4 2 2	0.181411234
1 1 2 3	0.053847407	1 4 2 3	0.029726932
1 1 2 4	0.010993479	1 4 2 4	0.006594159
1 1 3 3	0.251294784	1 4 3 3	0.136616235
1 1 3 4	0.090804270	1 4 3 4	0.048239462
1 1 4 4	0.290543372	1 4 4 4	0.15107^a
1 2 1 2	0.005846^a	2 2 2 2	1.687500000
1 2 1 3	0.01709^a	2 2 2 3	0.165598011
1 2 1 4	0.02044^a	2 2 3 3	0.323796847
1 3 1 3	0.07707^a	2 2 4 4	0.323796847
1 3 1 4	0.09246^a	2 3 2 3	0.021222527
1 4 1 4	0.11223^a	2 3 3 3	0.052500709
1 2 2 2	0.101562148	2 3 4 4	0.052500709
1 2 2 3	0.012383852	2 4 2 4	0.004501748
1 2 2 4	0.001347057	2 4 3 4	0.007470613
1 2 3 3	0.029757023	3 3 3 3	0.236132812
1 2 3 4	0.002777127	3 3 4 4	0.236132812
1 2 4 4	0.028909^a	3 4 3 4	0.052191840
1 3 2 2	0.167179098	4 4 4 4	0.254414062
1 3 2 3	0.027068280		

^{a} Estimated by the Mulliken approximation.

outlined in the 1956 paper of Roothaan et al. A few of the nastier integrals are estimated by means of the **Mulliken approximation** (Offenhartz, 1970), according to which

$$(ab|a'b') \simeq \tfrac{1}{4} S_{ab} S_{a'b'} \left[(aa|a'a') + (aa|b'b') + (bb|a'a') + (bb|b'b') \right]$$
$$(4.4\text{-}14)$$

In Fig. 4.6 we display plots of four different electron-repulsion integrals as a function of the internuclear distance.

With the integrals of Table 4.8, we can calculate a new **F** matrix according to eq. (4.4-13). This matrix is transformed as described in Step 2 and a new set of eigenvalues and eigenvectors obtained, as in Step 3. The process is repeated until successive sets of eigenvalues and eigenvectors differ insignificantly. At each stage the total molecular energy $\mathscr{E}$ is also computed,

$$\mathscr{E} = E + \frac{3}{R}$$

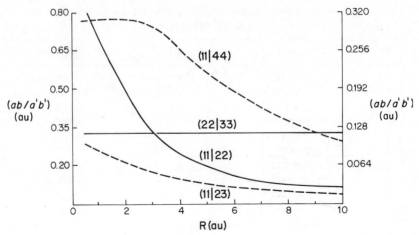

FIGURE 4.6. Plot of electron-repulsion integrals for LiH as a function of internuclear distance. Solid lines correspond to left and dashed line to right scales.

Table 4.9. Convergence of the Total Molecular Energy for Lithium Hydride

Cycle	$\mathscr{E}$ (au)
1	-11.50618^{a}
2	-7.97049
3	-7.97656
4	-7.97692
5	-7.97695
6	-7.97695
7	-7.97696

[a] Energy corresponding to the core Hamiltonian.

Table 4.10. Final Eigenvalues and Eigenvectors for Lithium Hydride

Basis Function[a] ε_i	1 -2.44661576 (au)	2 -0.318456186 (au)	3 $+0.0139237599$ (au)	4 $+0.526166267$ (au)
1	0.005726018	0.659521011	-0.128367785	-1.301124230
2	0.996549241	-0.127959412	-0.137857322	-0.021342984
3	0.016391155	0.321548854	0.812943084	0.862205423
4	-0.004897837	0.272216971	-0.593565168	1.110276520

[a] The basis set is the one listed in Table 4.3.

112

where E is given by eq. (4.1-5) and $3/R$ is the nuclear repulsion energy (in atomic units) at the distance R. Table 4.9 shows the value of $\mathscr{E}$ after each of several computational cycles and Table 4.10 gives the final eigenvalues and eigenvectors. These agree well with the ones reported by Ransil in his 1960 paper (see Exercise 4.3). Our total energy of -7.9770 au compares with one of the best values, -8.0600 au (Bender and Davidson, 1969).

4.5   COMPUTER HIGHLIGHT: CALCULATION OF THE INVERSE OF THE $S^{1/2}$ MATRIX

Familiarity with matrix manipulations is essential in physical chemistry, especially in computing. In this section we illustrate the programming for one such manipulation by determining the inverse of $S^{1/2}$ according to eq. (4.4-11).

We recall from elementary material on determinants that the **cofactor** of the ijth element is $(-1)^{i+j}$ times the determinant (**minor**) formed by striking out the ith row and jth column. In particular, the cofactor of the ijth element of $S^{1/2}$ is $(-1)^{i+j}$ times a determinant of the form

$$
\begin{bmatrix}
KN & KO & KP \\
LN & LO & LP \\
MN & MO & MP
\end{bmatrix}
\tag{4.5-1}
$$

For example, the cofactor of the $1, 2$ element of $S^{1/2}$ is $(-1)^{1+2}$ times the 3×3 determinant in which $K = 2$, $L = 3$, $M = 4$, and $N = 1$, $O = 3$, $P = 4$. Thus, element KN of the 3×3 determinant is the $2, 1$ element of $S^{1/2}$, element KO of the 3×3 determinant is the $2, 3$ element of $S^{1/2}$, element LN of the 3×3 determinant is the $3, 1$ element of $S^{1/2}$, and so on.

Equation (4.4-11) requires the transpose of the matrix of cofactors, but since S is symmetric, so is its square root. No actual transposition of cofactors is therefore needed in the present case. Figure 4.7 lists the program for calculation of the elements of $S^{-1/2}$, given the elements of $S^{1/2}$. Line 400 is the expansion of the general 3×3 determinant in eq. (4.5-1), while line 600 evaluates the denominator needed in eq. (4.4-11). A prudent programmer will verify that his $S^{-1/2}$ matrix, when multiplied by $S^{1/2}$, gives the unit matrix **1**. Such a check can be spliced onto the end of the program in Fig. 4.7.

4.6   BEYOND HARTREE–FOCK

Are the results in Section 4.4 the end of the story for lithium hydride? Certainly not, for we have used a very small basis set. However, even if a very large basis set had been used, significant error in the total energy would still have remained.

```
10    REM  PROGRAM TO CALCULATE THE
      INVERSE OF THE MATRIX (SQ)
25    DIM SQ(5,5)
50    INPUT SQ(1,1),SQ(2,2),SQ(3,3)
      ,SQ(4,4),SQ(1,2),SQ(1,3),SQ(
      1,4),SQ(2,3),SQ(2,4),SQ(3,4)

75    SQ(2,1) = SQ(1,2):SQ(3,1) = SQ
      (1,3):SQ(4,1) = SQ(1,4):SQ(3
      ,2) = SQ(2,3):SQ(4,2) = SQ(2
      ,4):SQ(4,3) = SQ(3,4)
100   DIM M(5,5),SI(5,5)
125   REM  M(I,J) IS AN ELEMENT OF
      THE COFACTOR OF THE IJ-TH EL
      EMENT OF THE 4X4 MATRIX (SQ)
      (THE SQUARE ROOT OF THE OVER
      LAP MATRIX (S))
150   FOR I = 1 TO 4
175   FOR J = 1 TO 4
200   IF I = 1 THEN K = 2:L = 3:M =
      4
225   IF I = 2 THEN K = 1:L = 3:M =
      4
250   IF I = 3 THEN K = 1:L = 2:M =
      4
275   IF I = 4 THEN K = 1:L = 2:M =
      3
300   IF J = 1 THEN N = 2:O = 3:P =
      4
325   IF J = 2 THEN N = 1:O = 3:P =
      4
350   IF J = 3 THEN N = 1:O = 2:P =
      4
375   IF J = 4 THEN N = 1:O = 2:P =
      3
400   M(I,J) =  - 1 ^ (I + J) * (SQ
      (K,N) * SQ(L,O) * SQ(M,P) +
      SQ(K,O) * SQ(L,P) * SQ(M,N) +
      SQ(K,P) * SQ(L,N) * SQ(M,O) -
      SQ(K,P) * SQ(L,O) * SQ(M,N) -
      SQ(K,O) * SQ(L,N) * SQ(M,P) -
      SQ(K,N) * SQ(M,O) * SQ(L,P))

450   NEXT J
500   NEXT I
550   REM  D=DET(SQ), AND (SI) IS
      THE INVERSE OF (SQ)
600   D = SQ(1,1) * M(1,1) + SQ(1,2
      ) * M(1,2) + SQ(1,3) * M(1,3
      ) + SQ(1,4) * M(1,4)
625   PRINT : PRINT
650   FOR I = 1 TO 4
700   FOR J = 1 TO 4
750   SI(I,J) = M(I,J) / D
800   PRINT  TAB( 4);I; TAB( 10);J
      ; TAB( 18);SI(I,J)
850   NEXT J
900   NEXT I
1000  END
```

FIGURE 4.7. Computer program for calculation of the inverse of a symmetric 4 × 4 matrix.

To understand why this is so, we must recall that the ground-state trial wave function for LiH was assumed to be a Slater determinant of molecular spin-orbitals. The results are therefore tied to use of the *orbital approximation*, by which we assert that the wave function can be expressed as a function only of the distances of the electrons from the nuclei. This cannot be a correct description of the molecule.

Motions of the electrons must be correlated with each other. There are two types of correlation: **spin correlation** and **Coulombic correlation**. Spin correlation is automatically taken into account by the use of an antisymmetrized Slater determinant. Two electrons with the same value of m_s are prevented from being in the same region of space. But what if two electrons have opposite values of m_s? The Pauli Antisymmetry Principle does not prohibit their approaching one another closely, but Coulombic repulsion should surely set in at close distances and cause them to avoid each other. The true wave functions for a molecule must, therefore, depend strongly on the distances of the electrons from each other, and this dependence has not been taken into account in our determinantal ground-state wave function for LiH.

The term **correlation energy** refers to the difference between the true (nonrelativistic) total energy of an atom or molecule and the energy calculated in the Hartree–Fock limit. The phrase is really a euphemistic expression for the penalty exacted when we insist upon using a procedure (the orbital approximation) that is inherently incorrect. It turns out to be rather hard to correct quantitatively for the correlation energy. One way in which the correction is attempted is by means of **configuration interaction**, a detailed discussion of which is outside the scope of this book. The underlying idea, however, is to take into account Coulombic correlation by forming linear combinations of the ground-state configuration (i.e., the Slater determinant from the original Hartree–Fock calculation) with several excited-state configurations (Slater determinants) formed by promoting one or more electrons into various of the virtual (unoccupied) molecular orbitals. Thus, if ϕ_1 is the ground-state configuration and ϕ_2 is a configuration in which two electrons have been promoted to a given virtual molecular orbital, we might write for the wave function of the ground state

$$\Psi = c_1\phi_1 + c_2\phi_2$$

The coefficients c_1 and c_2 are then to be found by means of the variation theorem.

In general, many excited configurational functions are needed (even for small molecules) if most of the Coulombic correlation energy is to be accounted for. For example, a configuration-interaction calculation on LiH (Bender and Davidson, 1969) employing over 900 configurations did not quite account for all of the correlation energy. The efficient calculation of atomic and molecular correlation energy is currently an active field of research.

Exercises

4.12.   Verify eqs. (4.4-3) and (4.4-6).

4.13.[C]   Prove the recursion relations in eqs. (4.4-5). Write a program that calculates all of the $A_n(x)$ and $B_n(x)$ for $n = 0$ up to some $n = N$ for an inputted nonzero value of x. How must $B_n(x)$ be evaluated when $x = 0$?

4.14.[C]   Work out the analytical expressions for H_{12} and H_{21} for lithium hydride with arbitrary values of the two Slater screening factors ξ. Using your program from Exercise 4.13, input the values of ξ and the value of R, and show numerically that $H_{12} = H_{21}$. You should get agreement with the entries in Table 4.6.

4.15.   Consider the matrix shown here.

$$\mathbf{A} = \begin{bmatrix} 3 & 2 & 1 \\ 2 & 4 & 0 \\ 1 & 0 & 1 \end{bmatrix}$$

Write out the characteristic equation and then determine the three eigenvalues.

4.16.[C]   Apply Sylvester's theorem to obtain a square root of the matrix $\mathbf{A}$ of the preceding exercise. Arrange your work by means of a program that prints out all the elements of $\mathbf{A}^{1/2}$. Build a check into your program verifying that $\mathbf{A}^{1/2}\mathbf{A}^{1/2} = \mathbf{A}$.

4.17.   Verify that eigenvector 2 in Table 4.10 is normalized, and that eigenvectors 1 and 2 are orthogonal.

4.18.   Are the qualitative trends of the curves in Fig. 4.4 what you would expect? How about the curves in Fig. 4.6?

4.19.   Interpret the behavior of the plot of H_{22} in Fig. 4.5. Calculate the limiting value of H_{22} as $R \to \infty$.

4.20.[C]   The elements of the matrix $\mathbf{S}^{-1/2}$ for LiH are shown here.

$$\begin{bmatrix} 1.34378 & -0.02465 & -0.37688 & -0.44272 \\ -0.02465 & 1.01111 & -0.07672 & 0.009248 \\ -0.37688 & -0.07673 & 1.15396 & 0.16841 \\ -0.44272 & 0.009248 & 0.16841 & 1.19779 \end{bmatrix}$$

With the aid of the program in Fig. 4.7, obtain the inverse of this matrix. Splice onto the program a set of statements to check that $\mathbf{S}^{1/2}$ obeys $\mathbf{S}^{1/2}\mathbf{S}^{-1/2} = \mathbf{1}$.

4.21.   Let $\mathbf{A}'$ be the unsymmetrical matrix obtained by changing the 2, 1 element of $\mathbf{A}$ in Exercise 4.15 from 2 to -2. Modify the program of Exercise 4.20 in order to calculate $\mathbf{A}'^{-1}$.

4.22. Test the Mulliken approximation on $(12|23)$ and $(23|44)$ for LiH.

4.23.^{C} Refer to eq. (4.2-12); then write a program to calculate the number of electrons in LiH from the data in Tables 4.5 and 4.10. Also have the program print out the electron populations in the distributions $\phi_1\phi_2$, $\phi_1\phi_3$, and $\phi_1\phi_4$.

REFERENCES

Bender, C. F. and Davidson, E. R., "Studies in Configuration Interaction: The First-Row Diatomic Hydrides," *Phys. Rev.*, **183**, 23 (1969).

Bessis, N. and Bessis, G., "Analytic Atomic Shielding Parameters," *J. Chem. Phys.*, **74**, 3628 (1981).

Blinder, S. M., "Basic Concepts of Self-Consistent-Field Theory," *Am. J. Phys.*, **33**, 431 (1965). A fine, detailed summary of the theory underlying the Hartree–Fock equations.

Boas, M., *Mathematical Methods in the Physical Sciences*, 2nd ed., Wiley, New York, 1983, Chap. 2. Nice introductory chapter on the calculus of variations.

Boys, S. F., "Electronic Wavefunctions. I. A General Method of Calculation for the Stationary States of any Molecular System," *Proc. Roy. Soc. London Ser. A*, **200**, 542 (1950).

Deif, A. S., *Advanced Matrix Theory for Scientists and Engineers*, Abacus Press, London, 1982, p. 98. Gives a proof of the statement that a real symmetric matrix has only real eigenvalues.

Dence, J. B., *Mathematical Techniques in Chemistry*, Wiley, New York, 1975, pp. 284–287, 333–345. The first set of pages discusses matrix inversion, and the second arrives at the Laplacian in generalized coordinate systems.

Frazer, R. A., Duncan, W. J., and Collar, A. R., *Elementary Matrices*, Cambridge University Press, London, 1963, pp. 78, 82. See these pages for a discussion of Sylvester's theorem.

Hall, G. G., "The Molecular Orbital Theory of Chemical Valency. XVIII. A Method of Calculating Ionization Potentials," *Proc. Roy. Soc. London, Ser. A*, **205**, 541 (1951). Equation (2.10) is our Roothaan eq. (4.2-2). Hall applies his method to methane and ethane, but in a semiempirical way by treating certain integrals as parameters.

Huber, K. P. and Herzberg, G., *Constants of Diatomic Molecules*, Van Nostrand Reinhold, New York, 1979.

Koopmans, T., "Über die Zuordnung von Wellenfunktionen und Eigenwerten zu den einzelnen Elektronen eines Atoms," *Physica*, **1**, 104 (1934). Exposition of the well-known approximate result connecting molecular orbital energies to ionization energies.

Miller, J., Gerhauser, J. M., and Matsen, F. A., *Quantum Chemistry Integrals and Tables*, University of Texas Press, Austin, 1959. Extensive tables of overlap, kinetic energy, potential energy, and electron-repulsion integrals.

Mulliken, R. S., Rieke, C. A., Orloff, D., and Orloff, H., "Formulas and Numerical Tables for Overlap Integrals," *J. Chem. Phys.*, **17**, 1248 (1949). Explicit formulas and some numerical values for the overlap integral between two Slater atomic orbitals.

Mulliken, R. S., "Electronic Population Analysis on LCAO–MO Molecular Wave Function. I," *J. Chem. Phys.*, **23**, 1833 (1955), and succeeding papers in the series. Detailed discussion of the breakdown of electronic populations into various subpopulations.

Offenhartz, P. O'D., *Atomic and Molecular Orbital Theory*, McGraw-Hill, New York, 1970, p. 329. Our eq. (4.4-14) is a particular case of the general Mulliken superposition approximation, $\phi_a\phi_b = \frac{1}{2}S_{ab}(\phi_a\phi_a + \phi_b\phi_b)$, discussed here.

Poirier, R., Kari, R., and Csizmadia, I. G., *Handbook of Gaussian Basis Sets*, Elsevier, Amsterdam, 1985.

Pople, J. A. and Beveridge, D. L., *Approximate Molecular Orbital Theory*, McGraw-Hill, New York, 1970, Chap. 2. This chapter is a nice introduction to self-consistent field molecular orbital theory, including open-shell systems.

Ransil, B. J., "Studies in Molecular Structure. II. LCAO–MO–SCF Wave Functions for Selected First-Row Diatomic Molecules," *Rev. Mod. Phys.*, **32**, 245 (1960). An early classic paper that gives molecular orbitals and their energies for several diatomic molecules (Li_2, C_2, N_2, F_2, LiH, HF, CO). This paper is one of several featured in a very worthwhile issue of *Reviews of Modern Physics*.

Roothaan, C. C. J., "New Developments in Molecular Orbital Theory," *Rev. Mod. Phys.*, **23**, 69 (1951a). Tough sledding, but this is *the* classic paper that revolutionized the handling of the Hartree–Fock equations for molecules.

Roothaan, C. C. J., "A Study of Two-Center Integrals Useful in Calculations on Molecular Structure. I.," *J. Chem. Phys.*, **19**, 1445 (1951b). Provides formulas for calculation of overlap, core Hamiltonian, and some of the simpler electron-repulsion integrals.

Roothaan, C. C. J., Ruedenberg, K., and Jaunzemis, W., "Study of Two-Center Integrals Useful in Calculations on Molecular Structure. III. A Unified Treatment of the Hybrid, Coulomb, and One-Electron Integrals," *J. Chem. Phys.*, **24**, 201 (1956). An alternative treatment to that in Part I of this series, which is useful for the electron-repulsion integrals.

Rosmus, P. and Meyer, W., "IV. Ionization Energies of First and Second Row Diatomic Hydrides and the Spectroscopic Constants of Their Ions," *J. Chem. Phys.*, **66**, 13 (1977).

Silver, D. M. and Nieuwpoort, W. C., "Universal Atomic Basis Sets," *Chem. Phys. Lett.*, **57**, 421 (1978). Authors show that a common or "universal" set of Slater-type orbitals with screening factors calculated in a manner independent of atomic identity works well in the calculation of ground-state energies of several light atoms. A related paper is D. L. Cooper and S. Wilson, "Universal Systematic Sequence of Even-tempered Exponential-type Functions in Electronic Structure Studies," *J. Chem. Phys.*, **77**, 5053 (1982).

Weinstock, R., *Calculus of Variations—With Applications to Physics and Engineering*, Dover, New York, 1974. Chapter 11 is on quantum mechanics and contains a few pages on the Hartree–Fock treatment of atoms.

CHAPTER 5

Calculations of Stationary Molecular Properties

Chapter 4 was concerned with the calculation of the ground-state wave function and energy of a molecule. In this chapter we focus on the use of the wave function and energy to calculate various molecular properties. We also discuss semiempirical procedures that can be applied to molecules too large to be handled by the more rigorous *ab initio* methods of Chapter 4.

5.1 DIPOLE MOMENTS

The **dipole moment** operator $\hat{\mu}$ of a closed-shell molecule of $2N$ electrons and M nuclei is given (in atomic units) by

$$\hat{\mu} = \sum_{\alpha=1}^{M} Z_\alpha \mathbf{R}_\alpha - \sum_{i=1}^{2N} \mathbf{r}_i \tag{5.1-1}$$

where the sum on α is over nuclei (of charge Z_α) and that on i is over electrons (of charge -1). The vector μ points from the center of charge of the electrons to that of the nuclei. Let us assume the nuclei are fixed and the electrons are described by the Born–Oppenheimer electronic wave function $\Psi(q)$. Then the expected value of $\hat{\mu}$ is, according to QM Postulate 5 (see Section 2.6),

$$\langle \mu \rangle = \int \Psi^*(q) \left(\sum_\alpha Z_\alpha \mathbf{R}_\alpha - \sum_i \mathbf{r}_i \right) \Psi(q) \, dq$$

$$= \sum_\alpha Z_\alpha \mathbf{R}_\alpha - \int \Psi^*(q) \left(\sum_i \mathbf{r}_i \right) \Psi(q) \, dq \tag{5.1-2}$$

where the second line follows from the fact that the nuclear contribution to $\hat{\mu}$ does not depend on the electronic configuration q. Note that the electronic contribution to $\hat{\mu}$ consists of a sum of one-electron operators $(\mathbf{r}_i)$ and thus is analogous to $\hat{H}_1$ in eq. (4.1-2). Further, since $\mathbf{r}_i = x\mathbf{e}_x + y\mathbf{e}_y + z\mathbf{e}_z$, the components of $\langle \boldsymbol{\mu} \rangle$ can be calculated separately. If $\Psi(q)$ is a single Slater determinantal ground-state wave function (see Section 3.3) compounded of N spatial orbitals $\{\chi_n(i)\}$, then the expected value of the z component of the electronic part of $\hat{\mu}$ is

$$
\langle \mu_z \rangle_e = \int \Psi^*(q) \sum_{i=1}^{2N} z_i \Psi(q) \, dq = 2 \sum_{n=1}^{N} \int \chi_n^*(i) z_i \chi_n(i) \, dq_i
$$

$$
= 2 \sum_{n=1}^{N} \int \sum_a c_{an}^* \phi_a^*(i) z_i \sum_b c_{bn} \phi_b(i) \, dq_i
$$

$$
= \sum_a \sum_b 2 \sum_n^{\text{occ}} c_{an}^* c_{bn} \int \phi_a^*(i) z_i \phi_b(i) \, dq_i
$$

$$
= \sum_a \sum_b P_{ab} \int \phi_a^*(i) z_i \phi_b(i) \, dq_i
$$

$$
= \sum_a \sum_b P_{ab} z_{ab} \tag{5.1-3}
$$

after replacement of each spatial molecular orbital by a linear combination of basis functions $\{\phi_a\}$ as in eq. (4.2-1) and after use of the definition of the electron-density matrix P_{ab} from eq. (4.2-9).

Equation (5.1-3) represents the only contribution to the ground-state dipole moment of a diatomic molecule (such as lithium hydride) if the molecule is

Table 5.1.   The 10 Independent Elements of the $P(I, J)$ and $Z1(I, J)$ Matrices Needed for a Hartree–Fock Calculation of the Dipole Moment of Lithium Hydride

I	J	$P(I, J)^a$	$Z1(I, J)$
1	1	0.870002	1.50771
1	2	−0.157369	−0.096076
1	3	0.424324	0.402492
1	4	0.359010	0.871599
2	2	2.018968	−1.50771
2	3	−0.049621	−0.250679
2	4	−0.079427	0.114619
3	3	0.207325	−1.50771
3	4	0.174902	2.220578
4	4	0.148252	−1.50771

^{a} Computed from the data in Table 4.10.

oriented along the z axis as in Fig. 4.3. The matrix elements z_{ab} in eq. (5.1-3) are again conveniently evaluated in confocal elliptical coordinates. From eq. (3.5-11) we have $z_i = R_0\xi\eta/2$ and

$$z_{ab} = \frac{R_0^4}{16} \int_0^{2\pi} d\theta \int_{-1}^1 \eta \, d\eta \int_1^\infty \xi(\xi^2 - \eta^2)\phi_a^*\phi_b \, d\xi \qquad (5.1\text{-}4)$$

For example, let $a = 1$ and $b = 3$ according to the basis set labeling in Table 4.3 for lithium hydride. The element z_{13} is then

$$z_{13} = \frac{R_0^5}{16}\left(\frac{0.65^5}{3}\right)^{1/2}$$

$$\times \left[A_4(\alpha)B_1(\beta) + A_3(\alpha)B_2(\beta) - A_2(\alpha)B_3(\beta) - A_1(\alpha)B_4(\beta) \right] \quad (5.1\text{-}5)$$

where the functions $A_n(\alpha)$ and $B_n(\beta)$ are defined in eqs. (4.4-4). The argu-

Table 5.2. SCF Calculations of the Dipole Moment of Selected Small Molecules

Molecule	$|\mu|$, Calculated[a]	$|\mu|$, Experimental[a, b]	Comments
LiH	6.426	$\left\{ 5.882^c \right.$	Subminimal Slater basis set[d]
	6.008		Extended Gaussian basis set[e]
H_2O	1.99	1.85	Extended Gaussian basis set[f]
NH_3	1.80	1.47	Minimal Slater basis set[g]
$H_2C=O$	2.587	$\left\{ 2.33 \right.$	Extended Gaussian basis set[h]
	0.983		Minimal Slater basis set[i]
$HC\equiv CF$	0.913	0.73	Slater ξ optimized; polarization functions present[j]
NaCl	9.243	9.00	Double-ξ basis set; polarization functions present[k]
$F-CN$	2.279	2.17	Same as for $HC\equiv CF$[j]

[a] The magnitude of the dipole moment vector, in units of debyes.
[b] R. D. Nelson, Jr., D. R. Lide, Jr., and A. A. Maryott, "Selected Values of Electric Dipole Moments for Molecules in the Gas Phase," in R. C. Weast and M. J. Astle (Eds.), *CRC Handbook of Chemistry and Physics*, 61st ed., CRC Press, Boca Raton, 1980, p. *E*-64.
[c] L. Wharton, L. P. Gold, and W. Klemperer, *J. Chem. Phys.*, **33**, 1255 (1960).
[d] This work.
[e] I. G. Csizmadia, *J. Chem. Phys.*, **44**, 1849 (1966).
[f] J. W. Moskowitz and M. C. Harrison, *J. Chem. Phys.*, **43**, 3550 (1965).
[g] U. Kaldor and I. Shavitt, *J. Chem. Phys.*, **45**, 888 (1966).
[h] N. W. Winter, T. H. Dunning, Jr., and J. H. Letcher, *J. Chem. Phys.*, **49**, 1871 (1968).
[i] M. D. Newton and W. E. Palke, *J. Chem. Phys.*, **45**, 2329 (1966).
[j] A. D. McLean and M. Yoshimine as quoted in M. Krauss (Ed.), *Compendium of ab initio Calculations of Molecular Energies and Properties*, NBS Technical Note 438, U.S. Dept. of Commerce, 1967, pp. 68–69.
[k] R. L. Matcha, *J. Chem. Phys.*, **48**, 335 (1968).

```
50    REM  PROGRAM TO CALCULATE THE
      DIPOLE MOMENT IN DEBYES OF
      LIH FROM A HARTREE-FOCK WAVE
      FUNCTION
75    DIM Z1(5,5),P(5,5),A(10),B(10
      ),AU(10),BU(10)
100   INPUT R
125   INPUT P(1,1),P(2,2),P(3,3),P
      (4,4)
150   INPUT P(1,2),P(1,3),P(1,4),P
      (2,3),P(2,4),P(3,4)
175   AL = R:BE =  - R:N = 3
200   GOSUB 1500
225   Z1(1,1) = 0.125 * R ^ 4 * (A(
      3) * B(1) - A(1) * B(3))
250   AL = 1.85 * R:BE = 0.85 * R:N
       = 3
275   GOSUB 1500
300   Z1(1,2) = 0.125 * 2.7 ^ 1.5 *
      R ^ 4 * (A(3) * B(1) - A(1) *
      B(3))
325   AL = 0.825 * R:BE =  - 0.175 *
      R:N = 4
350   GOSUB 1500
375   Z1(1,3) = (1 / 16) * (0.65 ^
      5 / 3) ^ 0.5 * R ^ 5 * (A(4)
      * B(1) + A(3) * B(2) - A(2)
      * B(3) - A(1) * B(4))
400   Z1(1,4) = (1 / 16) * 0.65 ^ 2
      .5 * R ^ 5 * (A(4) * B(2) -
      A(2) * B(4) + A(3) * B(1) -
      A(1) * B(3))
425   AL = 2.7 * R:BE = AL:N = 3
450   GOSUB 1500
475   Z1(2,2) = 0.125 * 2.7 ^ 3 * R
      ^ 4 * (A(3) * B(1) - A(1) *
      B(3))
500   AL = 1.675 * R:BE = AL:N = 4
525   GOSUB 1500
550   Z1(2,3) = (1 / 16) * 2.7 ^ 1.
      5 * (0.65 ^ 5 / 3) ^ 0.5 * R
      ^ 5 * (A(4) * B(1) + A(3) *
      B(2) - A(2) * B(3) - A(1) *
      B(4))
575   Z1(2,4) = (1 / 16) * (2.7 ^ 3
      * 0.65 ^ 5) ^ 0.5 * R ^ 5 *
      (A(4) * B(2) + A(3) * B(1) -
      A(2) * B(4) - A(1) * B(3))
600   AL = 0.65 * R:BE = AL:N = 5
625   GOSUB 1500
650   Z1(3,3) = (1 / 96) * 0.65 ^ 5
      * R ^ 6 * (A(5) * B(1) + 2 *
      A(4) * B(2) - 2 * A(2) * B(4
      ) - A(1) * B(5))
675   Z1(3,4) = (1 / 32) * 3 ^  - 0
      .5 * 0.65 ^ 5 * R ^ 6 * (A(5
      ) * B(2) + A(4) * B(3) + A(4
      ) * B(1) + A(3) * B(2) - A(3
      ) * B(4) - A(2) * B(3) - A(2
      ) * B(5) - A(1) * B(4))
700   Z1(4,4) = (1 / 32) * 0.65 ^ 5
      * R ^ 6 * (A(5) * B(3) + 2 *
      A(4) * B(2) + A(3) * B(1) -
      A(3) * B(5) - 2 * A(2) * B(4
      ) - A(1) * B(3))
715   PRINT : PRINT : PRINT
725   PRINT  TAB( 3);"DIPOLE MOMEN
      T OF LITHIUM HYDRIDE"
750   PRINT  TAB( 3);"------------
      --------------------"
775   PRINT : PRINT
800   PRINT  TAB( 5);"I"; TAB( 11)
      ;"J"; TAB( 22);"Z1(I,J)"
825   PRINT  TAB( 4);"---"; TAB( 1
      0);"---"; TAB( 22);"-------"
      : PRINT
875   FOR I = 1 TO 4
900   FOR J = 1 TO 4
920   Z1(J,I) = Z1(I,J):P(J,I) = P(
      I,J)
950   PRINT  TAB( 5);I; TAB( 11);J
      ; TAB( 20);Z1(I,J)
975   NEXT J
1000  NEXT I
1025  PRINT : PRINT
1050  RE = 0
1100  FOR J = 1 TO 4
1125  FOR I = 1 TO 4
1150  RE = RE + P(I,J) * Z1(I,J)
1175  NEXT I
1200  NEXT J
1250  DM =  - 2.541798 * (R + RE)
1275  PRINT "DIPOLE MOMENT IN DEB
      YES IS"; TAB( 28);DM
1300  END
1500  FOR L = 0 TO N
1525  A(0) = AL ^  - 1 * EXP ( -
      AL)
1550  A(L + 1) = A(L) * (L + 1) /
      AL + AL ^  - 1 * EXP ( - AL
      )
1575  AU(0) = BE ^  - 1 * EXP ( -
      BE)
1600  AU(L + 1) = AU(L) * (L + 1) /
      BE + BE ^  - 1 * EXP ( - BE
      )
1625  BU(0) =  - (BE ^  - 1) * EXP
      (BE)
1650  BU(L + 1) =  - BU(L) * (L +
      1) / BE - BE ^  - 1 * EXP (
      BE)
1675  B(L) = (( - 1) ^ (L + 1)) *
      BU(L) - AU(L)
1700  NEXT L
1725  RETURN
```

FIGURE 5.1. Computer program for the calculation of the dipole moment of lithium hydride.

ments of these functions in the present case are $\alpha = 0.825R_0$ and $\beta = -0.175R_0$. The values of z_{13} and the 9 other independent elements z_{ab} are given in Table 5.1 [where they are listed as $Z1(I, J)$, with $I \equiv a$ and $J \equiv b$] along with values of the 10 independent elements of the electron-density matrix as computed from the eigenvectors for LiH in Table 4.10.

Combination of eqs. (5.1-2) and (5.1-3) gives

$$\langle \mu \rangle = \left[\left(\frac{R_0}{2} - \frac{3R_0}{2} \right) - \langle \mu_z \rangle_e \right] \mathbf{e}_z$$

$$= \left[-R_0 - (-0.4873) \right] \mathbf{e}_z$$

$$= -2.5281 \mathbf{e}_z$$

where the value $R_0 = 3.0154$ is taken from Table 4.3. Conversion to debyes yields, finally, the value $\langle \mu \rangle = -6.426\mathbf{e}_z$ D. The experimental value is $-5.882\mathbf{e}_z$ D; the agreement with experiment is quite satisfactory for a wave function at the subminimal basis set level. Table 5.2 collects some Hartree–Fock calculations of the dipole moment of LiH and of other small molecules. We note that although none of the wave functions is of the highest quality and although the results are not strictly comparable, good agreement with experiment is found in most cases. Figure 5.1 shows the computer program that was used for our LiH calculation. Observe how the $A_n(\alpha)$ and $B_n(\beta)$ functions are computed recursively in a subroutine (statements 1525–1725) at the end of the program.

5.2 ATOMIC CHARGES

The property of "atomic charge" is perhaps not a molecular property at all! From a purist's viewpoint, once an atom A enters into a chemical combination with another atom, it is A no longer. A molecule is a new entity, distinct from its atomic constituents. However, chemists have found it profitable conceptually to carry over to the molecule certain properties associated with the atoms. Size is a good example of such a property. Another is the **charge q on an atom A in a molecule**. This ought to mean the difference between the electron population Q_A^0 of the neutral, free atom and the electron population Q_A assigned to A in the molecule.

$$q_A = Q_A^0 - Q_A \tag{5.2-1}$$

The problem is how to define the quantity Q_A.

Following Mulliken (see Section 4.2), we interpret a term P_{aa} in the electron-density matrix as the electron population belonging exclusively to atomic orbital a [see eq. (4.2-11b)]. A term $P_{ab}S_{ab}$ is the electronic population of the atomic overlap distribution $\phi_a^* \phi_b$. Mulliken extended these ideas by arbitrarily assigning as the electron population *of atom A* the sum of all those terms P_{aa}^A involving atomic orbitals on A, plus $\frac{1}{2}$ of the sum of all of those terms $P_{ab}^{AB}S_{ab}$ involving overlap of orbitals on A with orbitals on all other atoms B of the molecule. For a closed-shell molecule, all occupied molecular orbitals contain two electrons and we can express the charge on an atom A in a molecule as

$$q_A = Q_A^0 - \sum_a P_{aa}^A - \sum_{B \neq A} \sum_a \sum_b P_{ab}^{AB}S_{ab} \qquad (5.2\text{-}2)$$

$(a = \text{AO on atom A}, \; b = \text{AO on atom B})$

Let us now apply eq. (5.2-2) to lithium hydride. On the hydrogen atom there is only the $1s$ Slater orbital ($a = 1$). The index b, therefore, ranges from 2 to 4. We obtain

$$q_H = 1.000 - P_{11} - \sum_{b=2}^{4} P_{1b}^{HLi}S_{1b}$$

$$= 1.000 - (P_{11} + P_{12}S_{12} + P_{13}S_{13} + P_{14}S_{14})$$

$$= 1.000 - (1.256)$$

after making use of data in Tables 4.5 and 5.1. The charge on hydrogen is thus $-0.256e$, and that on lithium is $+0.256e$.

These charges are consistent with known physicochemical properties of lithium hydride. The material has a moderately high melting point (688°C), indicating some ionic character. It also can serve as a hydride donor, for example, to boron in the preparation of diborane.

$$6\text{LiH} + 8\text{BF}_3 \xrightarrow{\text{ether}} \text{B}_2\text{H}_6 + 6\text{LiBF}_4$$

On the other hand, considerable covalent character is indicated by the stability of LiH to oxygen below red heat (e.g., RbH, ignites spontaneously) and by its much less vigorous reaction with water than, say, NaH. Further, the apparent radius of H$^-$ in LiH is 0.1 Å less than that in NaH and KH.

Mulliken's definition of atomic charge can be criticized on two counts. First, the population P_{aa}^A is assigned entirely to atom A, even though the basis function a may have its maximum at a significant distance from A. Second, the total electron population $(2c_{an}^* c_{bn} S_{ab} + 2c_{bn}^* c_{an} S_{ba})$ in the overlap between orbital a on A and orbital b on B is divided equally between A and B; this would be unreasonable except in the special case where the two basis functions are identical. Various alternatives to Mulliken's definition have been proposed, but it is likely there is no truly satisfactory definition of atomic charge.

Mulliken himself has provided an interesting example of the weakness of population analysis (Politzer and Mulliken, 1971). Two very accurate wave functions for hydrogen fluoride were obtained from the literature; one used a basis set of 16 members, and the other a basis set of 18 members. Nearly

Table 5.3.   Calculated Atomic Charges for Some Linear Molecules

Molecule	Atom	Charge (Units of e)
LiH	Li	$+0.37^a$
HCCH	H	$+0.14^a$
HCCF	H, C	$+0.15, -0.19^a$
	C, F	$+0.09, -0.05$
HCCLi	C, Li	$-0.36, +0.49^a$
NNO	N	-0.08^b
	N, O	$+0.33, -0.25$
OCO	C	$+0.46^b$
HF	F	-0.27^c
NaF	F	-0.62^d

^{a}P. Politzer and R. R. Harris, *J. Am. Chem. Soc.*, **92**, 6451 (1970).
^{b}P. Politzer and P. H. Reggio, *J. Am. Chem. Soc.*, **94**, 8308 (1972).
^{c}P. Politzer and R. S. Mulliken, *J. Chem. Phys.*, **55**, 5135 (1971).
^{d}P. Politzer, *Theor. Chim. Acta* (*Berlin*), **23**, 203 (1971).

identical total energies, -100.0573 ± 0.0002 au, are predicted by the two wave functions. Yet, a population analysis based on one yields a charge on F of $-0.23e$ but a charge of $-0.48e$ with the other.

A recent example of an alternative definition is that of Politzer (Politzer and Harris, 1970). In it, the total volume of a molecule is partitioned into regions "belonging" to the constituent atoms, and the electron population associated with a particular atom is then ascertained by integrating the electronic density $\rho(\mathbf{r})$ over the region of space belonging to that atom. The atomic regions are defined such that in the limiting case of no interactions between the atoms, the electronic population associated with each one would be the same as for the free atom. Some results computed in this manner are summarized in Table 5.3. Noteworthy are the distinctly positive values for hydrogen in acetylene (consistent with the known acidity of this hydrocarbon), the strongly positive value on the central nitrogen in NNO, and the near neutrality of fluorine in fluoroacetylene. It is interesting that in an analysis of the pi-electron population in acetylene by his method, Politzer finds that 40% of the pi-electronic charge resides outside the C—C nuclear region, mainly in the adjacent bonding regions (Politzer and Harris, 1971). This provides a very direct explanation for the observed shortening of a single bond by the presence of an adjacent triple bond.

Exercises

5.1.   Verify the forms for $Z1(1,3)$ and $Z1(1,4)$ in the program of Fig. 5.1. Verify the numerical factor appearing in line 1250 for the conversion of atomic units to debyes.

5.2. Some of the integrals needed in eq. (5.1-3) can be written down readily in terms of certain previously calculated integrals. First, show that $z = r_H\cos\theta_H + \frac{1}{2}R = r_L\cos\theta_L - \frac{1}{2}R$. Also, note that if $1s_N = (\xi^3/\pi)^{1/2}e^{-\xi r}$ is a $1s$ Slater orbital centered on a nucleus N, then $(\xi r_N\cos\theta_N)1s_N$ is a $2p_z$ Slater orbital centered on that same nucleus. From these facts, show then that $Z1(2,3)$ is just $-\frac{1}{2}RS_{23}$, where S_{23} is the overlap integral for basis functions 2 and 3. Show also that $Z1(3,3)$ and $Z1(4,4)$ can be analogously obtained with very little work. Which overlap integral is needed to work out $Z1(2,4)$?

5.3.^{C} Write a modified version of the program in Fig. 5.1, and then run it with the coefficients given in Exercise 4.7 in order to calculate the dipole moment of HF. Compare with the experimental value of 1.82 D. Clementi (1962), using an extended basis set of Slater orbitals, calculated $\mu = 1.984$ D.

5.4.^{C} The overlap matrix for HF using the same basis set as in Exercise 5.3 is shown here.

	1	2	3	4	5	6
1	1					
2	0.0548	1				
3	0.4717	0.2377	1			
4	0.2989	0	0	1		
5	0	0	0	0	1	
6	0	0	0	0	0	1

The basis functions are numbered in the same order as in Exercise 4.7. Write a computer program to calculate the charge on fluorine according to a Mulliken population analysis, and compare with Politzer's value in Table 5.3.

5.5.^{C*} Ransil has reported a ground-state wave function for LiF from a minimal basis set of Slater orbitals with optimized screening factors (Ransil, 1960). The screening factors ξ are as follows:

	Basis function	ξ
1	$1s_F$	8.6501
2	$2s_F$	2.5639
3	$2pz_F$	2.5498
4	$1s_{Li}$	2.6865
5	$2s_{Li}$	0.6372
6	$2pz_{Li}$	0.6372

The occupied molecular orbitals have the following coefficients:

AO \ MO	1	2	3	4	5	6
1	-0.99710	-0.00781	0.24769	0.02166	-0.01185	-0.08247
2	-0.01280	0.02724	-0.99524	-0.11824	0.05940	0.59843
3	-0.00061	0.01715	-0.04596	0.98274	0.06751	0.19728
4	0.00017	0.99458	0.07925	-0.08815	0.15336	0.05399
5	0.00177	0.01610	-0.04296	0.10478	-0.89717	-0.55028
6	0.00250	-0.00357	-0.04834	0.08303	0.46436	-0.99248

After working out the necessary overlap integrals, use the program of Exercise 5.3 to calculate the dipole moment of LiF. The experimental value is 6.33 D; McLean (1964) calculated a value of 6.30 D with an extended basis set.

5.6. A ground-state wave function compounded from an extended Slater basis set with *polarization* functions for NaF was obtained by Matcha (1967). The eigenvector that is essentially a $1s$ orbital on fluorine has an energy of -26.0627 au. Politzer has proposed a charge-energy relationship (Politzer and Politzer, 1973) for fluorine in various molecules:

$$1.2717q_F + \sum_B \frac{q_B}{R_{FB}} = -1.6793\varepsilon_{1s,\,F} - 44.3643$$

Calculate the charge q_F on fluorine in NaF. Use $R_{NaF} = 3.628$ au. Calculate for comparison the charge on fluorine in HF; an accurate wave function obtained by Cade and Huo (1967) led to an orbital energy of -26.29428 au for the molecular orbital that is essentially a $1s$ orbital on fluorine. Use $R_{HF} = 1.7328$ au.

5.3 DISSOCIATION ENERGIES

Another molecular property of considerable interest is the **spectroscopic dissociation energy**, D_0. We may define this as the total energy of the products in whatever states they are formed minus the energy of the original molecule in its ground electronic and vibrational state (Herzberg, 1950). The particular case of diatomic molecules is easiest to discuss. In Fig. 5.2 we show a typical potential-energy curve for a diatomic molecule A—B in its ground electronic state. Addition of successive quanta of vibrational energy to the molecule ultimately results in the onset of a continuous spectrum of energies. The difference between the energy where the continuum begins and the energy at the minimum of the potential-energy curve is the **electronic dissociation energy**, D_e. This is not directly measurable; rather, if the zero-point energy $\hbar\omega_0/2$ (easily deduced from the infrared or Raman spectrum of the molecule) is subtracted from D_e, one obtains the spectroscopic dissociation energy D_0, which can be measured. For a diatomic molecule the products are usually the atoms in their *ground* electronic states. For polyatomic molecules the situation is more complex. First, the potential-energy plot is no longer a planar curve

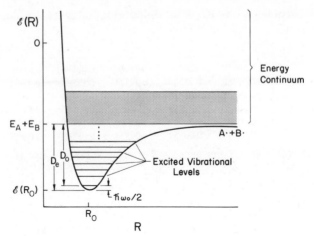

FIGURE 5.2. Illustrating the spectroscopic dissociation energy D_0 and the electronic dissociation energy D_e of a typical diatomic molecule AB. R is the internuclear distance.

but a multidimensional surface. Second, more than one pathway of dissociation is available to a polyatomic molecule. For example, NO_2 can dissociate in either of the following two ways:

$$NO_2 \rightarrow NO + O \qquad D_0 = 0.114 \text{ au}$$
$$NO_2 \rightarrow N + O_2 \qquad D_0 = 0.166 \text{ au}$$

Third, the dissociation products of a polyatomic molecule, unlike those of a diatomic molecule, may have various amounts of internal energy. The dissociation energies of far fewer polyatomic molecules than of diatomic molecules have been determined accurately, and so we shall only consider the latter here.

In principle, one can proceed with the calculation of D_0 for a diatomic molecule as follows: (a) invoke the Born–Oppenheimer approximation (see Section 3.4) and obtain $\mathscr{E}(R)$ as a function of the internuclear distance R; deduce the value $\mathscr{E}(R_0)$ at the equilibrium distance R_0; (b) fit $\mathscr{E}(R)$ to some analytic form and then solve the nuclear wave equation [eq. (3.4-3)]; deduce the zero-point energy $\hbar\omega_0/2$; (c) solve the energy-eigenvalue equation separately for the two atomic products and obtain their ground-state energies E_A and E_B. Then D_0 is given by

$$D_0 = E_A + E_B - \frac{1}{2}\hbar\omega_0 - \mathscr{E}(R_0) \qquad (5.3\text{-}1)$$

Table 5.4 gathers some typical results using the experimental values, however, for the zero-point energies. It is apparent that the Hartree–Fock calculations consistently do a poor job in estimating D_0, often only 50–60% of D_0 being accounted for. In the exceptional case of diatomic fluorine, not even the correct sign of D_0 is obtained! Thus, we see that whereas a good

**Table 5.4.   Calculated Spectroscopic Dissociation Energies D_0
for Selected Diatomic Molecules**

Molecule	$\hbar\omega_0/2$ (au)[a]	D_0(Calculated) (au)[b]	D_0(Experimental) (au)[d]
LiH	0.003	0.052	0.089
HF	0.009	0.152	0.215
CO	0.005	0.285	0.408
NH	0.007	0.070	0.138
CH	0.007	0.084	0.128
BF	0.003	0.189[c]	0.405[c]
F_2	0.002	-0.012[c]	0.059

[a] G. Herzberg, *Spectra of Diatomic Molecules*, 2nd ed., Van Nostrand Reinhold, New York, 1950.
[b] Calculated from data in W. Huo, *J. Chem. Phys.*, **43**, 624 (1965), and P. E. Cade and W. Huo, *J. Chem. Phys.*, **47**, 614 (1967), except where noted.
[c] B. J. Ransil, *Rev. Mod. Phys.*, **32**, 239 (1960).
[d] From D_0 values in P. G. Wilkinson, *Astrophys. J.*, **138**, 778 (1963), except where noted.

Hartree–Fock wave function may account for 99% of the total molecular energy, it may still yield a very poor dissociation energy. This is unfortunate because values of D_0 are of the same order of magnitude as the strengths of chemical bonds, and therefore are of chemical significance. It is generally agreed that the inability of Hartree–Fock wave functions to predict dissociation energies accurately is due largely to a failure to account for electron correlation (see Section 4.6) in both the molecule and its atomic constituents. When *extensive* configuration interaction is carried out, much improved results for the dissociation energies are usually obtained.

5.4   POLARIZABILITY—BACKGROUND

If a molecule is placed in a *weak* electric field **E**, a dipole moment μ^{ind} is induced in the molecule. The moment μ^{ind} is over and above any permanent moment μ_0 that the molecule may already possess (see Section 5.1). It is generally found that μ^{ind} is not in the same direction as **E** (Fig. 5.3). However, each component of μ^{ind} is found to be linear in the components of **E**. For example, if the field is parallel to the z axis,

$$\mu_x^{\text{ind}} = \alpha_{xz} E_z$$

$$\mu_y^{\text{ind}} = \alpha_{yz} E_z$$

$$\mu_z^{\text{ind}} = \alpha_{zz} E_z \tag{5.4-1}$$

The proportionality constant α_{iz} is associated with the ith component of the moment induced by the electric field along the z axis.

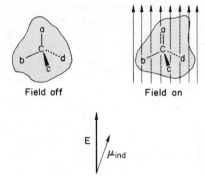

FIGURE 5.3. Illustrating the tensorial nature of the molecular polarizability.

A dipole moment is induced as charges are displaced. Imagine the electric field to be "turned on" gradually. If the charge q is displaced a distance dz by a field E_z, the work dw done on the charge is

$$dw = qE_z \, dz$$

$$= E_z \, d\mu_z^{\text{ind}}$$

$$= E_z \alpha_{zz} \, dE_z \tag{5.4-2}$$

from eq. (5.4-1). Recall from Section 1.1 that the negative of the work done by a conservative force on a particle as it moves from one point to another defines a difference of potential energy U between the two points. Let us take the zero of the potential energy to be the configuration before any displacement has occurred (i.e., at zero field). Then the potential energy of the charge in field E_z is

$$U = -\int_0^{E_z} E_z \alpha_{zz} \, dE_z$$

$$= -\frac{1}{2}\alpha_{zz}E_z^2 \tag{5.4-3}$$

Thus, U represents the additional energy the molecule has by virtue of the interaction of the induced moment with the electric field.

The relations (5.4-1)–(5.4-3) can be generalized to a field applied along an arbitrary direction.

$$\mu_x^{\text{ind}} = \alpha_{xx}E_x + \alpha_{xy}E_y + \alpha_{xz}E_z$$

$$\mu_y^{\text{ind}} = \alpha_{yx}E_x + \alpha_{yy}E_y + \alpha_{yz}E_z$$

$$\mu_z^{\text{ind}} = \alpha_{zx}E_x + \alpha_{zy}E_y + \alpha_{zz}E_z$$

or in matrix form,

$$
\begin{bmatrix} \mu_x^{ind} \\ \mu_y^{ind} \\ \mu_z^{ind} \end{bmatrix} = \begin{bmatrix} \alpha_{xx} & \alpha_{xy} & \alpha_{xz} \\ \alpha_{yx} & \alpha_{yy} & \alpha_{yz} \\ \alpha_{zx} & \alpha_{zy} & \alpha_{zz} \end{bmatrix} \begin{bmatrix} E_x \\ E_y \\ E_z \end{bmatrix}
\tag{5.4-4}
$$

Hence, in the integrand of eq. (5.4-3) we must replace E_z and dE_z by the vectors $\mathbf{E}$ and $d\mathbf{E}$, and α_{zz} must be replaced by the set of nine elements of the array $\boldsymbol{\alpha}$. The potential energy is then given by

$$
U = -\frac{1}{2}\sum_i\sum_j\alpha_{ij}E_iE_j
\tag{5.4-5a}
$$

The set of nine numbers α_{ij} constitutes what is called a **tensor of second rank**, and can be represented by the matrix $\boldsymbol{\alpha}$.[1] It has the unusual property that its product with a vector gives another vector, but generally in a different direction,

$$
\boldsymbol{\mu}^{ind} = \boldsymbol{\alpha} \cdot \mathbf{E}
$$

while multiplication once again by a vector gives a scalar

$$
-\frac{1}{2}\mathbf{E}\cdot\boldsymbol{\mu}^{ind} = -\frac{1}{2}\mathbf{E}\cdot\boldsymbol{\alpha}\cdot\mathbf{E} = U
\tag{5.4-5b}
$$

The tensor $\boldsymbol{\alpha}$ is called the **polarizability tensor**. It can be shown to be symmetric, that is, $\alpha_{ij} = \alpha_{ji}$ (Feynman et al., 1964). Therefore, to prescribe uniquely the polarizability tensor, we need specify only six elements. Further, by a suitable rotation of the xyz coordinate system, a new coordinate system can be found in which $\boldsymbol{\alpha}$ becomes diagonal. Thus, all information about $\boldsymbol{\alpha}$ is in fact specified by just the three numbers $\alpha_{xx}, \alpha_{yy}, \alpha_{zz}$.

$$
\boldsymbol{\alpha} = \begin{bmatrix} \alpha_{xx} & 0 & 0 \\ 0 & \alpha_{yy} & 0 \\ 0 & 0 & \alpha_{zz} \end{bmatrix}
\tag{5.4-6}
$$

We note that one is usually concerned with the **static polarizability**, the polarizability that obtains when the electric field $\mathbf{E}$ is time-independent. If the field oscillates (e.g., if $\mathbf{E} = \mathbf{E}_0\cos\omega_0 t$), then the response of the molecule varies with time and the polarizability, now called the **dynamic polarizability**, becomes a function of ω_0 and t.

---

[1] Other tensorial quantities of chemical interest include moment of inertia, electrical conductivity, and the nuclear magnetic shielding parameter (Dence, 1975).

We can write the following classical phenomenological expression for the total energy of a molecule in a static electric field (Buckingham, 1959):

$$\mathscr{E} = \mathscr{E}^{(0)} - \mu_0 \cdot \mathbf{E} - \frac{1}{2}\mathbf{E} \cdot \alpha \cdot \mathbf{E}$$

$$- \frac{1}{6}\sum_i \sum_j \sum_k \beta_{ijk} E_i E_j E_k + \cdots - \qquad (5.4\text{-}7)$$

In eq. (5.4-7), $\mathscr{E}^{(0)}$ is the energy of the molecule in the absence of the electric field. The term proportional to $\mathbf{E}$ arises from the interaction of the permanent moment with the field; the term proportional to the square of the field results from the coupling of the *induced* moment to the field. The next term in the energy expression involves elements β_{ijk} of the so-called **hyperpolarizability tensor**, a third-rank tensor. When the field strength is sufficiently low, only the first three terms of eq. (5.4-7) are significant. We see, therefore, from (5.4-7) that the components of the polarizability tensor α are given by

$$\alpha_{ij} = - \left[\frac{\partial^2 \mathscr{E}}{\partial E_i \, \partial E_j} \right]_{\mathbf{E}=0} \qquad (5.4\text{-}8)$$

Our next task is to calculate the α_{ij} according to quantum mechanics.

5.5   QUANTUM MECHANICAL TREATMENT OF MOLECULAR POLARIZABILITY—THE CLOSURE RELATION

Within the context of quantum mechanics, we interpret $\mathscr{E}$ in eq. (5.4-8) as the expected value of the Hamiltonian operator for the system. For a molecule in a static field we have

$$\hat{H} = \hat{H}^{(0)} - \left(\sum_{\alpha=1}^{M} Z_\alpha \mathbf{R}_\alpha - \sum_{i=1}^{2N} \mathbf{r}_i \right) \cdot \mathbf{E} \qquad (5.5\text{-}1)$$

Since $\mathbf{E}$ is assumed small, we can use perturbation theory to calculate the corrections to the ground-state energy $\mathscr{E}_0^{(0)}$ and then differentiate these with respect to the field components to obtain α_{ij} according to eq. (5.4-8). The first-order correction [see eq. (3.1-8)]

$$\mathscr{E}_0^{(1)} = - \int \Psi_0^{(0)*} \left(\sum_{\alpha=1}^{M} Z_\alpha \mathbf{R}_\alpha - \sum_{i=1}^{2N} \mathbf{r}_i \right) \cdot \mathbf{E} \Psi_0^{(0)} \, dq$$

$$= - \langle \mu \rangle \cdot \mathbf{E}$$

being proportional to **E**, cannot contribute to α. However, the second-order correction does.

Making use of Exercise 3.4, we can write for the second-order correction to the energy of the ground state

$$\mathscr{E}_0^{(2)} = -\sum_{k=1}^{\infty} \frac{\int \Psi_0^{(0)*}\hat{\mu} \cdot \mathbf{E}\Psi_k^{(0)} \, dq \int \Psi_k^{(0)*}\hat{\mu} \cdot \mathbf{E}\Psi_0^{(0)} \, dq}{E_k^{(0)} - E_0^{(0)}} \tag{5.5-2}$$

where $\{\Psi_k^{(0)}\}$ are the eigenfunctions of $\hat{H}^{(0)}$ and $\{E_k^{(0)}\}$ are the corresponding eigenvalues. Application of eq. (5.4-8) then yields

$$\alpha_{ij} = 2\sum_{k=1}^{\infty} \frac{\int \Psi_0^{(0)*}\hat{\mu}_i\Psi_k^{(0)} \, dq \int \Psi_k^{(0)*}\hat{\mu}_j\Psi_0^{(0)} \, dq}{E_k^{(0)} - E_0^{(0)}} \tag{5.5-3}$$

The value of α_{ij} calculated in this way depends on the coordinate system in which the molecule is described. Off-diagonal elements are not necessarily zero at this stage. Equation (5.5-3) shows that to calculate α_{ij}, one needs the eigenfunctions and eigenvalues of all the zero-order excited states. Rarely are these known, and so the use of eq. (5.5-3) necessitates approximations. The most severe of these is probably the "average energy" approximation.

To develop the "average energy" approximation we consider a quantity S of the form (Malta and Gouveia, 1983)

$$S = \sum_{k=1}^{\infty} \frac{\int \Psi_0^{(0)*}\hat{A}\Psi_k^{(0)} \, dq \int \Psi_k^{(0)*}\hat{B}\Psi_0^{(0)} \, dq}{\Delta E_k} \tag{5.5-4}$$

where $\hat{A}$ and $\hat{B}$ are Hermitian operators and $\Delta E_k \equiv E_k^{(0)} - E_0^{(0)}$. We assume that an "average value" $\overline{\Delta E}$ exists such that each ΔE_k can be replaced by $\overline{\Delta E}$

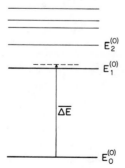

FIGURE 5.4. The "average energy" approximation.

(see Fig. 5.4) to give

$$S = \frac{1}{\Delta E} \sum_{k=1}^{\infty} \int \Psi_0^{(0)*} \hat{A} \Psi_k^{(0)} \, dq \int \Psi_k^{(0)*} \hat{B} \Psi_0^{(0)} \, dq$$

$$= \frac{1}{\Delta E} \sum_{k=0}^{\infty} \int \Psi_0^{(0)*} \hat{A} \Psi_k^{(0)} \, dq \int \Psi_k^{(0)*} \hat{B} \Psi_0^{(0)} \, dq$$

$$- \frac{1}{\Delta E} \int \Psi_0^{(0)*} \hat{A} \Psi_0^{(0)} \, dq \int \Psi_0^{(0)*} \hat{B} \Psi_0^{(0)} \, dq \qquad (5.5\text{-}5)$$

Introducing the dummy variable q' we can formally rewrite the first term of eq. (5.5-5) as

$$\frac{1}{\Delta E} \iint \Psi_0^{(0)*}(q) \hat{A}(q) \left(\sum_{k=0}^{\infty} \Psi_k^{(0)}(q) \Psi_k^{(0)*}(q') \right) \hat{B}(q') \Psi_0^{(0)}(q') \, dq \, dq'$$

According to the completeness relation, eq. (2.4-5), the expression in parentheses is the Dirac delta function, $\delta(q - q')$. Thus, formally integrating over q', we obtain

$$\frac{1}{\Delta E} \int \Psi_0^{(0)*}(q) \hat{A}(q) \hat{B}(q) \Psi_0^{(0)}(q) \, dq$$

and eq. (5.5-5) simplifies to

$$S = \frac{1}{\Delta E} \int \Psi_0^{(0)*} \hat{A} \hat{B} \Psi_0^{(0)} \, dq - \frac{1}{\Delta E} \int \Psi_0^{(0)*} \hat{A} \Psi_0^{(0)} \, dq \int \Psi_0^{(0)*} \hat{B} \Psi_0^{(0)} \, dq \quad (5.5\text{-}6)$$

We next introduce the quantities

$$F_k = \int \Psi_0^{*(0)} \hat{A} \Psi_k^{(0)} \, dq \int \Psi_k^{*(0)} \hat{B} \Psi_0^{(0)} \, dq$$

$$F = \sum_{k=0}^{\infty} F_k$$

$$x_k = 1 + \frac{\varepsilon}{\Delta E_k} \qquad (x_0 = 1)$$

where ε is an arbitrarily small energy of the same sign as ΔE_k. Since $0 < \Delta E_1 < \Delta E_2 < \cdots$, then $x_1 > x_2 > x_3 > \cdots 1$. Then we can rewrite

(5.5-4) as

$$S = \sum_{k=1}^{\infty} \frac{F_k}{\varepsilon/(x_k - 1)} = \frac{1}{\varepsilon} \sum_{k=1}^{\infty} F_k(x_k - 1)$$

$$\simeq \frac{1}{\varepsilon} \sum_{k=0}^{\infty} F_k \ln x_k$$

$$= \frac{1}{\varepsilon} \ln \prod_{k=0}^{\infty} (x_k)^{F_k}$$

$$= \frac{1}{\varepsilon} \left\{ \ln \left[(x_0)^F \left(\frac{x_1}{x_0} \right)^{F-F_0} \left(\frac{x_2}{x_1} \right)^{F-F_0-F_1} \cdots \right] \right\}$$

In the limit of very small ε one has $\lim_{\varepsilon \to 0} \{\varepsilon^{-1} \ln x_k\} \sim (x_k - 1)/\varepsilon$, and so

$$S = \lim_{\varepsilon \to 0} \left(\frac{F \ln x_0}{\varepsilon} + \frac{(F - F_0)\ln(x_1/x_0)}{\varepsilon} + \frac{(F - F_0 - F_1)\ln(x_2/x_1)}{\varepsilon} + \cdots \right)$$

$$(5.5-7)$$

It is now argued that the series in eq. (5.5-7) converges at least as rapidly as that in eq. (5.5-4). This is because $\lim_{k \to \infty}(x_{k+1}/x_k) = 1$ and $\lim_{k \to \infty}(F - F_0 - F_1 - \cdots - F_k) = 0$. Hence, taking only the first two terms in eq. (5.5-7), we have

$$S = \lim_{\varepsilon \to 0} \left[\frac{F \ln x_0}{\varepsilon} + \frac{(F - F_0)\ln(x_1/x_0)}{\varepsilon} \right]$$

$$= 0 + \lim_{\varepsilon \to 0} \frac{(F - F_0)\ln(x_1/x_0)}{\varepsilon}$$

$$= \lim_{\varepsilon \to 0} \frac{(F - F_0)\ln[1 + (\varepsilon/\Delta E_1)]}{\varepsilon}$$

$$= \frac{F}{\Delta E_1} - \frac{F_0}{\Delta E_1} \tag{5.5-8}$$

Comparison with eq. (5.5-6) then compels us to identify the "average energy" $\overline{\Delta E}$ as ΔE_1.

Application of the average energy approximation to eq. (5.5-3) yields the expression

$$\alpha_{ij} \simeq \frac{2}{E_1^{(0)} - E_0^{(0)}} \left(\int \Psi_0^{(0)*} \hat{\mu}_i \hat{\mu}_j \Psi_0^{(0)} \, dq \right.$$

$$\left. - \int \Psi_0^{(0)*} \hat{\mu}_i \Psi_0^{(0)} \, dq \int \Psi_0^{(0)*} \hat{\mu}_j \Psi_0^{(0)} \, dq \right) \tag{5.5-9}$$

This formula requires us to know (at least approximately) only the ground-state wave function and the energies of the ground and first-excited states.

Though the average-energy approximation may seem crude, yet another approximation is required in order to evaluate eq. (5.5-9). Generally one does not know $E_1^{(0)}$ very accurately, although as stated in Chapter 4 the virtual molecular orbitals of a Hartree–Fock, SCF calculation can form the basis for some of the excited states, and hence for the estimation of $E_1^{(0)}$. If an electron is promoted to a virtual orbital, the system becomes an open-shell system. The pertinent Hartree–Fock equations for an open-shell system no longer assume the form given in eq. (4.1-7); we have not treated this problem at all. As a rough approximation we take $E_1^{(0)} - E_0^{(0)}$ to be the excitation energy from the highest occupied molecular orbital to the lowest virtual orbital. That is, for a system of $2N$ paired electrons,

$$
\alpha_{ij} \simeq \underbrace{\frac{2}{\varepsilon_{N+1} - \varepsilon_N}\left(\int \Psi_0^{(0)*}\hat{\mu}_i\hat{\mu}_j\Psi_0^{(0)}\,dq \right.}_{A}
$$

$$
\underbrace{\left. - \int \Psi_0^{(0)*}\hat{\mu}_i\Psi_0^{(0)}\,dq \int \Psi_0^{(0)*}\hat{\mu}_j\Psi_0^{(0)}\,dq\right)}_{B} \tag{5.5-10}
$$

With this formula we are finally in a position to compute an approximate polarizability. As an example, we reconsider diatomic lithium hydride.

Imagine the LiH molecule to be oriented as in Fig. 4.3. Because of the cylindrical symmetry about the z axis, term B of eq. (5.5-10) is nonzero only for α_{zz}, for which $B = \langle\mu\rangle^2$. Term A involves a sum of two-particle operators, which can be written as

$$
\hat{\mu}_i\hat{\mu}_j = \underbrace{\sum_{\alpha=1}^{M}\sum_{\beta=1}^{M} Z_\alpha Z_\beta R_{\alpha i}R_{\beta j}}_{A_1} - \underbrace{\sum_{\alpha=1}^{M}\sum_{\lambda=1}^{2N} Z_\alpha\left(R_{\alpha i}r_{\lambda j} + R_{\alpha j}r_{\lambda i}\right)}_{A_2} + \underbrace{\sum_{\lambda=1}^{2N}\sum_{\mu=1}^{2N} r_{\lambda i}r_{\mu j}}_{A_3}
$$

$$
\tag{5.5-11}
$$

Again, because of the cylindrical symmetry, term A_1 makes no contribution to α_{ij} unless $i = j = z$. In this case, a simple calculation gives $\langle\mu_z\mu_z\rangle_{A1} = R_0^2$, where R_0 is the bond distance of LiH. Term A_2 also contributes to α_{ij} only when $i = j = z$, and the result is

$$
\langle\mu_z\mu_z\rangle_{A2} = 2(-Z_1 R_{1Z} - Z_2 R_{2Z})\langle z_1 + z_2 + z_3 + z_4\rangle
$$

$$
= 2\left(-\frac{R_0}{2} + \frac{3R_0}{2}\right)\langle z_1 + z_2 + z_3 + z_4\rangle
$$

$$
= 2R_0(-0.4873)
$$

$$
= -0.9746R_0
$$

where the last two lines follow from the calculation at the end of Section 5.1.

The term A_3 requires a more extensive analysis than do A_1 and A_2, because A_3 is a sum of *two-electron* operators. Contributions to A_3 involving cross terms such as $x_\lambda y_\mu$ or $y_\lambda z_\mu$ automatically vanish because of the x_λ or y_λ factor. However, squared terms such as $x_\lambda x_\mu$ or $z_\lambda z_\mu$ do not generally average to zero. The following terms give nonzero contributions to the diagonal components of $\langle \mu_i \mu_i \rangle$:

$$\hat{\mu}_x \hat{\mu}_x : \sum_{\lambda=1}^{4} x_\lambda^2 \qquad \hat{\mu}_y \hat{\mu}_y : \sum_{\lambda=1}^{4} y_\lambda^2 \qquad \hat{\mu}_z \hat{\mu}_z : \sum_{\lambda=1}^{4} \sum_{\mu=1}^{4} z_\lambda z_\mu$$

The averages of $\hat{\mu}_x \hat{\mu}_x$ and $\hat{\mu}_y \hat{\mu}_y$ are equal by symmetry.

Now let $\Psi(q)$ be a 4×4 Slater determinant of orthonormal molecular spin-orbitals $\Phi_p(j)$. Expansion of $\Psi(q)$ gives 24 terms.

$$\Psi(q) = (4!)^{-1/2} \{ [\Phi_1(1)\Phi_2(2) - \Phi_2(1)\Phi_1(2)]$$
$$\times [\Phi_3(3)\Phi_4(4) - \Phi_4(3)\Phi_3(4)] + 20 \text{ other terms} \} \quad (5.5\text{-}12)$$

Consider the operator x_1^2; then in $\Psi(q)$ electron 1 appears in six product terms as the argument of function Φ_1, in six other terms as the argument of function Φ_2, and so on. Because the Φ_p's are orthonormal, we therefore have

$$\int \Psi^*(q) x_1^2 \Psi(q) \, dq = \frac{6}{4!} \sum_{p=1}^{4} \int \Phi_p^*(1) x_1^2 \Phi_p(1) \, dq_1$$

and since the four electrons in LiH are indistinguishable,

$$\int \Psi^*(q) \sum_{\mu=1}^{4} x_\mu^2 \Psi(q) \, dq = \sum_{p=1}^{4} \int \Phi_p^*(\mu) x_\mu^2 \Phi_p(\mu) \, dq_\mu \quad (5.5\text{-}13)$$

For the $\hat{\mu}_z \hat{\mu}_z$ operator we can write

$$\sum_{\lambda=1}^{4} \sum_{\mu=1}^{4} z_\lambda z_\mu = \sum_{\lambda=1}^{4} z_\lambda^2 + \sum_{\lambda \neq \mu}^{4} \sum_{\mu=1}^{4} z_\mu z_\lambda \quad (5.5\text{-}14)$$

The first summation on the right leads to a formula analogous to eq. (5.5-13). For the double summation on the right we see from eq. (5.5-12) that electron λ appears in molecular spin-orbital p and electron μ appears simultaneously in spin-orbital p' in two product terms of the expansion of $\Psi(q)$. At the same time, electron λ appears in p' and electron μ appears in p in two other product terms. Thus, for a given pair of electrons we have

$$
\int \Psi^*(q) z_\lambda z_\mu \Psi(q)\, dq
$$

$$
= \frac{2}{4!} \sum_{p \neq p'} \sum_{p'=1}^{4} \left(\iint \Phi_p^*(\lambda) \Phi_{p'}^*(\mu) z_\lambda z_\mu \Phi_p(\lambda) \Phi_{p'}(\mu)\, dq_\lambda\, dq_\mu \right.
$$

$$
\left. - \iint \Phi_p^*(\lambda) \Phi_{p'}^*(\mu) z_\lambda z_\mu \Phi_{p'}(\lambda) \Phi_p(\mu)\, dq_\lambda\, dq_\mu \right) \quad (5.5\text{-}15)
$$

There are as many terms identical to eq. (5.5-15) as there are permutations of λ and μ, exclusive of the cases $\lambda = \mu$; the number of such permutations for LiH is $\frac{1}{2} \times 4!$. Hence, we have finally

$$
\int \Psi^*(q) z_\lambda z_\mu \Psi(q)\, dq
$$

$$
= \sum_{p \neq p'} \sum_{p'=1}^{4} \left(\iint \Phi_p^*(\lambda) \Phi_{p'}^*(\mu) z_\lambda z_\mu \Phi_p(\lambda) \Phi_{p'}(\lambda)\, dq_\lambda\, dq_\mu \right.
$$

$$
\left. - \iint \Phi_p^*(\lambda) \Phi_{p'}^*(\mu) z_\lambda z_\mu \Phi_{p'}(\lambda) \Phi_p(\mu)\, dq_\lambda\, dq_\mu \right) \quad (5.5\text{-}16)
$$

To summarize, combining all of the results from eqs. (5.5-11) to (5.5-16), we see that α for LiH is in diagonal form

$$
\alpha = \begin{bmatrix} \alpha_{xx} & 0 & 0 \\ 0 & \alpha_{yy} & 0 \\ 0 & 0 & \alpha_{zz} \end{bmatrix}
$$

where the diagonal components are given by the formulas

$$
\alpha_{xx} = \frac{2}{\varepsilon_3 - \varepsilon_2} \sum_{p=1}^{4} \int \Phi_p^*(\mu) x_\mu^2 \Phi_p(\mu)\, dq_\mu = \alpha_{yy} \quad (5.5\text{-}17a)
$$

$$
\alpha_{zz} = \frac{2}{\varepsilon_3 - \varepsilon_2} \left[R_0^2 - 0.9746 R_0 + \sum_{p=1}^{4} \int \Phi_p^*(\mu) z_\mu^2 \Phi_p(\mu)\, dq_\mu \right.
$$

$$
+ \sum_{p \neq p'} \sum_{p'=1}^{4} \left(\iint \Phi_p^*(\lambda) \Phi_{p'}^*(\mu) z_\lambda z_\mu \Phi_p(\lambda) \Phi_{p'}(\mu)\, dq_\lambda\, dq_\mu \right.
$$

$$
\left. - \iint \Phi_p^*(\lambda) \Phi_{p'}^*(\mu) z_\lambda z_\mu \Phi_{p'}(\lambda) \Phi_p(\mu)\, dq_\lambda\, dq_\mu \right)
$$

$$
\left. - (2.5281)^2 \right] \quad (5.5\text{-}17b)
$$

To simplify further the expressions for α_{xx} and α_{zz} and render them suitable for computer programming, we substitute the explicit forms for the spin-orbitals (i.e., $\Phi_1 = \chi_1\alpha$, $\Phi_2 = \chi_1\beta$, $\Phi_3 = \chi_2\alpha$, $\Phi_4 = \chi_2\beta$) and carry out the integrations over spin. The forms of the various terms in eq. (5.5-17) do not change. In the second of the two-electron terms we note that because of the orthonormality of the spin parts of the Φ_p's, the integrals are nonzero only if p and p' are either both odd or both even. Equations (5.5-17) then become

$$\alpha_{xx} = \frac{4}{\varepsilon_3 - \varepsilon_2} \sum_{n=1}^{2} \int \chi_n^*(\mu) x_\mu^2 \chi_n(\mu)\, dq_\mu = \alpha_{yy} \tag{5.5-18a}$$

$$\alpha_{zz} = \frac{2}{\varepsilon_3 - \varepsilon_2} \left[R_0^2 - 0.9746 R_0 + 2\sum_{n=1}^{2} \int \chi_n^*(\mu) z_\mu^2 \chi_n(\mu)\, dq_\mu \right.$$

$$+ \left(4 \sum_{n=1}^{2} \sum_{m=1}^{2} \iint \chi_n^*(\lambda) \chi_m^*(\mu) z_\lambda z_\mu \chi_n(\lambda) \chi_m(\mu)\, dq_\lambda\, dq_\mu \right.$$

$$\left. - 2 \sum_{n=1}^{2} \iint \chi_n^*(\lambda) \chi_n^*(\mu) z_\lambda z_\mu \chi_n(\lambda) x_n(\mu)\, dq_\lambda\, dq_\mu \right)$$

$$- 2 \sum_{n \neq m}^{2} \sum_{m=1}^{2} \iint \chi_n^*(\lambda) \chi_m^*(\mu) z_\lambda z_\mu \chi_m(\lambda) \chi_n(\mu)\, dq_\lambda\, dq_\mu$$

$$\left. - (2.5281)^2 \right] \tag{5.5-18b}$$

5.6   COMPUTER HIGHLIGHT: COMPUTATION OF α FOR LIH

To implement the calculation of α for LiH we now make the usual LCAO–MO approximation for both of the spatial molecular orbitals in eqs. (5.5-18). Reference to Table 4.10 shows that $\varepsilon_3 - \varepsilon_2 = 0.33238$. Component α_{xx}, for example, becomes

$$\alpha_{xx} = 12.0344 \sum_{n=1}^{2} \sum_{a,b} c_{an}^* c_{bn} \int \phi_a^*(\mu) x_\mu^2 \phi_b(\mu)\, dq_\mu = \alpha_{yy}$$

$$= 12.0344 \sum_{n=1}^{2} \sum_{a,b} c_{an}^* c_{bn} x_{ab}^2 \tag{5.6-1}$$

where the matrix element of x_μ^2 has been symbolized as x_{ab}^2. Values of this are obtained by carrying out the appropriate integration in confocal elliptical coordinates. The operator x_μ^2 is obtained from eq. (3.5-15). Equation (5.5-18b)

```
50  REM  PROGRAM TO CALCULATE POL
    ARIZABILITY TENSOR IN DIAGON
    AL FORM OF LITHIUM HYDRIDE
75  DIM C(5,5),X2(5,5),Z1(5,5),Z2
    (5,5),A(10),B(10),AU(10),BU(
    10),S(10)
100 INPUT R
125 INPUT C(1,1),C(2,2)
150 INPUT C(2,1),C(3,1),C(4,1),C
    (1,2),C(3,2),C(4,2)
175 INPUT Z1(1,1),Z1(2,2),Z1(3,3
    ),Z1(4,4)
200 INPUT Z1(1,2),Z1(1,3),Z1(1,4
    ),Z1(2,3),Z1(2,4),Z1(3,4)
275 AL = R:BE =  - R:N = 4
300 GOSUB 2000
325 X2(1,1) = (1 / 32) * R ^ 5 *
    (A(4) * B(0) - A(4) * B(2) +
    A(2) * B(4) - A(2) * B(0) -
    A(0) * B(4) + A(0) * B(2))
330 Z2(1,1) = (1 / 16) * R ^ 5 *
    (A(4) * B(2) - A(2) * B(4))
350 AL = 1.85 * R:BE = 0.85 * R:N
    = 4
375 GOSUB 2000
380 X2(1,2) = (1 / 32) * 2.7 ^ 1.
    5 * R ^ 5 * (A(4) * B(0) - A
    (4) * B(2) + A(2) * B(4) - A
    (2) * B(0) - A(0) * B(4) + A
    (0) * B(2))
385 Z2(1,2) = (1 / 16) * 2.7 ^ 1.
    5 * R ^ 5 * (A(4) * B(2) - A
    (2) * B(4))
400 AL = 0.825 * R:BE = - 0.175 *
    R:N = 5
425 GOSUB 2000
430 X2(1,3) = (0.65 ^ 5 / 3) ^ 0.
    5 * (1 / 64) * R ^ 6 * (A(5)
    * B(0) - A(5) * B(2) + A(3)
    * B(4) - A(3) * B(0) - A(1)
    * B(4) + A(1) * B(2) + A(4)
    * B(1) - A(4) * B(3) + A(2)
    * B(5) - A(2) * B(1) - A(0)
    * B(5) + A(0) * B(3))
435 Z2(1,3) = (0.65 ^ 5 / 3) ^ 0.
    5 * (1 / 32) * R ^ 6 * (A(5)
    * B(2) - A(3) * B(4) + A(4)
    * B(3) - A(2) * B(5))
450 X2(1,4) = (1 / 64) * 0.65 ^ 2
    .5 * R ^ 6 * (A(5) * B(1) -
    A(1) * B(5) + A(4) * B(0) -
    A(5) * B(3) + A(3) * B(5) -
    A(4) * B(2) + A(2) * B(4) -
    A(3) * B(1) + A(1) * B(3) -
    A(2) * B(0) + A(0) * B(2) -
    A(0) * B(4))

460 Z2(1,4) = (1 / 32) * 0.65 ^ 2
    .5 * R ^ 6 * (A(5) * B(3) -
    A(3) * B(5) + A(4) * B(2) -
    A(2) * B(4))
475 AL = 2.7 * R:BE = AL:N = 4
500 GOSUB 2000
525 X2(2,2) = (1 / 32) * 2.7 ^ 3 *
    R ^ 5 * (A(4) * B(0) - A(4) *
    B(2) + A(2) * B(4) - A(2) *
    B(0) - A(0) * B(4) + A(0) *
    B(2))
550 Z2(2,2) = (1 / 16) * 2.7 ^ 3 *
    R ^ 5 * (A(4) * B(2) - A(2) *
    B(4))
575 AL = 1.675 * R:BE = AL:N = 5
600 GOSUB 2000
625 X2(2,3) = (1 / 64) * 2.7 ^ 1.
    5 * (0.65 ^ 5 / 3) ^ 0.5 * R
    ^ 6 * (A(5) * B(0) - A(5) *
    B(2) + A(3) * B(4) - A(3) *
    B(0) - A(1) * B(4) + A(1) *
    B(2) + A(4) * B(1) - A(4) *
    B(3) + A(2) * B(5) - A(2) *
    B(1) - A(0) * B(5) + A(0) *
    B(3))
650 Z2(2,3) = (1 / 32) * 2.7 ^ 1.
    5 * (0.65 ^ 5 / 3) ^ 0.5 * R
    ^ 6 * (A(5) * B(2) - A(3) *
    B(4) + A(4) * B(3) - A(2) *
    B(5))
700 X2(2,4) = (1 / 64) * 2.7 ^ 1.
    5 * 0.65 ^ 2.5 * R ^ 6 * (A(
    5) * B(1) - A(1) * B(5) + A(
    4) * B(0) - A(5) * B(3) + A(
    3) * B(5) - A(4) * B(2) + A(
    2) * B(4) - A(3) * B(1) + A(
    1) * B(3) - A(2) * B(0) + A(
    0) * B(2) - A(0) * B(4))
725 Z2(2,4) = (1 / 32) * 2.7 ^ 1.
    5 * 0.65 ^ 2.5 * R ^ 6 * (A(
    5) * B(3) - A(3) * B(5) + A(
    4) * B(2) - A(2) * B(4))
750 AL = 0.65 * R:BE = AL:N = 6
775 GOSUB 2000
800 X2(3,3) = (1 / 384) * 0.65 ^
    5 * R ^ 7 * (A(6) * B(0) + 2
    * A(5) * B(1) - A(2) * B(4)
    - A(6) * B(2) - 2 * A(5) *
    B(3) + 2 * A(3) * B(5) + A(2
    ) * B(6) - A(4) * B(0) - 2 *
    A(3) * B(1) + 2 * A(1) * B(3
    ) + A(0) * B(4) + A(4) * B(2
    ) - 2 * A(1) * B(5) - A(0) *
    B(6))
```

FIGURE 5.5. Computer program for the calculation of the polarizability tensor of lithium hydride.

140

```
825 Z2(3,3) = (1 / 192) * 0.65 ^
    5 * R ^ 7 * (A(6) * B(2) + 2
    * A(5) * B(3) - 2 * A(3) *
    B(5) - A(2) * B(6))
850 D1 = A(6) * B(1) + A(5) * B(0
    ) - A(6) * B(3) - A(5) * B(4
    ) - A(4) * B(3) - A(3) * B(2
    ) + A(4) * B(5) + A(3) * B(6
    ) + A(3) * B(4) + A(2) * B(3
    ) - A(3) * B(0) - A(1) * B(6
    ) - A(2) * B(1) + A(1) * B(2
    ) + A(0) * B(3) - A(0) * B(5
    )
855 D2 = A(4) * B(0) - A(4) * B(2
    ) - A(2) * B(0) - A(0) * B(4
    ) - A(0) * B(2) + A(2) * B(4
    )
860 X2(3,4) = (1 / 384) * 3 ^ 0.5
    * 0.65 ^ 5 * R ^ 7 * (D1 +
    D2)
875 Z2(3,4) = 1 / 192) * 3 ^ 0.5
    * 0.65 ^ 5 * R ^ 7 * (A(6) *
    B(3) + A(5) * B(2) + A(5) *
    B(4) + A(4) * B(3) - A(4) *
    B(5) - A(3) * B(4) - A(3) *
    B(6) - A(2) * B(5))
900 X2(4,4) = (1 / 128) * 0.65 ^
    5 * R ^ 7 * (A(6) * B(2) + 2
    * A(5) * B(1) + A(4) * B(0)
    - 2 * A(3) * B(1) - A(2) *
    B(0) + 2 * A(1) * B(3) - A(6
    ) * B(4) + A(0) * B(2) - A(2
    ) * B(6) - 2 * A(1) * B(5) -
    A(0) * B(4) + 2 * A(2) * B(4
    ) - 2 * A(5) * B(3) - 2 * A(
    4) * B(2) + A(4) * B(6) + 2 *
    A(3) * B(5))
925 Z2(4,4) = (1 / 64) * 0.65 ^ 5
    * R ^ 7 * (A(6) * B(4) + 2 *
    A(5) * B(3) + A(4) * B(2) -
    A(4) * B(6) - 2 * A(3) * B(5
    ) - A(2) * B(4))
950 S1 = 0:S2 = 0:S3 = 0
975 FOR N = 1 TO 2
1000  S(N) = 0
1025  FOR I = 1 TO 4
1050  FOR J = 1 TO 4
1075  Z1(J,I) = Z1(I,J):Z2(J,I) =
      Z2(I,J):X2(J,I) = X2(I,J)
1100  S1 = S1 + C(I,N) * C(J,N) *
      X2(I,J)
1125  S2 = S2 + C(I,N) * C(J,N) *
      Z2(I,J)
1150  S(N) = S(N) + C(I,N) * C(J,N
      ) * Z1(I,J)
1175  NEXT J
1200  NEXT I
1225  S3 = S3 + S(N) ^ 2
1250  NEXT N
1275  S4 = 0: S5 = 0

1300  FOR N = 1 TO 2
1325  FOR M = 1 TO 2
1350  FOR I = 1 TO 4
1375  FOR J = 1 TO 4
1400  FOR K = 1 TO 4
1425  FOR L = 1 TO 4
1450  S4 = S4 + C(I,N) * C(J,N) *
      C(K,M) * C(L,M) * Z1(I,J) *
      Z1(K,L)
1475  S5 = S5 + C(I,2) * C(K,1) *
      C(L,1) * C(J,2) * Z1(I,K) *
      Z1(J,L) + C(I,1) * C(K,2) *
      C(L,2) * C(J,L) * Z1(I,K) *
      Z1(J,L)
1500  NEXT L
1525  NEXT K
1550  NEXT J
1575  NEXT I
1600  NEXT M
1625  NEXT N
1650  AX = 12.0344 * S1
1675  AZ = 6.0172 * (R ^ 2 - 0.974
      6 * R + 2 * S2 + 4 * S4 - 2 *
      S3 - 2 * S5 - 6.3913)
1700  PRINT : PRINT : PRINT  TAB(
      3);"POLARIZABILITY OF LITHIU
      M HYDRIDE": PRINT : PRINT
1725  PRINT  TAB( 2);"(I,J)"; TAB(
      12);"X2(I,J)"; TAB( 28);"Z2(
      I,J)": PRINT
1750  FOR I = 1 TO 4
1775  FOR J = 1 TO 4
1800  PRINT  TAB( 3);I; TAB( 4);"
      ,"; TAB( 5);J; TAB( 10);X2(I
      ,J); TAB( 26);Z2(I,J)
1825  NEXT J
1850  NEXT I
1875  PRINT : PRINT
1900  PRINT  TAB( 6);"ALPHA XX EQ
      UALS"; TAB( 22);AX: PRINT
1925  PRINT  TAB( 6);"ALPHA ZZ EQ
      UALS"; TAB( 22);AZ
1950  END
2000  FOR L = 0 TO N
2025  A(0) = AL ^  - 1 * EXP ( -
      AL)
2050  A(L + 1) = A(L) * (L + 1) /
      AL + AL ^  - 1 * EXP ( - AL
      )
2075  AU(0) = BE ^  - 1 * EXP ( -
      BE)
2100  AU(L + 1) = AU(L) * (L + 1) /
      BE + BE ^  - 1 * EXP ( - BE
      )
2125  BU(0) =  - (BE ^  - 1) * EXP
      (BE)
2150  BU(L + 1) =  - BU(L) * (L +
      1) / BE - BE ^  - 1 * EXP (
      BE)
2175  B(L) = (( - 1) ^ (L + 1)) *
      BU(L) - AU(L)
2190  NEXT L
2200  RETURN
```

FIGURE 5.5. Continued.

Table 5.5.   The 10 Independent Elements of Each of the $X2(I, J)$ and $Z2(I, J)$ Matrices Needed for a Calculation of the Polarizability of Lithium Hydride

I	J	$X2(I, J)$	$Z2(I, J)$
1	1	1	3.273189
1	2	0.040524	0.153768
1	3	1.158429	1.555056
1	4	0.993660	1.749634
2	2	0.137174	2.410364
2	3	0.0983769	0.476720
2	4	0	-0.345623
3	3	5.917160	8.190349
3	4	1.034614	-6.695975
4	4	3.550296	12.924077

is similarly transformed into

$$
\alpha_{zz} = 6.0172 \Bigg[R_0^2 - 0.9746 R_0 + 2 \sum_{n=1}^{2} \sum_{a,b} c_{an}^* c_{bn} z_{ab}^2
$$

$$
+ 4 \sum_{n=1}^{2} \sum_{m=1}^{2} \sum_{a,b} \sum_{a',b'} c_{an}^* c_{bn} c_{a'm}^* c_{b'm} z_{ab} z_{a'b'}
$$

$$
- 2 \sum_{n=1}^{2} \left(\sum_{a,b} c_{an}^* c_{bn} z_{ab} \right)^2
$$

$$
- 2 \sum_{n=1}^{2} \sum_{m \neq n} \sum_{a,b} \sum_{a',b'} c_{an}^* c_{bm} c_{a'm}^* c_{b'n} z_{ab} z_{a'b'} - 6.3913 \Bigg] \quad (5.6\text{-}2)
$$

where the matrix elements of z_μ^2 are designated z_{ab}^2, and the matrix elements of z_μ are the $Z1(I, J)$ given in Table 5.1. Figure 5.5 shows the listing of the program used to calculate α_{xx} and α_{zz}, and Table 5.5 presents the calculated values of the matrix elements x_{ab}^2 and z_{ab}^2, which are listed as $X2(I, J)$ and $Z2(I, J)$, respectively. The final polarizability tensor (with the components in atomic units, where 1 au $= 0.148184 \times 10^{-30}$ m^3) is found to be

$$
\alpha = \begin{bmatrix} 29.70 & 0 & 0 \\ 0 & 29.70 & 0 \\ 0 & 0 & 21.91 \end{bmatrix} \quad (5.6\text{-}3)
$$

There appears to be no experimental result for α in the literature. However, a recent theoretical calculation of α for LiH (Karlström et al., 1982) yielded values $\alpha_{xx} = 25.4$ au and $\alpha_{zz} = 22.1$ au. Our calculated components agree to within about 15%, but in view of the many approximations made in the

Table 5.6.   Comparison of Calculated and Experimental Polarizability Tensors for
Selected Molecules

Molecule		α_{xx}	α_{yy}	α_{zz}
$H_2{}^a$	calc	5.41		8.73
	exp	4.85		6.30
CO^a	calc	10.46		29.15
	exp	10.97		17.55
N_2	calc^b	9.48		14.49
	exp^c	10.20		14.82
H_2O^d	calc	7.87	8.37	6.68
	exp	9.91	10.32	9.55
HCCH	calc^d	14.96		28.57
	exp^e	19.37		31.92
CH_3F^d	calc	14.84		12.59
	exp	16.92		19.01

[a] H. J. Kolker and M. Karplus, *J. Chem. Phys.*, **39**, 2011 (1963).
[b] J. E. Gready, G. B. Bacskay, and N. S. Hush, *Chem. Phys.*, **22**, 141 (1977).
[c] G. D. Zeiss and W. J. Meath, *Mol. Phys.*, **33**, 1155 (1977).
[d] J. A. Hudis and R. Ditchfield, *Chem. Phys. Lett.*, **77**, 202 (1981).
[e] N. J. Bridge and A. D. Buckingham, *Proc. Roy. Soc. London, Ser. A*, **295**, 334 (1966).

development, the result of our calculation is satisfying. Table 5.6 shows some
typical results of polarizability calculations.

Exercises

5.7.   In this exercise we estimate the spectroscopic dissociation energy of
H_2. The ground-state wave function is a single Slater determinant,
and each molecular orbital is to be approximated as a linear combination of two $1s$ basis functions of the form

$$1s = \left(\frac{Z^3}{\pi} \right)^{1/2} e^{-rZ}$$

(a) We need first the core Hamiltonian integrals. Using the one-elec-
tron Hamiltonian $\hat{H}_1 = -\frac{1}{2}\nabla^2 - r_A^{-1} - r_B^{-1}$, show that the core
Hamiltonian integrals as defined in eq. (4.2-7) are (in atomic units)

$$H_{11} = \tfrac{1}{2}Z^2 - Z - R^{-1} + R^{-1}(1 + RZ)e^{-2RZ}$$

$$H_{12} = Z(Z - 2)(1 + RZ)e^{-RZ} - \tfrac{1}{2}S_{12}Z^2$$

$$S_{12} = e^{-RZ}\left(1 + RZ + \tfrac{1}{3}R^2Z^2\right)$$

(b) Next we need the electron-repulsion integrals, the matrix elements of the two-electron Hamiltonian $\hat{H}_2 = r_{12}^{-1}$. From formulas (Roothaan, 1951) we deduce

$$(11|11) = \frac{5}{8}Z$$

$$(11|22) = \frac{1}{R}\left\{1 - \left(1 + \frac{11}{8}RZ + \frac{3}{4}R^2Z^2 + \frac{1}{6}R^3Z^3\right)e^{-2RZ}\right\}$$

The integrals $(11|12)$ and $(12|12)$ require more work. We estimate them by means of the Mulliken approximation in eq. (4.4-14).

(c) Now evaluate the Hartree–Fock energy E according to eq. (4.1-5). The total energy $\mathscr{E}(R)$ is the energy E plus the energy of nuclear repulsion. Evaluate $\mathscr{E}(R)$, given that $R = 1.40153$ au; use $Z = 1.10$, instead of the value 1.00 indicated by Slater's rules.

(d) Apply eq. (5.3-1) to H_2; for E_H use the usual Bohr value of $-\frac{1}{2}$ au times the correction factor $[1 + (1836)^{-1}]^{-1}$, which allows for the reduced mass of the system. The zero-point vibrational frequency of H_2 is 4338 cm^{-1}. Compare your calculated D_0 with the experimental value of 0.1645 au (Herzberg, 1950).

5.8. An electric field is turned on in the x direction; what is the polarization energy of a molecule in the field? Next, a field in the y direction is turned on; what now is the total polarization energy of the molecule? The entire experiment is repeated, but the fields are turned on in the reverse order. What is the total potential energy in this case? Both experiments result in a molecule being subjected to a field with x and y components, and so the potential energy should be the same in the two cases. What does this imply about α_{xy} and α_{yx}? Draw a conclusion about the form of α.

5.9. The next several exercises probe some of the mathematical properties of second-rank tensors (Goodbody, 1982). Let $\{A_i\}$ be the components of a vector $\mathbf{A}$ in a (unprimed) coordinate system. If this system is now rotated about some axis to form a new (primed) coordinate system, then

$$A'_i = \sum_{j=1}^{3} \lambda_{ij}A_j$$

where λ_{ij} is the cosine of the angle between the i axis of the primed system and the j axis of the unprimed system. Prove that α transforms according to the law

$$\alpha'_{ij} = \sum_{k=1}^{3}\sum_{\ell=1}^{3} \lambda_{i\ell}\lambda_{jk}\alpha_{\ell k}$$

This result is frequently taken as *the definition* of a second-rank tensor in a Cartesian coordinate system.

5.10. In Exercise 5.9, let us define λ_{ij} as the cosine of the angle θ $(0 \le \theta \le \pi)$ needed to rotate the positive j axis of the unprimed system into the positive i axis of the primed system. Suppose LiH is initially oriented as in Fig. 4.3. The xyz coordinate system is then rotated $\frac{1}{4}\pi$ radians clockwise about the x axis. In the $x'y'z'$ coordinate system so formed, what are the components of α' if α is given by eq. (5.6-3)?

5.11. Show by actual multiplication that the transformation law for a second-rank Cartesian tensor can be expressed in matrix symbolism as

$$
\begin{bmatrix}
\alpha'_{11} & \alpha'_{12} & \alpha'_{13} \\
\alpha'_{21} & \alpha'_{22} & \alpha'_{23} \\
\alpha'_{31} & \alpha'_{32} & \alpha'_{33}
\end{bmatrix}
$$
$$
=
\begin{bmatrix}
\lambda_{11} & \lambda_{12} & \lambda_{13} \\
\lambda_{21} & \lambda_{22} & \lambda_{23} \\
\lambda_{31} & \lambda_{32} & \lambda_{33}
\end{bmatrix}
\begin{bmatrix}
\alpha_{11} & \alpha_{12} & \alpha_{13} \\
\alpha_{21} & \alpha_{22} & \alpha_{23} \\
\alpha_{31} & \alpha_{32} & \alpha_{33}
\end{bmatrix}
\begin{bmatrix}
\lambda_{11} & \lambda_{21} & \lambda_{31} \\
\lambda_{12} & \lambda_{22} & \lambda_{32} \\
\lambda_{13} & \lambda_{23} & \lambda_{33}
\end{bmatrix}
$$

5.12. If the direction cosines λ_{ij} are defined as in Exercise 5.10, then the determinant of the matrix λ can be shown to have the value $\det \lambda = 1$ for any rotation.

(a) What must therefore be true of the determinant of any second-rank tensor $\mathbf{T}$ under a rotation?

(b) Now let $\mathbf{S}$ be a new tensor defined as $\mathbf{S} = \mathbf{T} - k\mathbf{1}$, where k is a scalar and $\mathbf{1}$ is the unit tensor (i.e., the unit matrix). Prove that the **trace** of $\mathbf{T}$ must be invariant to a rotation:

$$
\text{Tr}\,\mathbf{T} = T_{11} + T_{22} + T_{33} = \text{constant}
$$

(c) Verify that the result of part (b) is true in Exercise 5.10. We note that in experiments where the components of a tensor $\mathbf{T}$ cannot be observed directly, one usually obtains the scalar average $\frac{1}{3}\text{Tr}\,\mathbf{T}$. For example, dielectric constant measurements of a dilute gas and use of the Debye equation yield $\frac{1}{3}\text{Tr}\,\alpha$, or $(\alpha_{11} + \alpha_{22} + \alpha_{33})/3$, for the polarizability.

5.13. Show that for a second-rank tensor $\mathbf{T}$ the quantity $I_2 = (T_{22}T_{33} - T_{23}T_{32} + T_{33}T_{11} - T_{13}T_{31} + T_{11}T_{22} - T_{12}T_{21})$ is invariant under a rotation. Let $I_1 = \text{Tr}\,\mathbf{T}$ and $I_3 = \det\mathbf{T}$, and let $\lambda_1, \lambda_2, \lambda_3$ be the three eigenvalues of $\mathbf{T}$ (obtained by solving the matrix equation $\mathbf{T} - \lambda\mathbf{1} = \mathbf{0}$). Prove that

$$
I_1 = \lambda_1 + \lambda_2 + \lambda_3
$$
$$
I_2 = \lambda_1\lambda_2 + \lambda_1\lambda_3 + \lambda_2\lambda_3
$$
$$
I_3 = \lambda_1\lambda_2\lambda_3
$$

5.14.   In view of the results of Exercise 5.13, determine the diagonal form of the tensor $\mathbf{T}$, where

$$\mathbf{T} = \begin{bmatrix} 2 & 1 & -3 \\ 1 & 0 & 2 \\ -3 & 2 & 4 \end{bmatrix}$$

5.15.   Work through the steps in the transformation of eq. (5.5-18b) into eq. (5.6-2).

5.16.   Spot-check the correctness of form of a couple of the $X2(I, J)$ integrals in the program of Fig. 5.5.

5.17.[C*]   Our object here is to estimate the polarizability of molecular hydrogen. You first need to deduce the molecular orbital energies (refer to Exercise 5.7 for pertinent information). Then work through the necessary polarizability equations in order to arrive at expressions for α_{xx}, α_{zz}. Work out the necessary matrix elements $Z1, Z2, X2$ as in eqs. (5.6-1) and (5.6-2), and then write the required program. Compare your answer with the experimental results given in Table 5.6.

5.7   NDDO THEORY

Our crude Hartree–Fock SCF calculation in Chapter 4 on the LiH molecule demonstrates that even in the case of a small molecule a great deal of labor is necessary to obtain the ground-state wave function and energy. Even more work would have been necessary if we had desired to optimize the screening factors ξ, to computationally determine the equilibrium internuclear distance R_0, and to improve on the energy $\mathscr{E}$ by means of configuration interaction. Clearly, a comparable calculation on a larger molecule such as ethane is feasible only on a powerful mainframe computer and requires the evaluation of many tens or hundreds of thousands of integrals.

A primary objective of quantitative chemistry is the development of a mathematical treatment of molecular properties that is sufficiently accurate, reliable, and inexpensive to answer questions of chemical interest on a routine basis. **Semiempirical** molecular orbital schemes try to do this by incorporating into the calculation data obtained from, or parameters that can be fitted to, experimental results. A number of monographs on semiempirical molecular orbital theory have appeared (Dewar, 1969; Pople and Beveridge, 1970; Murrell and Harget, 1972).

In fact, a plethora of semiempirical schemes have appeared in the primary literature. It would be impractical to discuss all of them. Nor can we describe in detail any one procedure. However, to give an indication of the nature of a typical semiempirical procedure, we discuss briefly one of the more recent ones and examine the kind of results it can yield.

Approximate molecular orbital schemes date back to 1931 when Erich Hückel in Germany first presented a crude treatment of π-electron systems. Modern treatments owe much to J. E. Lennard-Jones and C. A. Coulson in England, and more recently to J. A. Pople (Pople et al., 1965). These workers considered an orthogonal transformation among the atomic basis functions $\{\phi_a\}$ used in a Slater determinant. If the ϕ_a are real-valued functions, as we have assumed, then an orthogonal transformation is one that leads to a new set of functions $\{\chi_\mu\}$ according to

$$\chi = O\Phi \tag{5.7-1}$$

where χ and Φ are column matrices and O is any orthogonal matrix. By definition, matrix O is called **orthogonal** if

$$OO^T = 1 \tag{5.7-2}$$

where O^T is the transpose. For example, if O is the matrix

$$O = \begin{bmatrix} 0 & \tfrac{1}{2}\sqrt{3} & \tfrac{1}{2} \\ \tfrac{1}{2}\sqrt{2} & -\tfrac{1}{4}\sqrt{2} & \tfrac{1}{4}\sqrt{6} \\ \tfrac{1}{2}\sqrt{2} & \tfrac{1}{4}\sqrt{2} & -\tfrac{1}{4}\sqrt{6} \end{bmatrix}$$

then matrix multiplication of O and O^T yields

$$\begin{bmatrix} 0 & \tfrac{1}{2}\sqrt{3} & \tfrac{1}{2} \\ \tfrac{1}{2}\sqrt{2} & -\tfrac{1}{4}\sqrt{2} & \tfrac{1}{4}\sqrt{6} \\ \tfrac{1}{2}\sqrt{2} & \tfrac{1}{4}\sqrt{2} & -\tfrac{1}{4}\sqrt{6} \end{bmatrix} \begin{bmatrix} 0 & \tfrac{1}{2}\sqrt{2} & \tfrac{1}{2}\sqrt{2} \\ \tfrac{1}{2}\sqrt{3} & -\tfrac{1}{4}\sqrt{2} & \tfrac{1}{4}\sqrt{2} \\ \tfrac{1}{2} & \tfrac{1}{4}\sqrt{6} & -\tfrac{1}{4}\sqrt{6} \end{bmatrix} = \begin{bmatrix} 1 & 0 & 0 \\ 0 & 1 & 0 \\ 0 & 0 & 1 \end{bmatrix}$$

$$\qquad\qquad O \qquad\qquad\qquad\qquad O^T \qquad\qquad\qquad\quad 1$$

Matrix O is therefore orthogonal.

It can be proved that the molecular orbital energies and the total electronic energy are unchanged by an orthogonal transformation of the atomic orbital basis functions (Lennard-Jones, 1949). An example of such a transformation is one that mixes $2s$ and $2p$ orbitals on the same atom to produce a set of hybrid atomic orbitals (**hybridization**). Another example of an orthogonal transformation is one that mixes only the $2px, 2py, 2pz$ atomic orbitals on the same atom when the local axes on that atom are rotated. For example, if the local axes centered at a nucleus are rotated counterclockwise $\pi/4$ radians about the z axis, the following transformations result (Fig. 5.6):

$$2pz' = 2pz$$
$$2py' = [\cos(\pi/4)]2py - [\sin(\pi/4)]2px$$
$$2px' = [\sin(\pi/4)]2py + [\cos(\pi/4)]2px \tag{5.7-3a}$$

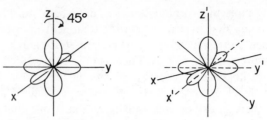

FIGURE 5.6. Illustrating the orthogonal transformation of a set of $2px, 2py, 2pz$ atomic orbitals by a rotation of the local axes.

This can be expressed by the matrix equation

$$
\underbrace{\begin{bmatrix} 1 & 0 & 0 \\ 0 & \frac{1}{2}\sqrt{2} & -\frac{1}{2}\sqrt{2} \\ 0 & \frac{1}{2}\sqrt{2} & \frac{1}{2}\sqrt{2} \end{bmatrix}}_{\mathbf{O}} \underbrace{\begin{bmatrix} 2pz \\ 2py \\ 2px \end{bmatrix}}_{\Phi} = \underbrace{\begin{bmatrix} 2pz \\ \frac{1}{2}\sqrt{2}\,(2py - 2px) \\ \frac{1}{2}\sqrt{2}\,(2py + 2px) \end{bmatrix}}_{\chi}
\qquad (5.7\text{-}3b)
$$

Then 3×3 rotation matrix **O** is easily verified to be orthogonal.

Any semiempirical molecular orbital scheme attempts to reduce the amount of work required in a full SCF treatment, primarily by neglecting certain integrals (i.e., setting them equal to zero) and by making approximations to certain others. However, such approximations may give final results that depend on the choice of coordinate system, especially for molecules of low symmetry and where there is no obvious or unique coordinate system. Since invariance to the rotation of local atomic axes is an inherent feature of molecular orbital energies obtained in a full Hartree–Fock SCF treatment, it must also be an essential feature of acceptable approximate semiempirical orbital energies. This required invariance places some restrictions on the allowable approximations.

The following set of approximations, which constitutes a semiempirical method that Pople calls **NDDO (Neglect of Diatomic Differential Overlap)**, obeys such restrictions:

Approximation 1. Only those electrons in the valence shell of each atom are treated explicitly, all other electrons being considered part of an unpolarizable core. An NDDO calculation on LiF, for example, is then an eight-electron calculation. Approximation 1 is called the **core approximation.**

Approximation 2. The atomic orbitals are treated as if they form an orthonormal set, that is, $S_{ab} = \int \phi_a^* \phi_b \, dq = \delta_{ab}$. The Roothaan equations [eq. (4.2-2)] now assume the form

$$
\sum_b F_{ab} c_{bn} = c_{an} \varepsilon_n
$$

Approximation 3.   All electron-repulsion integrals $(ab|a'b')$ in which either ϕ_a and ϕ_b are on different atoms or $\phi_{a'}$ and $\phi_{b'}$ are on different atoms, are set equal to zero.

With these approximations, the elements F_{ab} of the Fock matrix have the following form:

CASE 1. ϕ_a and ϕ_b are both on atom A.

$$F_{ab} = H_{ab} + \sum_{\substack{\text{all} \\ B \neq A}} \sum_{\substack{a', b' \\ \text{on A}}} P_{a'b'}(ab|a'b') - \frac{1}{2} \sum_{\substack{a', b' \\ \text{on A}}} P_{a'b'}(aa'|bb') \quad (5.7\text{-}4a)$$

Here,

$$H_{ab} = \int \phi_a^* \left\{ \frac{-\nabla^2}{2} + V_A \right\} \phi_b \, dq + \sum_{B \neq A} \int \phi_a^* V_B \phi_b \, dq \quad (5.7\text{-}4b)$$

where $V_N(q)$ is the effective potential energy of an electron in the field of the core of atom N. If ϕ_a, ϕ_b are Slater s, p, d functions, the first integral in eq. (5.7-4b) is 0 by symmetry unless $a = b$. The integrals in the summation generally do not vanish.

CASE 2. ϕ_a is on atom A and ϕ_b is on atom B.

$$F_{ab} = H_{ab} - \frac{1}{2} \sum_{\substack{b' \text{ on} \\ B \neq A}} \sum_{\substack{a' \\ \text{on A}}} P_{a'b'}(aa'|bb') \quad (5.7\text{-}5)$$

In Case 2, it is clear that Approximation 3 greatly reduces the number of electron-repulsion integrals that need to be taken into account.

In Case 1, the integrals H_{ab} are relatively easy to evaluate (or approximate semiempirically); in Case 2, they are less so. A final approximation is therefore invoked by Pople.

Approximation 4.   The off-diagonal matrix elements H_{ab} between atomic orbitals on different atoms are estimated by the formula

$$H_{ab} = \beta_{AB}^0 S_{ab}$$

where β_{AB}^0 is a parameter depending only on the nature of the atoms A and B and not on the orbitals a and b. The restriction on the proportionality constant is necessary if the calculations are to be invariant under transformation of the atomic basis set. The proportionality is one form of the class of approximations known collectively as **Mulliken approximations**, and is similar in spirit to the particular approximation given in eq. (4.4-14). It should be noted that here S_{ab} is not set equal to zero in spite of Approximation 2. In an NDDO calculation all overlap integrals must therefore be calculated, but they do not appear explicitly in the Roothaan equations.

5.8 THE MNDO METHOD

The MNDO method (**Modified Neglect of Diatomic Overlap**) is a particular modification of the NDDO theory (Dewar and Thiel, 1977a). The only integrals evaluated explicitly are the overlap integrals S_{ab}; all others are either neglected or are parametrized. All parameters are atomic rather than molecular since then their number can be kept to a minimum. Using a set of standard molecules, Dewar obtained optimum values for the set of parameters that minimize the sum of the squares of the weighted errors in the calculated values of a number of reference properties of the standard molecules. Some of these reference properties are the following: (a) standard molar enthalpy of formation, $\Delta \tilde{H}_f^0$; (b) ionization potential; (c) dipole moment; (d) bond lengths and bond angles.

Reference properties (a–d) are calculable in a straightforward way from the ground-state wave function. For property (a), we write the total energy $\mathscr{E}(Q)$ (see Section 4.1) of the molecule in a fixed nuclear configuration as

$$\mathscr{E}(Q) = E + \sum_{A<B} \sum_{B} E_{AB}^{core} \qquad (5.8\text{-}1)$$

where E is the ground-state electronic energy for the valence-electron system, and E_{AB}^{core} is the energy of interaction between the core of atom A (nucleus plus inner-shell electrons) and the core of atom B. By means of a thermodynamic cycle (see Fig. 5.7), the standard molar enthalpy of formation of the gaseous molecule is related to the electronic energies E_A of the component gaseous atoms and the experimental enthalpies of formation $\Delta \tilde{H}_f^0(A)$ of these

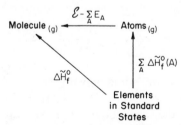

FIGURE 5.7. Thermodynamic cycle connecting total molecular energy $\mathscr{E}(Q)$ and standard molar enthalpy of formation of molecule.

atoms by the equation

$$\Delta \tilde{H}_f^0 = \mathscr{E}(Q) - \sum_A E_A + \sum_A \Delta \tilde{H}_f^0(A)$$

$$= E + \sum_{A<B} \sum_B E_{AB}^{core} - \sum_A E_A + \sum_A \Delta \tilde{H}_f^0(A) \qquad (5.8\text{-}2)$$

The quantities E_{AB}^{core} are treated as parameters.

The complete set of MNDO parameters can now be listed:

1. ξ_a The screening factor for an atomic orbital ϕ_a, obtained by optimization rather than by Slater's rules.

2. U_{aa} The one-electron, one-atom energy, which is the kinetic energy of an electron in atomic orbital a on atom A plus its potential energy of interaction with the core of A, is obtained by fitting calculated values to experimental spectroscopic energies of several valence states of the atom and of some of its ions.

3. H_{ab} The one-electron, two-atom (a and b on different atoms) matrix element is formulated as in Approximation 4 of Section 5.7, with $\beta_{AB}^0 = (\beta_a^A + \beta_b^B)/2$, where β_a^A, β_b^B are atomic orbital parameters of atoms A and B.

4. $(ab|a'b')$ The two-electron, one-, or two-atom repulsion integral represents the interaction of one charge distribution $e\phi_a\phi_b$ another $e\phi_{a'}\phi_{b'}$. Each distribution of charge is treated classically as a composite of monopolar, dipolar, and quadrupolar distributions of charges of fixed orientation and location in space (Dewar and Thiel, 1977b).

5. $V_{ab,B}$ The one-electron, two-atom (a and b both on atom A) attraction between the charge distribution $e\phi_a\phi_b$ and the core of atom B. The action of the core is mimicked by means of the spherical valence-shell charge distribution $s^B s^B$ at atom B, where s^B means a $1s$ basis function on B. The quantity $V_{ab,B}$ is formulated as $-Z_B(ab|s^B s^B)$.

6. E_{AB}^{core} The repulsion between the cores of atoms A and B is formulated as

$$E_{AB}^{core} = Z_A Z_B \left(s^A s^A | s^B s^B \right) \left[1 + \left(e^{-\alpha_A R_{AB}} + e^{-\alpha_B R_{AB}} \right) \right]$$

where both cores are simulated by spherical valence-shell distributions, and α_A, α_B are atomic parameters. Values of α_A, α_B are obtained by the matching of experimental enthalpies of formation with calculated ones according to eq. (5.8-2).

Table 5.7. MNDO Parameters for the Elements H and Be to F

Parameter[a]	H	Be[b]	B	C	N	O	F[c]
$\xi\,(a_0^{-1})$	1.331967	1.004210	1.506801	1.787537	2.255614	2.699905	2.848487
U_{ss} (eV)	−11.906276	−16.602378	−34.547130	−52.279745	−71.932122	−99.643090	−131.071548
U_{pp} (eV)	—	−10.703771	−23.121690	−39.205558	−57.172319	−77.797472	−105.782137
β_s (eV)	−6.989064	{−4.017096	{−8.252054	{−18.985044	{−20.495758	{−32.688082	{−48.29046
β_p (eV)							−36.50854
ρ_0 (Å)	0.560345	0.799924	0.679822	0.588660	0.529751	0.466882	0.425492
ρ_1 (Å)		0.788356	0.539446	0.430254	0.337322	0.275822	0.243849
ρ_2 (Å)		0.684928	0.476128	0.395734	0.324853	0.278678	0.255793
α (Å^{-1})	2.544134	1.669434	2.134993	2.546380	2.861342	3.160604	3.419661

[a]From M. J. S. Dewar and W. Thiel, *J. Am. Chem. Soc.*, **99**, 4899 (1977), except where noted.
[b]M. J. S. Dewar and H. S. Rzepa, *J. Am. Chem. Soc.*, **100**, 777 (1978).
[c]M. J. S. Dewar and H. S. Rzepa, *J. Am. Chem. Soc.*, **100**, 58 (1978).

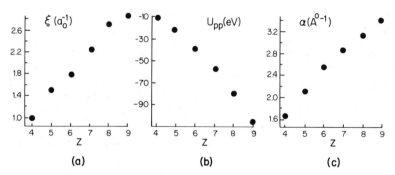

FIGURE 5.8. The MNDO parameters show a regular variation with atomic number across the second row.

Table 5.7 lists the pertinent parameters for the elements H, and Be to F. Some parameters are available for certain of the third-row elements, but have not been as thoroughly investigated as those for the second-row elements (Dewar et al., 1978). No parameters are available for lithium. Considerable regularity in the values of the parameters is apparent upon proceeding across the second row from Be to F. Figures 5.8(a)–(c) show the behavior of three of the parameters as a function of atomic number.

Of particular interest are the difficult two-electron, two-atom, electron-repulsion integrals. Dewar has performed some comparison calculations between the approximate MNDO formulas and the exact analytical expressions (Dewar and Thiel, 1977b). It is generally found that the absolute value of each semiempirical integral is less than that of the analytical integral. However, this is no cause for alarm because the Hartree–Fock theory neglects Coulombic correlation (see Section 4.6). Consequently, the electronic repulsions are

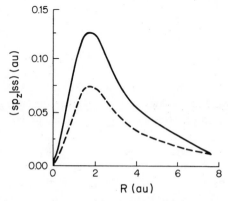

FIGURE 5.9. Two-electron, two-atom repulsion integrals plotted as a function of internuclear distance R for a C—C fragment with valence-shell Slater atomic orbitals ($\xi = 1.787537$) and calculated by semiempirical (---) and analytical (—) means.

Table 5.8.   MNDD Calculations of Standard Molar Enthalpy of Formation, Electron Affinity, Dipole Moment, and Molecular Geometry of Selected Molecules[a–c]

Molecule	$\Delta \tilde{H}_f^0$ (kcal mol^{-1})	EA (eV)	μ (D)	Geometry
C_2H_4	15.3	−1.02	—	123.2°
	12.5	−1.55		121.2°
C_6H_6	21.2	−0.07	—	1.407 Å (CC)
	19.8	−1.15		1.397 Å
Pyridine	28.7	0.22	1.97	1.353 Å (NC)
	34.6	−0.62	2.22	1.338 Å
CO_2	−75.4	0.18	—	1.186 Å (CO)
	−94.1	−0.60		1.162 Å
N_2O	30.9	1.12	0.76	1.181 Å (NO)
	19.6	0.22	0.17	1.162 Å
O_3	48.5	1.85	1.18	117.6° (∠OOO)
	34.2	2.0	0.53	116.8°
CF_4	−223.0	−0.71	—	1.347 Å (CF)
	−214.3			1.321 Å
$\underset{\text{H}\,\overset{\text{O}}{\overset{\|}{\text{C}}}\,-\text{OH}}{}$	−92.7		1.49	1.227 Å (C=O)
	−90.7		1.41	1.202 Å
F_2	7.3	0.75	—	1.266 Å (FF)
	0	3.08		1.418 Å
BeO	38.2	1.78	5.32	1.335 Å (BeO)
	31	1.77		1.331 Å
Cyclopropane	68.2		0.48	151.6° (∠HC=C)
	66.2		0.45	149.9°

[a] M. J. S. Dewar and H. S. Rzepa, *J. Am. Chem. Soc.*, **100**, 58, 777, 784 (1978).
[b] M. J. S. Dewar and W. Thiel, *J. Am. Chem. Soc.*, **99**, 4907 (1977).
[c] For each molecule the upper number is the calculated value and the lower number is the experimental value.

overestimated; this overestimation is thus partially compensated in the MNDO treatment by deriving the values of the repulsion integrals from various properties of the constituent atoms. Values found in this way are therefore smaller than the analytical values. Fig. 5.9 shows a typical comparison.

Several molecular properties have been calculated by the MNDO procedure. Table 5.8 displays the results of four such properties for a selection of molecules. No simple generalizations seem possible; for some properties and some molecules good results are obtained, while for other properties and other molecules, the results are not so good. Some molecules fare badly in essentially all properties, for example, N_2O and F_2. However, it may be pointed out that comparative calculations show MNDO to be at least as good as, if not superior to, other semiempirical procedures, and often to be as good as rigorous *ab initio* methods, which themselves frequently give poor agreement with experiment (Dewar and Ford, 1979). Semiempirical procedures, therefore, will continue to be useful tools for understanding molecular properties.

Exercises

5.18. A rotation counterclockwise by θ about the z axis followed by a rotation counterclockwise by ϕ about the y' axis can be represented by the single rotation matrix **R**.

$$\mathbf{R} = \begin{bmatrix} \cos\theta\cos\phi & \sin\theta\cos\phi & \sin\phi \\ -\sin\theta & \cos\theta & 0 \\ -\cos\theta\sin\phi & -\sin\theta\sin\phi & \cos\phi \end{bmatrix}$$

Is **R** an orthogonal matrix?

5.19. A set of four equivalent sp^3 hybrid orbitals using the $n = 2$ valence-shell atomic orbitals is

$$\Psi_1 = \tfrac{1}{2}\chi_{2s} + \tfrac{1}{2}\sqrt{3}\,\chi_{2pz}$$

$$\Psi_2 = \tfrac{1}{2}\chi_{2s} + \tfrac{1}{3}\sqrt{6}\,\chi_{2px} - \tfrac{1}{6}\sqrt{3}\,\chi_{2pz}$$

$$\Psi_3 = \tfrac{1}{2}\chi_{2s} - \tfrac{1}{6}\sqrt{6}\,\chi_{2px} + \tfrac{1}{2}\sqrt{2}\,\chi_{2py} - \tfrac{1}{6}\sqrt{3}\,\chi_{2pz}$$

$$\Psi_4 = \tfrac{1}{2}\chi_{2s} - \tfrac{1}{6}\sqrt{6}\,\chi_{2px} - \tfrac{1}{2}\sqrt{2}\,\chi_{2py} - \tfrac{1}{6}\sqrt{3}\,\chi_{2pz}$$

Write the matrix that converts the vector of χ functions into the vector of Ψ functions, and determine whether it is orthogonal.

5.20. Suppose $\{\phi_\mu\}$ is a basis set and **A** is a square matrix of the form

$$\mathbf{A} = \begin{bmatrix} \int\phi_1\hat{A}\phi_1\,dq & \int\phi_1\hat{A}\phi_2\,dq & \int\phi_1\hat{A}\phi_3\,dq & \cdots \\ \int\phi_2\hat{A}\phi_1\,dq & \int\phi_2\hat{A}\phi_2\,dq & \int\phi_2\hat{A}\phi_3\,dq & \cdots \\ \int\phi_3\hat{A}\phi_1\,dq & \int\phi_3\hat{A}\phi_2\,dq & \int\phi_3\hat{A}\phi_3\,dq & \cdots \\ \vdots & \vdots & \vdots & \end{bmatrix}$$

where $\hat{A}$ is any operator. Now let **O** be an orthogonal matrix that converts the set $\{\phi_\mu\}$ into the set $\{\chi_\sigma\}$ according to $\chi = \mathbf{O}\Phi$. Then the effect on **A** is to convert it to a new matrix **A'**, where

$$\mathbf{A'} = \mathbf{OAO}^T$$

Verify this relation.

5.21. In view of the result in Exercise 5.20, show that the Roothaan equations (4.2-5) become in the new basis set $\{\chi_\sigma\}$

$$\mathbf{F'(Oc)} = \mathbf{S'(Oc)}\varepsilon$$

This shows that the orbital coefficients for the solution of the Roothaan equations in the new basis are $(\mathbf{Oc})$ and the orbital energies $\{\varepsilon_\mu\}$ are unchanged by the orthogonal transformation.

5.22. (a) In Hartree–Fock–Roothaan theory, prove that the total electronic energy E is expressible as

$$E = \sum_i^{occ} \varepsilon_i + \frac{1}{2} \sum_a \sum_b P_{ab} H_{ab}$$

where the P_{ab} are elements of the electron density matrix.
(b) If the orthogonal transformation $\mathbf{O}$ is made, what is the effect on $\mathbf{P}$?
(c) What then is the effect on E of the orthogonal transformation $\mathbf{O}$?

5.23. If lithium hydride is treated in the NDDO approximation using the basis set of Table 4.3, write symbolically the secular determinantal equation [see below eq. (4.2-5)]. Now consider the element $F_{11} - \varepsilon$ in the secular determinant. According to eq. (5.7-4a), how many distinct repulsion integrals $(ab|a'b')$ and $(aa'|bb')$ survive? How many such integrals are there in the full Hartree–Fock treatment? Consider the Fock element F_{13}. Now how many of the repulsion integrals in eq. (5.7-5) survive in the NDDO procedure?

REFERENCES

Buckingham, A. D., "Molecular Quadrupole Moments," *Quart. Rev.*, **13**, 183 (1959). Very interesting article, although most of it is not pertinent to the present chapter; our eq. (5.4-7) is taken from Buckingham's eq. (24).

Cade, P. E. and Huo, W. M., "Electronic Structure of Diatomic Molecules. VI. A. Hartree-Fock Wavefunctions and Energy Quantities for the Ground States of the First-Row Hydrides, AH," *J. Chem. Phys.*, **47**, 614 (1967).

Clementi, E., "SCF-MO Wave Functions for the Hydrogen Fluoride Molecule," *J. Chem. Phys.*, **36**, 33 (1962).

Dence, J. B., *Mathematical Techniques in Chemistry*, Wiley, New York, 1975, pp. 345–360. Brief introduction to second-rank Cartesian tensors.

Dewar, M. J. S., *The Molecular Orbital Theory of Organic Chemistry*, McGraw-Hill, New York, 1969. See especially Chapter 10 on "Calculations Including σ Electrons."

Dewar, M. J. S. and Thiel, W., "Ground States of Molecules. 38. The MNDO Method. Approximations and Parameters," *J. Am. Chem. Soc.*, **99**, 4899 (1977a). Outline of the basic approximations of the MNDO method.

Dewar, M. J. S. and Thiel, W., "A Semiempirical Model for the Two-Center Repulsion Integrals in the NDDO Approximation," *Theor. Chim. Acta (Berlin)*, **46**, 89 (1977b).

Dewar, M. J. S., McKee, M. L., and Rzepa, H. S., "MNDO Parameters for Third Period Elements," *J. Am. Chem. Soc.*, **100**, 3607 (1978). Values of parameters for Si, P, S, Cl; no calculations are reported.

Dewar, M. J. S. and Ford, G. P., "An Addendum to a Recent Paper by Halgren, Lipscomb, and Their Co-workers Concerning the Relative Accuracies of Several Current MO Methods," *J. Am. Chem. Soc.*, **101**, 5558 (1979). The results of MNDO calculations are compared with results obtained by CNDO/2, PRDDO, and minimum basis-set SCF procedures on a set of test molecules.

Feynman, R. P., Leighton, R. B., and Sands, M., *The Feynman Lectures on Physics*, Vol. 2, Addison-Wesley, Reading, 1964, Chap. 31. This chapter talks about tensors, mainly the polarizability, moment of inertia, and stress tensors. Our Exercise 5.8 on the symmetry of α is suggested by Feynman's remarks on p. 31-4.

Goodbody, A. M., *Cartesian Tensors*, Ellis Horwood, Chichester, 1982. A fine, accessible book on the mathematics of Cartesian tensors, with applications drawn from mechanics and elasticity.

Herzberg, G., *Spectra of Diatomic Molecules*, 2nd ed., Van Nostrand Reinhold, New York, 1950, pp. 363 and 437. Brief discussion of dissociation energy, with reference to H_2^+.

Karlström, G., Roos, B. O., and Sadlej, A. J., "Ground-State Dipole Polarizability of Lithium Hydride. Accurate SCF and CAS–SCF Calculations," *Chem. Phys. Lett.*, **86**, 374 (1982). Quite similar results are reached ($\alpha_{xx} = 25.67$ au, $\alpha_{zz} = 23.12$ au) for the polarizability tensor of LiH in J. E. Gready, G. B. Bacskay, and N. S. Hush, "Finite-Field Method Calculations of Molecular Polarisabilities," *Chem. Phys.*, **22**, 141 (1977).

Lennard-Jones, J., "The Molecular-Orbital Theory of Chemical Valency. I, II," *Proc. Roy. Soc. London Ser. A*, **198**, 1, 14 (1949). Two classic papers on modern molecular orbital theory by one of the pioneers in the field.

Malta, O. L. and Gouveia, E. A., "Comment on the Average Energy Denominator Method in Perturbation Theory," *Phys. Lett.*, **97A**, 333 (1983). The discussion immediately following eq. (5.5-4) is based on this short paper.

Matcha, R. L., "Theoretical Analysis of the Electronic Structure and Molecular Properties of the Alkali Halides. II. Sodium Fluoride," *J. Chem. Phys.*, **47**, 5295 (1967).

McLean, A. D., "Optimization of Molecular Wavefunctions by Scaling," *J. Chem. Phys.*, **40**, 2774 (1964).

Murrell, J. N. and Harget, A. J., *Semi-empirical Self-consistent-field Molecular-orbital Theory of Molecules*, Wiley, New York, 1972.

Politzer, P. and Harris, R. R., "Properties of Atoms in Molecules. I. A Proposed Definition of the Charge on an Atom in a Molecule," *J. Am. Chem. Soc.*, **92**, 6451 (1970). Charges associated with the atoms in a molecule are assigned by integrating the molecular electronic density over regions "belonging" to the individual atoms.

Politzer, P. and Harris, R. R., "The Distribution of the Pi Electronic Charge of the Carbon-Carbon Triple Bond," *Tetrahedron*, **27**, 1567 (1971). Must reading for organic chemists. A related paper is P. Politzer and S. D. Kasten, "Analysis of the Charge Distributions in Molecules of the Types XCCH and XCN," *J. Phys. Chem.*, **80**, 283 (1976).

Politzer, P. and Mulliken, R. S., "Comparison of Two Atomic Charge Definitions, as Applied to the Hydrogen Fluoride Molecule," *J. Chem. Phys.*, **55**, 5135 (1971).

Politzer, P. and Politzer, A., "Properties of Atoms in Molecules. V. An Easy Procedure for Estimating Atomic Charges from Calculated Core-Electron Energies," *J. Am. Chem. Soc.*, **95**, 5450 (1973). Development of useful linear relationships for C, N, O, F, Cl between the charge on the atom in a molecule and the calculated orbital energy of its $1s$ electrons.

Pople, J. A., Santry, D. P., and Segal, G. A., "Approximate Self-Consistent Molecular Orbital Theory. I. Invariant Procedures," *J. Chem. Phys.*, **43**, S129 (1965). Seminal paper on semiempirical molecular orbital theory, upon which our Section 5.7 is based.

Pople, J. A. and Bevridge, D. L., *Approximate Molecular Orbital Theory*, McGraw-Hill, New York, 1970. Somewhat similar in spirit to the book by Dewar.

Ransil, B. J., "L.C.A.O.-M.O.-S.C.F. Wave Functions for Selected First-Row Diatomic Molecules," *Rev. Mod. Phys.*, **32**, 245 (1960).

Roothaan, C. C. J., "A Study of Two-Center Integrals Useful in Calculations on Molecular Structure. I," *J. Chem. Phys.*, **19**, 1445 (1951).

CHAPTER 6

Principles of Statistical Mechanics

A **macroscopic state** of a chemical system of n independent components at equilibrium is specified by $(n + 1)$ state variables, which may consist of the temperature (T), the volume (V), and $(n - 1)$ mole fractions $(X_1, X_2, \ldots, X_{n-1})$. In classical mechanics, a **microscopic state** of the system is determined by giving the $2\sum_{i=1}^{n} f_i N_i$ coordinates and momenta at an arbitrary instant $t = t_0$, where f_i is the number of classical degrees of freedom of a molecule of component i, and there are N_i molecules of this component present. On the other hand, in quantum mechanics the microscopic state is completely specified by the state function $\Psi(q, t_0)$, and this is determined by a set of quantum numbers comparable in size to the number of classical quantities just indicated. In either case, an enormous number of microscopic states must correspond to any particular macroscopic state, so that any macroscopic observable results from an appropriate average over the microscopic states. **Statistical mechanics** is the apparatus by which such an average is obtained; it provides the bridge between the microscopic and the macroscopic worlds.

6.1 PRINCIPLE OF EQUAL *A PRIORI* PROBABILITY

Suppose that the stationary quantum states of the system of interest are known. Since any real system cannot be truly isolated, weak "external" perturbations that cause transitions among these (approximately) stationary states are always present. For example, if the system is an ideal gas, the stationary states are taken to be the usual particle-in-a-box eigenstates (see

Table 2.1). In reality, the walls of the box are rough on a molecular scale and therefore induce transitions among the stationary states. If the Hamiltonian for the particle-in-a-box model is regarded as the unperturbed Hamiltonian and the roughness as a perturbation, the results of Section 2.10 can be applied. According to eq. (2.10-14), only transitions between degenerate states are allowed. Moreover, the rates of transition between these states must satisfy the principle of detailed balance [eq. (2.10-15)].

Although these conclusions derive from consideration of a special case, we take them to be generally valid.

SM POSTULATE 1. Let a and b be any two stationary states of a degenerate energy level of an isolated system. Then transitions between a and b are always possible, even if it is necessary to go via other states c, d, and so on, of that energy level.

This postulate, known as the **quantum ergodic hypothesis**, says that over a sufficiently long time all of the states of a degenerate level will be visited. The ergodic hypothesis has never been proved rigorously, but the fact that it has not so far led to discrepancies with experiment acts as a justification.

Now let p_a stand for the probability that an isolated system with fixed total energy is in stationary state a. From simple kinetic considerations the rate of change of p_a ought to be given by

$$\frac{dp_a}{dt} = \dot{p}_a(t) = \sum_b w_{ab} p_b(t) - \sum_b w_{ba} p_a(t) \qquad (6.1\text{-}1)$$

where the w's are time-independent transition rates. Thus, w_{ab} is the probability per unit time of transition from state b to state a and is given, in principle, by eq. (2.10-14). The first term on the right side of eq. (6.1-1) represents transitions from all other (degenerate) states b into state a; the second term includes transitions out of state a into all other (degenerate) states b (Fig. 6.1). The first term increases the probability of the system's being in state a, whereas the second term decreases it. The form of eq. (6.1-1) can be justified rigorously, but we shall not do this here.

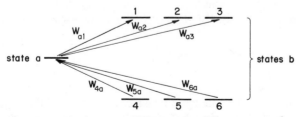

FIGURE 6.1. Rates of transition (w_{ab}, w_{ba}) between states (all seven states shown belong to the same degenerate energy level).

An equation of type (6.1-1) can be written for each degenerate state. At equilibrium, the properties of the system become time-independent. In particular, $\dot{p}_a = 0$ and we have from eq. (6.1-1)

$$\sum_b w_{ab} p_b = \sum_b w_{ba} p_a$$

$$= \sum_b w_{ab} p_a \qquad (6.1\text{-}2)$$

where the second line follows from eq. (2.10-15). Since eq. (6.1-2) must hold for *all* states a and the w_{ab} depend, in general, on a and b, the only nontrivial solution is

$$p_a = p_b \qquad (6.1\text{-}3)$$

for all a and b. That is, *for an isolated system at equilibrium, every stationary state of a degenerate level is equally likely to be occupied.* This result is known as the **Principle of Equal a Priori Probability.**

6.2   BOLTZMANN'S h-THEOREM

Imagine a system having a fixed number N of molecules in a fixed volume V. The stationary states depend, in general, upon N and V. Let $E(N, V)$ denote a given energy *level*, to which belong $\Omega(N, V, E)$ distinct stationary states (i.e., Ω is the degeneracy). If the system is *isolated*, then according to SM Postulate 1, over a sufficiently long time it will pass through all $\Omega(N, V, E)$ stationary states consistent with the constraints of fixed N, V, and E, eventually reaching equilibrium. Now suppose the system is not initially at equilibrium and consider the quantity

$$h = \sum_a p_a(t) \ln p_a(t) \qquad (6.2\text{-}1)$$

where $p_a(t)$ is the probability that the system is in stationary state a of energy level $E(N, V)$. The summation in eq. (6.2-1) extends over all $\Omega(N, V, E)$ degenerate states. The rate of change of h with respect to time is

$$\dot{h}(t) = \sum_a \dot{p}_a(t) \ln p_a(t) + \sum_a \dot{p}_a(t) \qquad (6.2\text{-}2)$$

Since the system must be in some linear combination of the stationary states, $\sum_a p_a(t) = 1$ and, therefore, the second summation in eq. (6.2-2) is zero. Substituting eq. (6.1-1) into the first summation of eq. (6.2-2) and invoking

$w_{ab} = w_{ba}$, we obtain

$$\dot{h} = \sum_a \sum_b w_{ab}[p_b(t) - p_a(t)]\ln p_a(t) \tag{6.2-3a}$$

$$= -\sum_a \sum_b w_{ab}[p_b(t) - p_a(t)]\ln p_b(t) \tag{6.2-3b}$$

after interchanging the dummy indices a and b. Addition of eqs. (6.2-3) yields

$$\dot{h} = -\frac{1}{2}\sum_a \sum_b w_{ab}[p_b(t) - p_a(t)][\ln p_b(t) - \ln p_a(t)] \tag{6.2-4}$$

We now argue that the right-hand side of eq. (6.2-4) must be negative. By definition, the transition rate w_{ab} must be nonnegative [see eq. (2.10-14)]. Moreover, it is easy to show that the quantity $(x - y)(\ln x - \ln y)$ is also nonnegative for all real, nonzero x, y. It follows from eq. (6.2-4) that $\dot{h}(t) \leq 0$, and therefore *the function $h(t)$ decreases monotonically with time*. This is the celebrated ***h*-Theorem**, proposed in 1872 by the Austrian physicist Ludwig Boltzmann (1844–1906).[1] The theorem has often been discussed from the standpoint of the kinetic theory of collisions or of classical statistical mechanics (Fitts and Mucci, 1962), but the preceding outline seems more direct. Of course, quantal concepts were not available to Boltzmann in 1872.

The minimum value of h is attained when the system reaches equilibrium and all p_a are equal. We conclude that if an isolated system is prepared in a nonequilibrium state, its approach to equilibrium is tracked by a continuously decreasing value of h. From the principle of equal *a priori* probability we have $p_a = \Omega^{-1}$ at equilibrium, and insertion of this into eq. (6.2-1) gives

$$-h = \ln \Omega \tag{6.2-5}$$

That is, $-h$ achieves a maximum value of $\ln \Omega(N, V, E)$ at equilibrium. But from the Second Law of Thermodynamics we know that an isolated system attains maximal entropy at equilibrium. This strongly suggests the following relation between the thermodynamic entropy and the degeneracy:

$$S_{eq} = C \ln \Omega + B \tag{6.2-6}$$

where C and B are constants to be determined. If the Third Law of Thermodynamics is to hold for a perfectly ordered system, then S_{eq} must approach 0 as T approaches 0 K. From a quantum mechanical standpoint, a perfectly ordered system occupies only the nondegenerate ground state at 0 K. Thus, $\Omega = 1$ and we conclude from eq. (6.2-6) that B must vanish. Hence, eq.

---

[1] Boltzmann, who tragically committed suicide, is a fascinating historical character in science. A recent biography (Broda, 1983) is one of the few available.

(6.2-6) reduces to

$$S_{eq} = C \ln \Omega(N, V, E) \tag{6.2-7}$$

which with C suitably identified (see Section 6.9), is Boltzmann's formula for the absolute entropy of the system. This relation provides a fundamental link between the microscopic world of quantum states and the macroscopic world of classical thermodynamics.

It is emphasized that the thermodynamic entropy is defined only for systems at equilibrium. The quantity

$$S = -C\sum_{a} p_a(t)\ln p_a(t) \tag{6.2-8}$$

is sometimes referred to as the **generalized entropy**. The thermodynamic entropy eq. (6.2-7) is a special case of this.

6.3   THE MICROCANONICAL ENSEMBLE

Over a sufficiently long period of time an isolated system at equilibrium continues to sample all of the $\Omega(N, V, E)$ degenerate stationary states available to it. A measurable property is generally a function of time. The pressure, for example, is the time rate of change of momentum per unit area due to collisions of molecules with the walls of the container or with the surfaces of a pressure-measuring device. If the observation is of very short duration, say the time required for a molecular collision ($\approx 10^{-14}$ s), then the pressure appears to fluctuate wildly about some mean value. The ordinary time-independent pressure measured on a macroscopic scale over a period of, say, minutes must be a time average,

$$P_{obs}(\Delta t) = \frac{1}{\Delta t} \int_{t_0}^{t_0 + \Delta t} P(t)\, dt \tag{6.3-1}$$

where $P(t)$ is the instantaneous pressure when the system is in a certain quantum state. The quantity P_{obs} does not depend on the initial time t_0, nor upon the interval of observation Δt if this is sufficiently long. We take the experimentally observed value ($\overline{P}$) to be the limit

$$\overline{P} = \lim_{\Delta t \to \infty} P_{obs}(\Delta t) \tag{6.3-2}$$

For a truly macroscopic system containing typically 10^{23} molecules, it is practically impossible to determine either the quantum mechanical or the classical microstate of the system and carry out the averaging indicated by eqs.

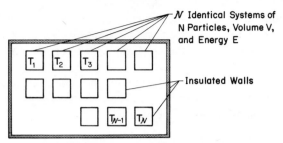

FIGURE 6.2. A microcanonical ensemble.

(6.3-1) and (6.3-2).[2] It was the suggestion of the American physical chemist Josiah Willard Gibbs (1839–1903)[3] to examine a virtual (imagined) collection, or **ensemble**, of replicas of the system of interest instead of the time evolution of a single system. Recall that the concept of ensemble was introduced in Section 2.6. The ensemble contains an astronomically large number $\mathcal{N}$ of systems, each one identical with the system of interest and isolated from all the others (Fig. 6.2). That is, each system of the ensemble has precisely the same values of N, V, and E, but the temperature and pressure, for example, may vary from system to system. All possible stationary states of the system consistent with these constraints are represented in the ensemble. The systems are distributed so that there are N_a replicas in state a, where $N_a/\mathcal{N} = p_a = \Omega^{-1}$. That is, all the accessible states are equally represented in the ensemble. Gibbs called this the **microcanonical ensemble**.

Note that the ensemble is static; the systems can be viewed as "trapped" in their idealized stationary states. The properties of each system of the ensemble are time-independent and depend only on the quantum numbers that specify the state. We know that every system has the same energy $E(N, V)$. Therefore, a given system can occupy only one of the $\Omega(N, V, E)$ allowed stationary states. Since we have no additional knowledge about the system that would permit us to favor any particular states over the others, every state must be equally represented in the ensemble.

The **ensemble average** $\langle O \rangle$ of an observable O is then given by

$$\langle O \rangle = \sum_a O_a \left(\frac{N_a}{\mathcal{N}} \right) = \sum_a O_a p_a = \Omega^{-1} \sum_a O_a \qquad (6.3\text{-}3)$$

where O_a is the observed value of O when the system is in state a. Note that

---

[2]We note, however, that high-speed electronic computers make it feasible to calculate classical trajectories for systems containing as many as 1000 particles. From these trajectories the properties can be computed approximately by equations similar to eq. (6.3-2). This technique is known as **molecular dynamics** (see Section 8.10).

[3]Gibbs, together with Boltzmann, is one of the founding fathers of statistical mechanics; see (Wheeler, 1951) for a biography.

eq. (6.3-3) is analogous to eq. (2.6-6) and has a similar interpretation: The ensemble average of an observable is simply the sum over the accessible states of the system (as dictated by the constraints defining the ensemble) of the value of the observable in each state, weighted by the probability that the system is in that state. The relation between the time average $\overline{O}$ [calculated from an equation similar to eq. (6.3-2)] and the ensemble average $\langle O \rangle$ is contained in the following crucial postulate.

SM POSTULATE 2. The ensemble average of an observable O is equal to the time average of O for a single system.

We shall assume SM Postulate 2 holds for all other ensembles characterizing different types of systems.

6.4 THE CANONICAL ENSEMBLE

Since we never deal with truly isolated systems in practice, it is convenient to introduce ensembles corresponding to systems under more realistic conditions. A common example is a *closed* system immersed in a thermal bath. The system contains a fixed number N of molecules, fixed volume V, and fixed temperature T. To construct the ensemble we assemble (mentally) an enormous collection of systems with the same fixed values of N and V and with walls that are permeable to heat. The entire collection is placed in an exceedingly large heat bath at a fixed temperature T until thermal equilibrium is reached. The collection is then insulated as a whole from the rest of the universe so that it becomes an isolated supersystem. Gibbs called this virtual collection a **canonical ensemble**. In contrast to the microcanonical ensemble, the temperature but not the energy is the same throughout the canonical ensemble.

Let the number of systems in the canonical ensemble be $\mathcal{N}$ and the total energy be E. The number of systems in stationary state i with the energy E_i is denoted N_i. Then the ensemble must satisfy the following constraints:

$$\sum_i N_i = \mathcal{N} \qquad (6.4\text{-}1a)$$

$$\sum_i N_i E_i = E \qquad (6.4\text{-}1b)$$

It is clear from eq. (6.4-1b) that all possible stationary states of the system of interest are not equally represented in the canonical ensemble. This is so simply because the energy is not the same for all the systems. The absolute temperature of a system can be shown to be proportional to the *average* kinetic energy of the molecules. However, two systems of identical composition and at the same temperature may well have equal mean kinetic energies but different potential energies and, therefore, different total energies.

Group1	Group 2	Group 3
①	② ③ ④	⑤
	② ③ ⑤	④
	② ④ ⑤	③
	③ ④ ⑤	②
②	① ③ ④	⑤
	① ③ ⑤	④
	① ④ ⑤	③
	③ ④ ⑤	①
③	① ② ④	⑤
	① ② ⑤	④

Group1	Group 2	Group 3
③	① ④ ⑤	②
	② ④ ⑤	①
④	① ② ③	⑤
	① ② ⑤	③
	① ③ ⑤	②
	② ③ ⑤	①
⑤	① ② ③	④
	① ② ④	③
	① ③ ④	②
	② ③ ④	①

FIGURE 6.3. Twenty ways of distributing five objects so that one is in Group 1, three are in Group 2, and one is in Group 3 ($W = 5!/(1!3!1!) = 20$).

We now turn to calculating the fraction of systems in the ensemble that occupy a given stationary state. We shall identify this fraction with the probability p_i that a system drawn "at random" from the ensemble is in the ith quantum state. Our first step is to calculate the number of ways $W(N_1, N_2, \ldots, N_i, \ldots)$ that the ensemble can be constituted with N_1 systems in state 1, N_2 in state 2, and so on. That is, $W(N_1, N_2, \ldots, N_i, \ldots)$ is the number of ways the *distribution* $\{N_1, N_2, \ldots, N_i, \ldots\}$ can be realized. This is equivalent to the number of ways $\mathcal{N}$ objects can be divided into K groups (the number of different stationary states) such that there are N_1 objects in group 1, N_2 in group 2, and so on, irrespective of the ordering of the objects. This problem is illustrated in Fig. 6.3. The total number of ways of arranging the $\mathcal{N}$ objects if permutations are significant in each of the K groups is

$$\underset{\substack{\text{First} \\ \text{group}}}{[\mathcal{N}(\mathcal{N}-1)\cdots(\mathcal{N}-N_1+1)]}$$

$$\times \underset{\substack{\text{Second} \\ \text{group}}}{[(\mathcal{N}-N_1)(\mathcal{N}-N_1-1)\cdots(\mathcal{N}-N_1-N_2+1)]} \times \cdots \times$$

$$\times \underset{\substack{K\text{th} \\ \text{group}}}{[(N-N_1-N_2-\cdots-N_{K-1})(\mathcal{N}-N_1-N_2-\cdots-N_{K-1}-1)\ldots1]}$$

$$= \mathcal{N}! \tag{6.4-2}$$

But if permutations within each group are not to be important, then all $N_1!$ ways of forming group 1 are equivalent, all $N_2!$ ways of forming group 2 are equivalent, and so on. The result is that the number of combinatorially distinct ways of dividing the $\mathcal{N}$ objects according to the specified distribution and without regard for ordering within each group is

$$W = \frac{\mathcal{N}!}{\prod_{i=1}^{k} N_i!} \tag{6.4-3}$$

Each of these ways of dividing the $\mathcal{N}$ objects represents a state of the *ensemble*. Hence, viewing the ensemble as an isolated supersystem, we can write

$$\Omega(\mathcal{N}, E) = \sum W(N_1, N_2, \ldots, N_K) \tag{6.4-4}$$

where the summation runs over all distributions $\{N_1, N_2, \ldots, N_K\}$ compatible with relations (6.4-1). By SM Postulate 1 all of the $\Omega(\mathcal{N}, E)$ states are equally likely. The probability of any particular distribution is therefore

$$p(N_1, N_2, \ldots, N_K) = \frac{W(N_1, N_2, \ldots, N_K)}{\Omega(\mathcal{N}, E)} \tag{6.4-5}$$

The number of systems that are in the ith stationary state is obtained by averaging N_i with the probability distribution function (6.4-5).

$$\langle N_i \rangle = \sum N_i p(N_1, N_2, \ldots, N_K) \tag{6.4-6}$$

The summation extends over all possible distributions, not over the subscript i. Making use of eqs. (6.4-4) and (6.4-5), we obtain

$$\langle N_i \rangle \Omega(\mathcal{N}, E) = \mathcal{N} \sum \frac{(\mathcal{N}-1)!}{N_1! N_2! \cdots (N_i - 1)! \cdots N_K!} \tag{6.4-7}$$

To simplify eq. (6.4-7) somewhat, we make the substitutions $\mathcal{N} - 1 = M$, $N_i - 1 = M_i$, and $N_j = M_j$ (for $j \neq i$). Equations (6.4-1) become

$$\sum_j M_j = \mathcal{N} - 1 \tag{6.4-8a}$$

$$\sum_j M_j E_j = E - E_i \tag{6.4-8b}$$

Equation (6.4-7) may then be put in the form

$$\langle N_i \rangle \Omega(\mathcal{N}, E) = \mathcal{N} \sum \frac{M!}{M_1! M_2! \cdots M_K!}$$

$$= \mathcal{N} \Omega(\mathcal{N} - 1, E - E_i) \tag{6.4-9}$$

in view of the definition in eq. (6.4-4). Therefore, the fraction of systems that we expect to find in state i is

$$\frac{\langle N_i \rangle}{\mathcal{N}} = \frac{\Omega(\mathcal{N} - 1, E - E_i)}{\Omega(\mathcal{N}, E)} \tag{6.4-10}$$

We now recall that $\mathcal{N}$, N_i, E, and Ω are all extremely large since the ensemble itself is assumed astronomical in size. This suggests that we work with the logarithm of eq. (6.4-10).

$$\ln \frac{\langle N_i \rangle}{\mathcal{N}} = \ln \Omega(\mathcal{N} - 1, E - E_i) - \ln \Omega(\mathcal{N}, E) \tag{6.4-11}$$

Since $\Omega(\mathcal{N} - 1, E - E_i)$ is so close to $\Omega(\mathcal{N}, E)$, let us expand the first term on the right-hand side of eq. (6.4-11) in a Taylor series about $(\mathcal{N}, E)$ and retain terms only through first order (Courant and John, 1974).

$$\ln \Omega(\mathcal{N} - 1, E - E_i) \simeq \ln \Omega(\mathcal{N}, E) - \left(\frac{\partial \ln \Omega(\mathcal{N}, E)}{\partial \mathcal{N}} \right)_E$$

$$- E_i \left(\frac{\partial \ln \Omega(\mathcal{N}, E)}{\partial E} \right)_{\mathcal{N}}$$

$$= \ln \Omega(\mathcal{N}, E) - \alpha - \beta E_i \tag{6.4-12}$$

where the second line defines the constants α, β. Substituting eq. (6.4-12) into eq. (6.4-11) and taking the antilog, we obtain

$$\langle N_i \rangle = \mathcal{N} e^{-\alpha} e^{-\beta E_i} \tag{6.4-13}$$

Condition (6.4-1a), combined with eq. (6.4-13), yields

$$\sum_i \langle N_i \rangle = \mathcal{N} = e^{-\alpha} \sum_i e^{-\beta E_i} \mathcal{N} \tag{6.4-14}$$

from which it follows that

$$e^{\alpha} = \sum_i e^{-\beta E_i} = Q$$

Finally, the desired probability p_i of finding the system in stationary state i is

$$p_i = \frac{\langle N_i \rangle}{\mathcal{N}} = \frac{e^{-\beta E_i}}{Q} \tag{6.4-15}$$

Equation (6.4-15) is the **canonical distribution function** and Q is called the **canonical partition function**. The partition function is the bridge between the microscopic world of stationary states and the macroscopic world of thermodynamic properties. As we shall see presently, all of the thermodynamic properties can be calculated from the partition function.[4]

Exercises

6.1. Explain how Boltzmann's formula, eq. (6.2-7), depends on SM Postulate 1.

6.2. Consider the system consisting of a single, spin-free hydrogen atom. Write an explicit expression for the degeneracy $\Omega(E_n)$. At large values of the principal quantum number n, what is the density of states in an interval of energy of width dE? Evaluate this in atomic units for $n = 100$.

6.3. Without using the method of Lagrangian multipliers, show that the function $h = \sum_a p_a \ln p_a$, subject to the restrictions $0 < p_a < 1, \sum_a p_a = 1$, has a local extremum at $p_1 = p_2 = \cdots = $, and so on. Verify that the extremum of h is not a maximum.

6.4. A hypothetical system consists of eight independent particles. Each particle has available to it the nondegenerate energy levels shown here. Assume further that the particles are indistinguishable, that is, placement of one particle in, say, the $E = 2$ level does not constitute a state distinct from that in which another particle is placed in that level instead. At equilibrium, let the system in isolation have an energy of 12 units. Calculate the entropy in units of C [see eq. (6.2-7)].

E

5   ___

4   ___

3   ___

2   ___

1   ___

6.5. If in eq. (6.4-3) the N_i's are assumed to be so large that they are essentially continuous, show that an extremum in W, subject to con-

---

[4] In early works the canonical partition function is symbolized by Z, standing for the German *Zustandssumme* (sum over states). The term "partition function" was introduced by the English workers C. G. Darwin and R. H. Fowler in 1922.

straint (6.4-1a), occurs when all N_i's are equal. [*Hint*: Make use of Stirling's approximation (see Section 6.6).]

6.6. Consider 10 distinguishable objects that are to be distributed in 4 groups. Write out all the possible distributions; the occupancy of each group is integral, and some may be permitted to contain no objects. Show that the result of Exercise 6.5 is roughly true in this discrete case. Assume that the groups are distinguishable, even when some contain equal numbers of objects. How many distributions are possible?

6.7. Consider the groups of Exercise 6.6 to be states with the energies shown here. We impose the additional constraint on the possible distributions: $\sum_{i=1}^{4} N_i E_i = 16$. Now determine how many different distributions are permissible, and compare with your answer in Exercise 6.6.

Group

	4		$E_3 = 3$
2		3	$E_2 = 2$
	1		$E_1 = 1$

6.8. For each allowed distribution in Exercise 6.7, calculate the number of *ensemble states*. What is the total number of accessible ensemble states? Determine the average number of systems in each of the four system states (remember, there are 10 systems in the ensemble).

6.9. One particular distribution in Exercise 6.7 leads to the greatest number of ensemble states. Identify it and compare its population of states with the average population of states deduced in Exercise 6.8.

6.10. Refer to eq. (6.4-13) and determine to what extent a consistent set of (α, β) parameters can be found that fits the average populations found in Exercise 6.8. Why is a perfect fit not to be expected?

6.5   FLUCTUATIONS IN THE CANONICAL ENSEMBLE

Recall that we arrived at the canonical distribution eq. (6.4-15) by averaging N_i over the probability distribution function $p(N_1, N_2, \ldots, N_K)$ [see eq. (6.4-6)]. In general, we expect that p as a function of $\{N_1, N_2, \ldots, N_K\}$ will have some spread about a **mean distribution** $\{\langle N_1 \rangle, \langle N_2 \rangle, \ldots, \langle N_K \rangle\}$. We now demonstrate, however, that the spread can be made negligible simply by taking a sufficiently large ensemble. Thus, the only important distribution is the mean one, and for a very narrow distribution the mean distribution coincides with the **most probable distribution**.

As a measure of the width of the spread about the mean, we look at the **mean-square deviation**, $\sigma_{N_i}^2$ (Yourgrau et al., 1982).

$$\sigma_{N_i}^2 \equiv \left\langle (N_i - \langle N_i \rangle)^2 \right\rangle$$
$$= \langle N_i^2 \rangle - \langle 2N_i \langle N_i \rangle \rangle + \langle N_i \rangle^2$$
$$= \langle N_i^2 \rangle - \langle N_i \rangle^2 \qquad (6.5\text{-}1)$$

The first quantity on the right side of eq. (6.5-1) is by definition

$$\langle N_i^2 \rangle = \sum N_i^2 p(N_1, N_2, \ldots, N_K) \qquad (6.5\text{-}2)$$

where the summation is over all allowed distributions. By a sequence of steps parallel to those leading from eqs. (6.4-6) to (6.4-9), eq. (6.5-2) can be converted into

$$\langle N_i^2 \rangle \Omega(\mathcal{N}, E) = \mathcal{N} \sum \frac{M_i M!}{M_1! M_2! \cdots M_i! \cdots M_K!} + \mathcal{N} \Omega(\mathcal{N} - 1, E - E_i) \qquad (6.5\text{-}3)$$

To simplify the first term on the right-hand side of eq. (6.5-3), we again make the substitutions $M - 1 = M'$, $M_i - 1 = M_i'$, and $M_j = M_j'$ (for $j \neq i$). One now has the constraints

$$\sum_j M_j' = \mathcal{N} - 2 \qquad (6.5\text{-}4a)$$

$$\sum_j M_j' E_j = E - 2E_i \qquad (6.5\text{-}4b)$$

and, accordingly, eq. (6.5-3) becomes

$$\langle N_i^2 \rangle \Omega(\mathcal{N}, E) = \mathcal{N}(\mathcal{N} - 1)\Omega(\mathcal{N} - 2, E - 2E_i) + \mathcal{N}\Omega(\mathcal{N} - 1, E - E_i)$$
$$= \mathcal{N}(\mathcal{N} - 1)\Omega(\mathcal{N} - 2, E - 2E_i) + \langle N_i \rangle \Omega(\mathcal{N}, E) \qquad (6.5\text{-}5)$$

where the second line follows from eq. (6.4-9).

Continuing the development along the same line as before, we expand the logarithm of $\Omega(\mathcal{N} - 2, E - 2E_i)$ in a Taylor series about the point $(\mathcal{N}, E)$. This gives

$$\ln \Omega(\mathcal{N} - 2, E - 2E_i) \simeq \ln \Omega(\mathcal{N}, E) - 2\alpha - 2\beta E_i$$

and therefore to first order one has

$$\mathcal{N}(\mathcal{N} - 1)\Omega(\mathcal{N} - 2, E - 2E_i) \simeq \mathcal{N}^2 \Omega(\mathcal{N}, E) e^{-2(\alpha + \beta E_i)} \qquad (6.5\text{-}6)$$

Combination of eqs. (6.5-5) and (6.5-6) yields

$$\langle N_i^2 \rangle = \mathcal{N}^2 e^{-2(\alpha + \beta E_i)} + \langle N_i \rangle$$
$$= \langle N_i \rangle^2 + \langle N_i \rangle \tag{6.5-7}$$

from eq. (6.4-13). Finally, from eq. (6.5-1) we have simply

$$\sigma_{N_i}^2 = \langle N_i \rangle$$

or, equivalently,

$$\frac{\sigma_{N_i}}{\langle N_i \rangle} = \frac{1}{\sqrt{\langle N_i \rangle}} \tag{6.5-8}$$

This last equation tells us that the fractional root-mean-square deviation of the number of systems in state i tends to zero as the number of systems $\mathcal{N}$ in the ensemble, and hence also as the number of systems N_i in state i, increases without bound. Thus, in this sense the probability distribution $p(N_1, N_2, \ldots, N_K)$ is very strongly peaked about the mean distribution $\{\langle N_1 \rangle, \langle N_2 \rangle, \ldots, \langle N_K \rangle\}$. In computing properties of the system, we may safely assume that only the mean distribution need be considered. We shall illustrate this point shortly by computing the entropy.

6.6   CALCULATION OF MACROSCOPIC PROPERTIES

Equation (6.4-15) gives the probability that a system chosen at random from the canonical ensemble is in the ith stationary state. By SM Postulate 2 we identify this probability with the probability that the system in equilibrium (with N, V, T fixed) is found in state i at any particular instant. It is emphasized that we need only know the stationary-state energies E_i in order to compute p_i. Once the number of molecules N and V are fixed, these energy eigenvalues are completely determined in principle. Questions concerning the statistics (i.e., Bose–Einstein or Fermi–Dirac) obeyed by the particles composing the system arise only in connection with the determination of the E_i and not in their use to evaluate p_i.

We now deduce an expression for the entropy of a closed, isothermal system (i.e., a system with fixed N, V, T) at equilibrium. The ensemble as a whole is regarded as an isolated supersystem, whose entropy is given by

$$S_{\text{ens}} = C \ln\left[\sum W(N_1, N_2, \ldots, N_K)\right] \tag{6.6-1}$$

from eqs. (6.2-7) and (6.4-4). The analysis of Section 6.5, as well as the result of Exercise 6.7, makes it clear that the average distribution far outweighs all others when the number of systems in the ensemble increases without limit. Only the average distribution need be included in the summation in eq. (6.6-1), and the entropy *per system* is then

$$\langle S \rangle = \frac{S_{\text{ens}}}{\mathcal{N}} = \frac{C}{\mathcal{N}} \ln W(\langle N_1 \rangle, \langle N_2 \rangle, \langle N_3 \rangle, \ldots, \langle N_K \rangle) \qquad (6.6\text{-}2)$$

The N_i's are all large numbers. Making use of eq. (6.4-3) and Stirling's approximation for large factorials,[5] we can write eq. (6.6-2) as

$$\langle S \rangle = \frac{C}{\mathcal{N}} \left[(\mathcal{N} \ln \mathcal{N} - \mathcal{N}) - \sum_i \{ \langle N_i \rangle \ln \langle N_i \rangle - \langle N_i \rangle \} \right]$$

$$= -C \sum_i \frac{\langle N_i \rangle}{\mathcal{N}} \ln \left(\frac{\langle N_i \rangle}{\mathcal{N}} \right)$$

$$= -C \sum_i p_i \ln p_i \qquad (6.6\text{-}3)$$

from eq. (6.4-15). This formula is identical in form to eq. (6.2-8) at equilibrium. The difference is that p_a in eq. (6.2-8) refers to any stationary state having a *fixed energy*, whereas p_i in eq. (6.6-3) refers to any stationary state whatsoever, regardless of its energy, E_i.

Now substituting eq. (6.4-15) into eq. (6.6-3), we obtain

$$\langle S \rangle = -C \left(-\beta \sum_i E_i \frac{e^{-\beta E_i}}{Q} - \ln Q \sum_i \frac{e^{-\beta E_i}}{Q} \right)$$

$$= C \left(\beta \sum_i p_i E_i + \ln Q \sum_i p_i \right)$$

$$= \beta C \langle E \rangle + C \ln Q \qquad (6.6\text{-}4)$$

where we have invoked SM Postulate 2 in reaching the last line. In Section 6.8 it is shown that $\beta CT = 1$, and in Section 6.9 that C is, in fact, Boltzmann's constant k_B. For convenience, we temporarily take these for granted and write eq. (6.6-4) as $\langle S \rangle = \langle E \rangle / T + k_B \ln Q$. Comparing this with the thermodynamic relation $S = (E - A)/T$, where A is the Helmholtz free energy, and again making use of SM Postulate 2, we conclude

$$\langle A \rangle = -k_B T \ln Q \qquad (6.6\text{-}5)$$

Equation (6.6-5) is the most fundamental connection between macroscopic

---

[5]**Stirling's asymptotic series** (after the Scottish mathematician James Stirling, 1692–1770) is $\ln(x!) \sim (x + \frac{1}{2}) \ln x - [x - 1/12x - \frac{11}{12} + \cdots -]$, and for large x this is nearly equal to $x(\ln x - 1)$ (Dence, 1975).

thermodynamic properties of a closed, isothermal system and its quantum states.

Although the entropy has already been expressed in eq. (6.6-3) in terms of the stationary-state probabilities p_i, it is more useful to relate it to the partition function Q. This is accomplished by using the relation

$$dA = -S\,dT - P\,dV + \mu\,dN \tag{6.6-6}$$

where P is the pressure and μ is the chemical potential. This relation expresses the infinitesimal change in the Helmholtz free energy under a reversible change in which the system does only mechanical work. Regarding A as a function of N, V, and T, we can also write

$$dA = \left(\frac{\partial A}{\partial T}\right)_{N,V} dT + \left(\frac{\partial A}{\partial V}\right)_{N,T} dV + \left(\frac{\partial A}{\partial N}\right)_{V,T} dN \tag{6.6-7}$$

for any arbitrary infinitesimal change. Therefore, comparing eqs. (6.6-6) and (6.6-7), we find after invoking SM Postulate 2 once again

$$\langle S \rangle = -\left(\frac{\partial \langle A \rangle}{\partial T}\right)_{N,V}$$

or

$$\langle S \rangle = k_B \ln Q + k_B T \left(\frac{\partial \ln Q}{\partial T}\right)_{N,V} \tag{6.6-8}$$

Finally, we can now calculate $\langle E \rangle$ through identification with the thermodynamic expression $E = A + TS$.

$$\langle E \rangle = k_B T^2 \left(\frac{\partial \ln Q}{\partial T}\right)_{N,V} \tag{6.6-9a}$$

It follows immediately from the thermodynamic relation $C_v = (\partial E/\partial T)_{N,V}$ and SM Postulate 2 that

$$\langle C_v \rangle = \left(\frac{\partial \langle E \rangle}{\partial T}\right)_{N,V} \tag{6.6-9b}$$

Similarly, expressions for the thermodynamic pressure and the chemical potential can be obtained since $P = -(\partial A/\partial V)_{N,T}$ and $\mu = (\partial A/\partial N)_{T,V}$. It should be noticed that although Q and also $\langle A \rangle$ might appear to be functions only of T and not of V and N also, this is not so since the energies E_i are themselves functions of the volume of the system and of the number of molecules.

Note that the right side of eq. (6.6-6) involves the differentials of the parameters N, V, T that characterize the canonical ensemble. This indicates that A is the *natural* or *characteristic* thermodynamic function and explains the simple relationship between A and Q in eq. (6.6-5). Similarly, for the microcanonical ensemble, the natural parameters are N, V, E and the characteristic thermodynamic function is the entropy, since for a reversible process

$$dS = \frac{1}{T} dE + \frac{P}{T} dV - \frac{\mu}{T} dN \qquad (6.6\text{-}10)$$

Still other ensembles are conceivable. In Chapter 8 we shall introduce the **grand canonical ensemble**, the ensemble in which T, V, μ are held fixed. For this ensemble, the natural thermodynamic function is PV, since for a reversible process

$$d(PV) = S\, dT + P\, dV + N\, d\mu \qquad (6.6\text{-}11)$$

Exercises

6.11. Work through the steps leading to eq. (6.5-3).

6.12. Apply eq. (6.5-1) to one of the states in Exercise 6.7, and calculate $\sigma_{N_i}/\langle N_i \rangle$. Compare this with the value you would get using eq. (6.5-8). Comment on the difference.

6.13. To help get a feel for the accuracy of Stirling's approximation, use all of the terms given in footnote 5 and estimate the value of 15! Compute the percentage error from the exact value of 1,307,674,368,000. How well does the two-term expression for $\ln(x!)$ work on 15! ?

6.14.[C] A hypothetical system consists of five indistinguishable, noninteracting particles. Energy levels of $E = 0$, $2E_0$, $3E_0$ are available to each particle. Write a computer program that will calculate values of $\langle E \rangle$ in units of E_0 as a function of the variable $x = k_B T/E_0$. Let x range from 0.1 to 9.7 in jumps of 0.4.

6.15.[C] For the system described in Exercise 6.14, calculate also with the aid of a program values of $\langle C_v \rangle$ in units of k_B as a function of $k_B T/E_0$.

6.16.[C] Plot the entropy of the system in Exercise 6.14 in units of k_B versus $x = k_B T/E_0$. To what value should the entropy tend as $x \to 0$?

6.17. Show that the same expression results for the statistical mechanical pressure $\langle P \rangle$, whether one formulates it as $-(\partial \langle A \rangle/\partial V)_{N,T}$ or as

$$-\sum_i \left(\frac{\partial E_i}{\partial V} \right)_N Q^{-1} \exp\left(\frac{-E_i}{k_B T} \right)$$

where $-(\partial E_i/\partial V)_N$ is the pressure in the ith stationary state.

6.18. Derive eq. (6.6-10). (*Hint:* Use the First and Second Laws of Thermodynamics to suggest appropriate independent variables for dE.) Also, derive eq. (6.6-11).

6.7 FLUCTUATIONS IN MACROSCOPIC OBSERVABLES

In Section 6.5 we examined fluctuations in the distribution $\{N_1, N_2, \ldots, N_K\}$ of systems in the canonical ensemble. We found that the fraction of distributions differing from the average one $\{\langle N_1 \rangle, \langle N_2 \rangle, \ldots, \langle N_K \rangle\}$ becomes vanishingly small as the size of the ensemble increases without bound. Analogously, we now investigate the fluctuations of *observables* associated with a closed, isothermal system at equilibrium. We shall see that for macroscopic systems these fluctuations are negligible, so that it is highly unlikely to measure a value of any property that differs significantly from the mean value.

Consider first the internal energy, the mean value of which is given by

$$\langle E \rangle = \sum_i p_i E_i$$

$$= \sum_i \frac{e^{-E_i/k_B T}}{Q} E_i \tag{6.7-1}$$

Differentiation of eq. (6.7-1) with respect to T yields

$$\left(\frac{\partial \langle E \rangle}{\partial T} \right)_{N,V} = \frac{1}{k_B T^2} \sum_i \frac{e^{-E_i/k_B T}}{Q} E_i^2 - \sum_i E_i \frac{e^{-E_i/k_B T}}{Q} \left(\frac{1}{Q} \frac{\partial Q}{\partial T} \right)_{N,V}$$

$$= \frac{1}{k_B T^2} \langle E^2 \rangle - \langle E \rangle \left(\frac{\partial \ln Q}{\partial T} \right)_{N,V} \tag{6.7-2}$$

Solving (6.7-2) for $\langle E^2 \rangle$, we obtain

$$\langle E^2 \rangle = k_B T^2 \left(\frac{\partial \langle E \rangle}{\partial T} \right)_{N,V} + k_B T^2 \langle E \rangle \left(\frac{\partial \ln Q}{\partial T} \right)_{N,V} \tag{6.7-3}$$

From eqs. (6.6-9) and (6.7-3) it follows that

$$\langle E^2 \rangle = k_B T^2 \langle C_v \rangle + \langle E \rangle^2 \tag{6.7-4}$$

Thus, the mean-square deviation of the internal energy is

$$\sigma_E^2 = \langle E^2 \rangle - \langle E \rangle^2 = k_B T^2 \langle C_v \rangle$$

or

$$\frac{\sigma_E}{\langle E \rangle} = \frac{T(k_B \langle C_v \rangle)^{1/2}}{\langle E \rangle} \tag{6.7-5a}$$

To get a notion of the magnitude of σ_E, we compute it for one mole of an ideal gas. In this case, $\langle \tilde{E} \rangle = 3N_0 k_B T/2$ and $\langle \tilde{C}_v \rangle = 3N_0 k_B/2$, where N_0 is Avogadro's number. Hence,

$$\frac{\sigma_E}{\langle \tilde{E} \rangle} = \left(\tfrac{2}{3}\right)^{1/2} N_0^{-1/2} \simeq 10^{-12} \tag{6.7-5b}$$

which tells us that the root-mean-square deviation of the internal energy from the mean value is a minute fraction of that value. As long as the sample is macroscopic, then the deviation is negligible. However, if we consider a very small sample, say just a few tens of molecules, then the deviation may be of the same order as the mean value.

It is possible to present the information contained in eq. (6.7-5) in a graphic, more illuminating way. We begin by observing that the probability of the system's having energy E_i, identical to the ith stationary-state eigenvalue, is

$$p(E_i) = \frac{e^{-E_i/k_B T}}{Q} \Omega(N, V, E_i) \tag{6.7-6}$$

where $\Omega(N, V, E_i)$ is the degeneracy defined in Section 6.2. For a large system, the eigenvalues are very closely spaced. We may then regard E as a continuous variable, in which case we must define the probability of the system's having an energy lying between E and $E + dE$. This probability, $dp(E)$, is given generally by the relation

$$dp(E) = f(E)\, dE$$

where $f(E)$ is a probability density function (i.e., probability per unit interval of energy). The form of $f(E)$ suggested by eq. (6.7-6) is

$$f(E) = Ae^{-E/k_B T} \Omega(N, V, E) \tag{6.7-7}$$

where A is a normalization constant determined by the requirement

$$\int f(E)\, dE = 1 \tag{6.7-8}$$

Since $\Omega(N, V, E)$ increases with E, whereas the exponential in eq. (6.7-7) decreases, we expect $f(E)$ to possess a maximum, say at $E = E^*$. Furthermore, E^* should be very nearly equal to $\langle E \rangle$, since the spread about $\langle E \rangle$ is very narrow. Let us then expand $\ln f(E)$ about $\langle E \rangle$ in a Taylor series.

$$\ln f(E) = \ln f(\langle E \rangle) + \left(\frac{\partial \ln f}{\partial E}\right)_{\langle E \rangle} (E - \langle E \rangle)$$

$$+ \frac{1}{2}\left(\frac{\partial^2 \ln f}{\partial E^2}\right)_{\langle E \rangle} (E - \langle E \rangle)^2 + \cdots \tag{6.7-9}$$

Because $f(E)$, and consequently $\ln f(E)$, have maxima at $E = \langle E \rangle$, it follows that $(\partial \ln f/\partial E)_{\langle E \rangle} = 0$. From eq. (6.7-7) we have explicitly

$$\left(\frac{\partial \ln f}{\partial E} \right) = \frac{1}{k_B T} + \frac{\partial \ln \Omega}{\partial E}$$

and a second differentiation then gives

$$\left(\frac{\partial^2 \ln f}{\partial E^2} \right)_{\langle E \rangle} = \left(\frac{\partial^2 \ln \Omega}{\partial E^2} \right)_{\langle E \rangle} \tag{6.7-10}$$

To evaluate the right side of eq. (6.7-10), we recall from eq. (6.4-12) that $\beta = (\partial \ln \Omega/\partial E)_{\mathcal{N}}$, and in Section 6.6 we agreed to accept the identification $\beta = 1/k_B T$. It follows that

$$\left(\frac{\partial^2 \ln \Omega}{\partial E^2} \right)_{\langle E \rangle} = \left(\frac{\partial\left(k_B^{-1} T^{-1}\right)}{\partial E} \right)_{\langle E \rangle}$$

$$= \frac{-1}{k_B T^2} \left(\frac{\partial T}{\partial E} \right)_{\langle E \rangle}$$

$$= -\frac{1}{k_B T^2 \langle C_v \rangle} \tag{6.7-11}$$

Inserting eq. (6.7-11) into eq. (6.7-9) and taking antilogs, we obtain

$$f(E) = f(\langle E \rangle) \exp\left[-\frac{1}{2} \left(\frac{E - \langle E \rangle}{\sigma_E} \right)^2 \right] \tag{6.7-12}$$

Finally, the normalization requirement of eq. (6.7-8) must be met; the result is

$$f(E) = \left\{ 2\pi k_B C_v T^2 \right\}^{-1/2} \exp\left[-\frac{1}{2} \left(\frac{E - \langle E \rangle}{\sigma_E} \right)^2 \right] \tag{6.7-13}$$

Equation (6.7-13) is of the form of a normal (Gaussian) distribution of a random variable E with mean $\langle E \rangle$ and variance σ_E^2. Figure 6.4 shows a qualitative plot of $f(E)$. Taking the case of 1 mol of a monatomic gas at 25°C we find that if E lies $10^{-12} \langle E \rangle$ units on either side of $\langle E \rangle$, the value of $f(E)$ drops to roughly 0.6 of its maximum value. When E lies $10^{-11} \langle E \rangle$ units on either side of $\langle E \rangle$, the value of $f(E)$ drops to 10^{-20} of its maximum value! Clearly, $f(E)$ appears as an extremely high, extremely narrow spike centered at $\langle E \rangle$. In fact, the probability density function is so narrow that the canonical ensemble is equivalent to the microcanonical ensemble in the sense that nearly all systems in the canonical ensemble have an identical energy,

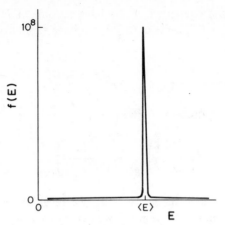

FIGURE 6.4. Probability density function for the distribution of energy in a canonical ensemble.

which is equal to the mean energy $\langle E \rangle$. Note that this is only true if the system contains a sufficiently large number of particles.

Equation (6.7-5) refers to a fluctuation at constant N and V (i.e., at constant density). As long as the density is kept constant, fluctuations in other properties are determined by those of the energy through the equilibrium thermodynamic relations. For example, the mean-square deviation of the entropy would be calculated by averaging $(S - \langle S \rangle)^2$ over the states of the system in analogy to eq. (6.7-5). Let us, therefore, consider S to be a function of N, V, E, and if N, V are kept fixed, then S is just a function of E. When E departs slightly from its mean $\langle E \rangle$, S departs slightly from its mean $\langle S \rangle$. To first order, the deviation in S is given by

$$S - \langle S \rangle \simeq \left(\frac{\partial S}{\partial E} \right)_{N,V} (E - \langle E \rangle) \qquad (6.7\text{-}14)$$

But from eq. (6.6-10) we have $(\partial S / \partial E)_{N,V} = T^{-1}$, and hence

$$\sigma_S^2 = \langle (S - \langle S \rangle)^2 \rangle = T^{-2} \langle (E - \langle E \rangle)^2 \rangle$$
$$= k_B \langle C_v \rangle \qquad (6.7\text{-}15)$$

from eq. (6.7-5). The relative deviation of the entropy is therefore

$$\frac{\sigma_S}{\langle S \rangle} = \frac{(k_B \langle C_v \rangle)^{1/2}}{\langle S \rangle} \qquad (6.7\text{-}16)$$

Again, using the example of a mole of a monatomic gas, for which $\langle \tilde{S} \rangle$ at

25°C is typically about 20 R, we calculate

$$\frac{\sigma_S}{\langle \tilde{S} \rangle} \simeq \frac{1}{20}\left(\frac{3}{2}\right)^{1/2} N_0^{-1/2} \simeq 10^{-13}$$

On a relative basis, fluctuations in the entropy are very small.

6.8 DETERMINATION OF THE CANONICAL PARAMETERS

In connection with eq. (6.2-7), where the statistical-mechanical definition of entropy is introduced, we implicitly take C (and B, there) to be temperature-independent constants. We need not do this; both B and C could be assumed to approach zero with the temperature. However, that B and C are strictly constant is the simpler hypothesis. If this is valid, then the following argument shows that C is a universal constant.

Imagine two closed systems A and B immersed in a thermal reservoir so that they are at a common temperature T (Fig. 6.5). The systems are now withdrawn and temporarily isolated. The combined entropy is, from eq. (6.2-7)

$$S_{AB} = C_{AB} \ln \Omega_{AB}(N, V, E) \tag{6.8-1}$$

where $N = N_A + N_B$, $V = V_A + V_B$, and $E = E_A + E_B$. The degeneracy of the combined system is simply the product of the degeneracies of the separate systems,

$$\Omega_{AB}(N, V, E) = \Omega_A(N_A, V_A, E_A)\Omega_B(N - N_A, V - V_A, E - E_A) \tag{6.8-2}$$

because the separate systems are isolated from each other. Substitution of eq. (6.8-2) into eq. (6.8-1) yields

$$S_{AB} = C_{AB}(\ln \Omega_A + \ln \Omega_B) \tag{6.8-3}$$

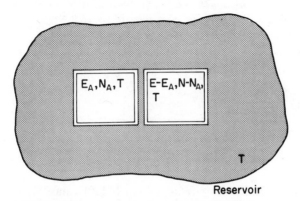

FIGURE 6.5. Two systems in thermal but not material contact.

Next, imagine a differential amount of energy dE_A to be transferred out of A and into B. Since AB is itself isolated, S_{AB} is maximal at equilibrium and we have

$$\left(\frac{\partial S_{AB}}{\partial E_A}\right)_{N,V} = C_{AB}\left[\left(\frac{\partial \ln \Omega_A}{\partial E_A}\right)_{N_A,V_A} + \left(\frac{\partial \ln \Omega_B}{\partial E_B}\right)_{N_B,V_B}\frac{dE_B}{dE_A}\right] = 0 \quad (6.8\text{-}4)$$

But $-dE_A = dE_B$, so $dE_B/dE_A = -1$ and eq. (6.8-4) leads to

$$\left(\frac{\partial \ln \Omega_A}{\partial E_A}\right)_{N_A,V_A} = \left(\frac{\partial \ln \Omega_B}{\partial E_B}\right)_{N_B,V_B} \quad (6.8\text{-}5)$$

The entropies of the separate systems are also given by expressions of the form of eq. (6.2-7), for example,

$$\left(\frac{\partial \ln \Omega_A}{\partial E_A}\right)_{N_A,V_A} = \frac{1}{C_A}\left(\frac{\partial S_A}{\partial E_A}\right)_{N_A,V_A}$$

Recalling once again from thermodynamics that $(\partial S/\partial E)_{N,V} = T^{-1}$, then we can simplify eq. (6.8-4) to $(C_A T_A)^{-1} = (C_B T_B)^{-1}$. But systems A and B are both initially at a temperature T, and remain so upon separation and isolation. We conclude that $C_A = C_B$, and since systems A and B are entirely arbitrary, C must be a universal constant, independent of the system.

Return now to eq. (6.6-4), where we consider the parameter β. Differentiation of this equation with respect to T gives

$$\left(\frac{\partial \langle S \rangle}{\partial T}\right)_{N,V} = \beta C\left(\frac{\partial \langle E \rangle}{\partial T}\right)_{N,V} + C\langle E \rangle\left(\frac{\partial \beta}{\partial T}\right)_{N,V} + \frac{C}{Q}\left(\frac{\partial Q}{\partial T}\right)_{N,V} \quad (6.8\text{-}6)$$

By SM Postulate 2 we make the usual identification of thermodynamic and statistical mechanical quantities, and rewrite eq. (6.8-6) as

$$\frac{C_v}{T} = \beta C C_v + C\langle E \rangle\left(\frac{\partial \beta}{\partial T}\right)_{N,V} + \frac{C}{Q}\left(\frac{\partial Q}{\partial T}\right)_{N,V} \quad (6.8\text{-}7)$$

It may be shown that

$$\frac{1}{Q}\left(\frac{\partial Q}{\partial T}\right)_{N,V} = -\langle E \rangle\left(\frac{\partial \beta}{\partial T}\right)_{N,V}$$

and, therefore, the last two terms of eq. (6.8-7) cancel. The final result is $\beta = (CT)^{-1}$, and so β itself is the product of a universal constant and the reciprocal of the absolute temperature. Equation (6.6-4) then reduces to

$$\langle A \rangle = \langle E \rangle - T\langle S \rangle = -CT \ln Q \qquad (6.8\text{-}8)$$

and it remains only to discover the identity of the universal constant C.

6.9 DETERMINATION OF THE PARAMETER C

The parameter C is now to be evaluated by choosing a particular system for which the mathematics is tractable. We select for this purpose the ideal gas, which consists of N identical, independent particles. The stationary states of *any* system of *identical, independent* particles have general features that influence the form of the partition function. In order to express Q, we need first to examine in more detail the effects of identity and independence.

From the discussion of separation of variables in Section 2.8, we know that the wave function for a system of independent particles can be written as a product of one-particle functions, the energy-eigenvalue equation is separable into several one-particle equations, and the energy of the system is a simple sum of one-particle energies.

$$E_{ij\ldots\ell} = \varepsilon_{i1} + \varepsilon_{j2} + \varepsilon_{k3} + \cdots + \varepsilon_{\ell N} \qquad (6.9\text{-}1)$$

The indices $i, j, k, \ldots, \ell$ on the separate ε's may vary independently, so that the canonical partition function becomes

$$Q = \sum_{i,\,j\ldots\ell}\sum\cdots\sum e^{-E_{ij\ldots\ell}/CT}$$

$$= \sum_{i} e^{-\varepsilon_{i1}/CT}\sum_{j} e^{-\varepsilon_{j2}/CT} \cdots \sum_{\ell} e^{-\varepsilon_{\ell N}/CT}$$

$$= q_1 q_2 \cdots q_N \qquad (6.9\text{-}2)$$

Since the same set of one-particle energy states is available to each particle, we have $Q = q^N$, where q is called the **molecular partition function**.

The equation $Q = q^N$ has been derived on the assumption that the particles can be labeled. Classical mechanics recognizes that identical particles, once they are labeled, can always be distinguished during the course of their motion. In quantum mechanics this is not so, and indistinguishability may be viewed as a consequence of the fact that the wave function must be either symmetric or antisymmetric upon exchange of the labels of any two identical particles (see Section 3.3). Since this symmetry is correlated with the fact that all known particles possess either integral or half-integral spins, the explanation of indistinguishability is thus pushed back to the explanation of spin itself and of its restriction in nature to only certain values.

The result is that we overcount the number of states of any multicomponent system (whether the particles are independent or not) if we assume that the particles can be labeled. The correction factor needed to give the proper number of states and hence the proper partition function depends in a complex way on the number of particles and on what restrictions there are on the populations of the single-particle states. A situation that is easy to handle and fortunately occurs for many systems of interest is where the number of one-particle energy states is much larger than the number of particles. States with an energy, for example, of

$$E_{ij\ldots\ell} = \varepsilon_{i1} + \varepsilon_{j2} + \varepsilon_{j3} + \cdots + \varepsilon_{\ell N}$$

are then extremely rare, that is, it is most unlikely for two particles to be in the same state. Under these conditions, the correction factor is simply the number of permutations of N objects, or just $N!$ and the corrected partition function is

$$Q_{\text{corr}} = \frac{q^N}{N!}$$

N particles

independent, indistinguishable

number of states $\gg N$

(6.9-3)

No real systems obey eq. (6.9-3) exactly, but it is nevertheless convenient to refer to those particles that are approximately described by it as **boltzons** (or by some authors, as **corrected boltzons**). Generally, eq. (6.9-3) is a good approximation for atoms and molecules, but not for bound electrons.

The ideal gas fits the requirements of eq. (6.9-3). We note from Table 2.1 that the energy levels of a particle of mass m in a box $L \times W \times H$ are

$$\varepsilon_{npq} = \frac{h^2}{8m}\left(\frac{n^2}{L^2} + \frac{p^2}{W^2} + \frac{q^2}{H^2}\right)$$

where n, p, q are positive integers. The molecular partition function is then

$$q = \sum_n \exp\left(\frac{-h^2}{8mCT}\frac{n^2}{L^2}\right)\sum_p \exp\left(\frac{-h^2}{8mCT}\frac{p^2}{W^2}\right)\sum_q \exp\left(\frac{-h^2}{8mCT}\frac{q^2}{H^2}\right) \quad (6.9\text{-}4)$$

if the particles are presumed structureless, that is, without rotational, vibrational, and electronic energy. Since translational energy levels are generally very densely packed, little harm is done mathematically if each summation in eq. (6.9-4) is replaced by an integration over the semi-infinite range $(0, \infty)$. If we do this, then the corrected partition function becomes

$$Q_{corr} = \frac{\left(\int_0^\infty e^{-h^2 n^2/8mCTL^2}\, dn \int_0^\infty e^{-h^2 p^2/8mCTW^2}\, dp \int_0^\infty e^{-h^2 q^2/8mCTH^2}\, dq\right)^N}{N!}$$

$$= \frac{\left[L\left(\dfrac{2\pi mCT}{h^2}\right)^{1/2} \times W\left(\dfrac{2\pi mCT}{h^2}\right)^{1/2} \times H\left(\dfrac{2\pi mCT}{h^2}\right)^{1/2}\right]^N}{N!}$$

$$= \left(\frac{2\pi mCT}{h^2}\right)^{3N/2} \frac{V^N}{N!} \qquad (6.9\text{-}5)$$

where $V = LWH$. Equation (6.9-5) may be shown, with the aid of *classical statistical mechanics* (some elements of which are covered in Section 8.3), to be valid for a container of any shape (Rushbrooke, 1949).

The Helmholtz free energy is now obtained easily from eq. (6.9-5) by making use of eq. (6.8-8).

$$\langle A \rangle = -CT\left[\frac{3}{2}N \ln\left(\frac{2\pi mCT}{h^2}\right) + N \ln V - \ln(N!)\right] \qquad (6.9\text{-}6)$$

The ensemble-averaged pressure is given by $\langle P \rangle = -(\partial\langle A\rangle/\partial V)_{N,T}$, which from eq. (6.9-6) is just NCT/V. Since P for the ideal gas is Nk_BT/V, then by SM Postulate 2 we conclude $C = k_B$.

Exercises

6.19. Refer back to Exercise 6.15. When $x = 8.1$, compute the relative deviation of the energy, $\sigma_E/\langle E \rangle$. How does this compare with what you would obtain from an equation like eq. (6.7-5b), if such an equation were valid for a system of the size of the present one?

6.20. Prove that the expression in eq. (6.7-13) is properly normalized. Note that the interval of integration is $(-\infty < E < \infty)$.

6.21.[C] To get a feel for the behavior of the distribution function in eq. (6.7-13), write a computer program that calculates values of $f(t)$ for the Gaussian distribution

$$f(t) = \frac{1}{\sigma\sqrt{2\pi}} e^{-[(t-m)/\sigma]^2/2}$$

where t is the random variable, m is the mean, and σ^2 is the variance. Prepare different plots of $f(t)$ versus $(t - m)$, corresponding to values of $\sigma = 1$, 0.2, 0.1, 0.05. Where have you seen behavior like this previously?

6.22. Show that

$$\frac{1}{Q}\left(\frac{\partial Q}{\partial T}\right)_{N,V} = -\langle E \rangle \left(\frac{\partial \beta}{\partial T}\right)_{N,V}$$

as claimed in the text following eq. (6.8-7).

6.23. A theory for a real gas yields the expression

$$Q_{\text{real}} = Q_{\text{ideal}} f(V, T)$$

According to the theory, how is the real gas enthalpy related to the ideal gas enthalpy?

REFERENCES

Broda, E., *Ludwig Boltzmann: Man, Physicist, Philosopher* (translated by L. Gay and E. Broda), OxBow Press, Woodbridge, CT, 1983.

Courant, R. and John, F., *Introduction to Calculus and Analysis*, Vol. 2, Wiley, New York, 1974, p. 68. Discussion of Taylor's theorem for functions of more than one variable.

Dence, J. B., *Mathematical Techniques in Chemistry*, Wiley, New York, 1975, p. 195. Derivation of Stirling's asymptotic series from the Euler–Maclaurin summation formula.

Fitts, D. D. and Mucci, J. F., "The Boltzmann *H*-Function," *J. Chem. Educ.*, **39**, 515 (1962). The Boltzmann *h*-theorem is discussed from the standpoint of classical statistical mechanics.

Rushbrooke, G. S., *Introduction to Statistical Mechanics*, Oxford University Press, London, 1949, p. 62. Derivation of the canonical, translational partition function by means of classical statistical mechanics, valid for a vessel of arbitrary shape.

Wheeler, L. P., *Josiah Willard Gibbs: The History of a Great Mind*, Yale University Press, New Haven, 1951. Gibbs was insufficiently appreciated in his lifetime; here is one of the very few biographies of this great figure in American chemistry and physics.

Yourgrau, W., van der Merwe, A., and Raw, G., *Treatise on Irreversible and Statistical Thermophysics*, Dover, New York, 1982, pp. 104–106. Our argument in Section 6.5 is taken from this excellent monograph.

CHAPTER 7

The Calculation of Equilibrium
Properties of Macroscopic Systems

Chapter 6 has presented all the principles needed to calculate the thermodynamic functions of substances at equilibrium. In practice, severe difficulties
stand in the way unless one makes certain physical approximations. The basic
problem is that it is (apparently) no easier to write down the exact canonical
partition function Q for a system of interacting molecules than it is to write
down the exact energy states of the system. Many-body interactions must still
be reckoned with in statistical mechanics, even though the quantities of
interest are obtained by statistical averaging. However, two ideal states of
matter to which analogous real states approach and for which the canonical
partition function is tractable are the dilute gas of noninteracting molecules
and the harmonic crystal. In fact, the dilute gas has already been addressed in
eq. (6.9-5). Most of this chapter consists of sample calculations on the dilute
gas; the chapter closes with a brief treatment of crystals.

7.1 MONATOMIC GASES

We begin by using the canonical partition function to compute the molar
entropy of a structureless gas from eq. (6.6-8).

$$\langle S \rangle = k_B \ln Q + k_B T \left(\frac{\partial \ln Q}{\partial T} \right)_{N,V}$$

If the gas is assumed to be ideal, then Q is given by eq. (6.9-5), and

$$\langle \tilde{S} \rangle = k_B \ln\left[\left(\frac{2\pi m k_B T}{h^2} \right)^{3N_0/2} \frac{\tilde{V}^{N_0}}{N_0!} \right] + k_B T \left\{ \frac{\partial}{\partial T} \ln\left[\left(\frac{2\pi m k_B T}{h^2} \right)^{3N_0/2} \frac{\tilde{V}^{N_0}}{N_0!} \right] \right\}_V$$

or

$$\langle \tilde{S} \rangle = R\left[\frac{5}{2} + \frac{3}{2}\ln\left(\frac{2\pi m k_B}{h^2} \right) + \frac{3}{2}\ln T + \ln\left(\frac{\tilde{V}}{N_0} \right) \right] \qquad (7.1\text{-}1)$$

Equation (7.1-1), historically important in the development of quantum mechanics, was given by O. Sackur and H. Tetrode in 1912–1913. Note that eq. (7.1-1) gives the *absolute* entropy; in contrast, thermodynamics gives the entropy only to within an additive constant.

If we apply eq. (7.1-1) to krypton at 298.15 K and 1 atm and take $\tilde{V}$ to be that given by the ideal gas law, then we obtain a value for the standard molar entropy $\langle \tilde{S}^\circ \rangle$ of 163.97 J mol^{-1} K^{-1}. The Handbook[1] value is 163.97 J mol^{-1} K^{-1}, in superb agreement. However, the Handbook value is taken from the extensive compilations of the National Bureau of Standards, and further checking shows that for the monatomic gases the tabulated values of $\tilde{S}^\circ$ are actually obtained by computation with eq. (7.1-1) on the ideal gas reference states instead of being measured thermochemically and corrected for nonideality. We have merely confirmed some arithmetic.

A substance that is chemically more interesting than krypton is rubidium, the alkali metal nearest to krypton in the periodic table. We expect the entropy of rubidium to be almost the same as that of krypton since the atomic masses differ but slightly. Application of eq. (7.1-1) to rubidium vapor at 298.15 K and 1 atm yields $\langle \tilde{S}^\circ \rangle = 164.22$ J mol^{-1} K^{-1}; the Handbook value, however, is 169.99 J mol^{-1} K^{-1}, for a difference of 5.77 J mol^{-1} K^{-1}. What is the problem here? Could eq. (7.1-1) fail to apply to rubidium for some reason? As we shall now see, the fault lies not in eq. (7.1-1) per se, but in our failure to take into account a new factor not present in krypton.

Any monatomic species must have a nuclear and an electronic, as well as a translational, contribution to its molecular partition function. The three contributions are separable to a high degree of accuracy.

$$Q = \frac{(q_{\text{trans}} q_{\text{nuc}} q_{\text{elec}})^N}{N!} \qquad (7.1\text{-}2)$$

The energy levels of a nucleus are widely spaced, on the order of 10^{11} J mol^{-1}. Except at exceedingly high temperatures, a nucleus will remain in its ground

---

[1]Here and henceforth, "Handbook" refers to either the *CRC Handbook of Chemistry and Physics* or *Lange's Handbook of Chemistry*.

level. If the nuclear spin is I (see footnote 4 near the end of Section 7.3), then the nuclear partition function is just

$$q_{\text{nuc}} \simeq g_1 e^{-\varepsilon_1/k_B T}$$
$$= (2I + 1)e^{-\varepsilon_1/k_B T} \qquad (7.1\text{-}3)$$

It is customary to regard the ground level as the reference point for the excited nuclear states. This amounts to setting $\varepsilon_1 = 0$. Further, since nuclei are left invariant in chemical processes, the residual entropy of $R \ln(2I + 1)$ that persists at absolute zero always cancels out in the difference of entropy between reactants and products. This residual entropy due to nuclear spin is therefore customarily set equal to 0, that is, we set $q_{\text{nuc}} = 1$ by convention. The problem with rubidium must lie in the q_{elec} factor.

The lowest-energy electronic configuration of rubidium is given by the term symbol $^2S_{1/2}$, and is doubly degenerate.[2] If the ground electronic level is taken as the reference point for the excited electronic states, then the electronic molecular partition function is given approximately by

$$q_{\text{elec}} \simeq 2 e^{-\varepsilon_1/k_B T} + g_2 e^{-\varepsilon_2/k_B T}$$
$$\simeq 2$$

where ε_1, ε_2 now refer to electronic energies; the second line assumes $(\varepsilon_2 - \varepsilon_1)/k_B T \gg 1$, which is true in the case of Rb at moderate temperatures. Insertion of $q_{\text{elec}} = 2$ into eq. (7.1-2) and then substitution of the result into eq. (6.6-8) leads to an additional contribution to $\langle \tilde{S} \rangle$ of just $R \ln 2$, or 5.76 J mol^{-1} K^{-1}. This is precisely the amount by which the Sackur–Tetrode formula (7.1-1) is in error. It is because rubidium contains an unpaired electron and krypton does not that an additional entropy contribution is obtained.

7.2   VIBRATION OF DIATOMIC MOLECULES

The next level of application of the canonical partition function is to the calculation of thermodynamic properties of diatomic molecules in the gas phase. We present a somewhat less rigorous discussion in this section and the next, reserving for Section 7.4 a more careful treatment.

Let us recall from eq. (1.3-8) the classical equation for the relative motion of a two-atom system AB. We consider as a special case strictly vibratory

---

[2]An **atomic term symbol** denotes an electronic energy level and has the form $^{2S+1}L_J$. The states corresponding to the $(2J + 1)$ allowed values of the total angular momentum along some fixed axis are generally very nearly equal in energy for atoms with $Z < 40$, and so we may take $2J + 1$ as the degeneracy of the level indicated by the term symbol. The degeneracy $2J + 1$ is not to be confused with the quantity $2S + 1$, which is called the **multiplicity** of the level (Pauling, 1960).

motion, for which $\ell = 0$. The simplest potential curve with any credence for a diatomic molecule is

$$V(R_{AB} - R_{AB}^{\circ}) = \tfrac{1}{2}\mu\omega_0^2(R_{AB} - R_{AB}^{\circ})^2 \qquad (7.2\text{-}1)$$

where R_{AB}° is the equilibrium internuclear distance. This is just the potential energy for the linear harmonic oscillator, which was treated quantum mechanically in Section 2.7 and summarized in Table 2.1. The energy eigenvalues are

$$\varepsilon_{\text{vib}} = \left(n + \tfrac{1}{2}\right)\hbar\omega_0 \qquad (n = 0, 1, 2, 3, \ldots,) \qquad (7.2\text{-}2)$$

In evaluating the partition function Q_{vib} for a collection of harmonic oscillators (diatomic molecules), a correction for indistinguishability is not needed since this has already been made in connection with translation. Equation (6.9-2) therefore applies. Let us measure the energies of the excited states relative to the bottom (i.e., $R_{AB} = R_{AB}^{\circ}$) of the potential well $V(R_{AB} - R_{AB}^{\circ})$. Then the vibrational partition function becomes

$$Q_{\text{vib}} = \left(\sum_{n=0}^{\infty} e^{-n\hbar\omega_0/k_B T}\right)^N e^{-N\hbar\omega_0/2k_B T}$$

Summing the infinite geometric series, we obtain

$$Q_{\text{vib}} = \left(\frac{e^{-\hbar\omega_0/2k_B T}}{1 - e^{-\hbar\omega_0/k_B T}}\right)^N \qquad (7.2\text{-}3)$$

Substitution of eq. (7.2-3) into eq. (6.6-8) yields

$$\langle \tilde{S}_{\text{vib}} \rangle = R\left[\frac{\Theta_v}{T}\frac{e^{-\Theta_v/T}}{1 - e^{-\Theta_v/T}} - \ln(1 - e^{-\Theta_v/T})\right] \qquad (7.2\text{-}4)$$

where the **characteristic vibrational temperature** Θ_v is defined as $\Theta_v = \hbar\omega_0/k_B$. The vibrational contribution to the thermodynamic internal energy may be found from eq. (6.6-9), and from this the heat capacity at constant volume, $\langle C_{v,\text{vib}} \rangle$, is obtained by differentiation with respect to the temperature.

$$\langle \tilde{C}_{v,\text{vib}} \rangle = \left(\frac{\partial \langle \tilde{E}_{\text{vib}} \rangle}{\partial T}\right)_v$$

$$= k_B T^2 \frac{\partial^2 \ln Q_{\text{vib}}}{\partial T^2} + 2k_B T \frac{\partial \ln Q_{\text{vib}}}{\partial T} \qquad (7.2\text{-}5a)$$

or from eq. (7.2-3)

$$\langle \tilde{C}_{v,\text{vib}} \rangle = R\left(\frac{\Theta_v}{T}\right)^2 e^{-\Theta_v/T}(1 - e^{-\Theta_v/T})^{-2} \qquad (7.2\text{-}5b)$$

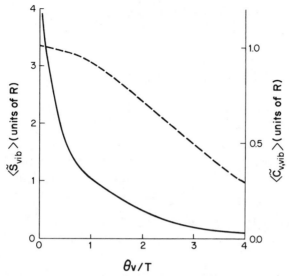

FIGURE 7.1. Graphs of the vibrational contribution to the molar entropy (—) and the molar heat capacity (---) as a function of Θ_v/T.

Plots of $\langle \tilde{S}_{\text{vib}} \rangle$ and $\langle \tilde{C}_{v,\text{vib}} \rangle$ are presented in Fig. 7.1. Note that $\langle \tilde{C}_{v,\text{vib}} \rangle$ is defined at $(\Theta_v/T) = 0$, but $\langle \tilde{S}_{\text{vib}} \rangle$ is not. Extensive tables of $\langle \tilde{S}_{\text{vib}} \rangle$ and $\langle \tilde{C}_{v,\text{vib}} \rangle$ are available (Abramowitz and Stegun, 1965).

Using characteristic vibrational temperatures Θ_v, deduced from infrared or Raman (in the case of a homonuclear diatomic) spectra, we have applied eqs. (7.2-4) and (7.2-5) to some representative diatomic molecules in Table 7.1. Since thermochemical measurements yield only *total* entropies and heat capac-

Table 7.1.   Vibrational Contribution to the Entropy and the Heat Capacity at 298.15 K of Some Diatomic Molecules

Molecule	$\tilde{\nu}_0{}^{a,\,b}$	Θ_v (K)	$\langle \tilde{S}_{\text{vib}}^{\circ} \rangle^c$	$\langle \tilde{C}_{v,\text{vib}} \rangle^d$
Na_2^{23}	159.23	229.09	10.71	7.92
I_2^{127}	214.57	308.71	8.39	7.61
S_2^{32}	725.68	1044.07	1.16	3.27
$N^{14}O^{16}$	1904.03	2739.40	0.00866	0.0718
N_2^{14}	2359.61	3394.87	0.00116	0.0122
H^1Br	2649.67	3812.19	0.00032	0.00381
			0.106^e	0.586^e

[a] Data taken from G. Herzberg, *Spectra of Diatomic Molecules*, 2nd ed., Van Nostrand Reinhold, New York, 1950, pp. 502–580.
[b] These are *wavenumbers* (in cm^{-1}) of the fundamental absorptions; hence, $\Theta_v = hc\tilde{\nu}_0/k_B$.
[c] Equation (7.2-4), in units of $J\ mol^{-1}\ K^{-1}$.
[d] Equation (7.2-5b), in units of $J\ mol^{-1}\ K^{-1}$.
[e] At 600 K.

ities, we cannot judge how good the values in Table 7.1 are until we have computed the other contributions to $\tilde{S}^\circ$ and $\tilde{C}_v^\circ$.

Exercises

7.1. Verify the form of eq. (7.1-1). Then verify the calculated value of $\langle \tilde{S}^\circ \rangle$ for krypton.

7.2. Let $f = f(T, P)$ be the fugacity of a pure gas at some specified temperature and pressure. Prove that the difference between the actual molar entropy of the gas at T, P and the entropy for the ideal gas is

$$S(T, P) - S(T, P)_{\text{ideal}} = -R\left[\ln\left(\frac{f}{P}\right) + T\left(\frac{\partial \ln f}{\partial T}\right)_P\right]$$

7.3. Assume that $P = 1$ atm is sufficiently low that the van der Waals equation of state may be solved *approximately* for $\tilde{V}$ as a function only of T and P. Find the form of $\tilde{V}$ and of the fugacity f. Then estimate the magnitude of the correction $\tilde{S}(T, P) - \tilde{S}(T, P)_{\text{ideal}}$ in units of joules per mol per degree kelvin for krypton at 298.15 K and 1 atm, given that $a = 2.318$ L^2 atm mol^{-2} and $b = 0.3978$ L mol^{-1}.

7.4. The first three energy levels of gaseous silicon are as follows (Moore, 1949):

$$^3P_0, 0 \text{ cm}^{-1} \qquad ^3P_1, 77.15 \text{ cm}^{-1} \qquad ^3P_2, 223.31 \text{ cm}^{-1}$$

Compute the standard molar entropy of silicon and compare with the tabulated value of 167.86 J mol^{-1} K^{-1}.

7.5. Compare the calculated standard molar entropy of argon vapor at its boiling point of 87.29 K with the experimentally determined value of 129.75 J mol^{-1} K^{-1} (Landolt–Börnstein, 1961).

7.6. The absolute entropy of a substance at T, P means (if $S_0^\circ = 0$) the change in entropy upon heating the material at P from 0 K to T. In view of Exercise 7.3, why should eq. (7.1-1) work so well in computing the *actual* entropy of krypton at 298.15 K if for a significant part of the interval [0, 298.15] krypton is far from an ideal gas?

7.7. Verify the formulas in eqs. (7.2-4) and (7.2-5b). Compare the value of $\langle \tilde{S}_{\text{vib}}^\circ \rangle$ for Cl$_2$ using eq. (7.2-4) versus using a three-term series expansion of the molecular vibrational partition function. Refer to Table 7.5 for Θ_v.

7.8. The atoms of an ideal gas have two internal energy levels (each nondegenerate) separated by $\Delta\varepsilon$. Define the dimensionless variable $u = k_B T/\Delta\varepsilon$.

Prepare a plot of $\langle \tilde{C}_p^{\circ} \rangle$ versus u, where $\langle \tilde{C}_p^{\circ} \rangle$ contains both the translational and the internal contributions.

7.9. A dilute monolayer of gas adsorbed on a solid surface can be modeled crudely as particles constrained to move in a two-dimensional box of dimensions a by b. Derive expressions for the translational contribution to the entropy and Helmholtz free energy of the gas.

7.3   ROTATION OF DIATOMIC MOLECULES

To consider the rotation of a *rigid* diatomic AB, return to eq. (1.3-7),

$$\mu R_{AB}^{\circ 2} \omega = \ell$$

where the angular velocity ω is defined as $\omega = d\phi/dt$, the time rate of change of the angle of rotation ϕ relative to some arbitrary orientation. Define $I = \mu R_{AB}^{\circ 2}$ as the **moment of inertia** of AB about an axis passing through its center of mass and perpendicular to the internuclear bond. Since the rotational kinetic energy of AB is given by $T = \frac{1}{2} I \omega^2$ [see the expression for L_{rel} just preceding eq. (1.3-6a)], we can rewrite this equation as

$$T = \frac{\ell^2}{2I} \tag{7.3-1}$$

In the quantum mechanical version of eq. (7.3-1), T is replaced by the kinetic energy operator and ℓ^2 is replaced by the operator for the square of the total orbital angular momentum. The result is just the (θ, ϕ) equation of the hydrogen atom (see Table 2.1), whose eigenvalues are given by

$$\varepsilon_{\text{rot}} = \frac{J(J+1)\hbar^2}{2I} \qquad (J = 0, 1, 2, 3, \ldots,) \tag{7.3-2a}$$

Spectroscopists usually write eq. (7.3-2a) as

$$\frac{\varepsilon_{\text{rot}}}{hc} = BJ(J+1) \tag{7.3-2b}$$

where the **rotational constant** B (in cm^{-1}) of the molecule is $h/8\pi^2 cI$. The value of B may be inferred from microwave spectra. From eq. (7.3-2b), the separations between the lower rotational levels is of the order hcB. Taking HCl as an example, we find

$$\Delta \tilde{E} \simeq \frac{(6.626 \times 10^{-34})^2 (6.023 \times 10^{23}) \text{ J}^2 \text{ s}^2 \text{ mol}^{-1}}{8\pi^2 (1.673 \times 10^{-27} \text{ kg})(1.275 \times 10^{-10} \text{ m})^2}$$

$$= 123 \text{ J mol}^{-1}$$

At room temperature, thermal energy ($\simeq k_B T$) is of the order 8.314 J mol^{-1} K^{-1} × 298.15 K, or 2480 J mol^{-1}. Rotational energy levels are, therefore, closely spaced in comparison to thermal energy.

We are now ready to set up the molecular rotational partition function for a heteronuclear diatomic molecule. Note that each level J is $(2J + 1)$-fold degenerate, corresponding to the $2J + 1$ possible projections of the angular momentum upon the laboratory z axis (see Table 2.1). Thus, we have

$$q_{\text{rot}} = \sum_{J=0}^{\infty} (2J + 1)e^{-hcBJ(J+1)/k_B T} \tag{7.3-3}$$

The series in eq. (7.3-3) cannot be summed analytically, but can be approximated closely by replacing the summation by an integration over J, since the levels are closely spaced. This replacement leads to

$$q_{\text{rot}} = \int_0^{\infty} (2J + 1)e^{-hcBJ(J+1)/k_B T}\, dJ$$

$$= \frac{T}{\Theta_r} \qquad (\Theta_r \ll T) \tag{7.3-4}$$

where $\Theta_r = hcB/k_B$ is the **characteristic rotational temperature**. It is emphasized that eq. (7.3-4) is valid only for $T \gg \Theta_r$.

The rotational partition function of a system of N independent, identical heteronuclear diatomic molecules is then

$$Q_{\text{rot}} = \left(\frac{T}{\Theta_r}\right)^N$$

and substitution into eqs. (6.6-8) and (6.6-9), followed by differentiation, yields the molar rotational contributions to the entropy and heat capacity.

$$\langle \tilde{S}_{\text{rot}} \rangle = R\left(1 + \ln\frac{T}{\Theta_r}\right) \qquad \langle \tilde{C}_{v,\text{rot}} \rangle = R \tag{7.3-5}$$

Table 7.2 gives the rotational contribution to the standard molar entropy of some heteronuclear diatomic molecules. The values are generally two or more orders of magnitude larger than typical vibrational contributions at room temperature.

To see how well the calculations have gone so far, we now assemble the translational, vibrational, and rotational contributions to $\tilde{S}^{\circ}$ and $\tilde{C}_p^{\circ}$ for a few molecules in Table 7.3, where comparison with experiment is made. The agreement is generally excellent except for NO, which appears suspiciously

Table 7.2. Standard Molar Rotational Entropies of Some Heteronuclear Diatomic Molecules at 298.15 K

Molecule	$R_{AB}^{\circ\,a}$	B (cm^{-1})^a	Θ_r (K)^b	$\langle \tilde{S}_{rot}^{\circ} \rangle^c$
I^{127}Cl^{35}	2.32088	0.11416	0.1642	70.71
P^{31}N^{14}	1.49087	0.78649	1.1316	54.66
N^{14}O^{16}	1.15077	1.67195	2.4055	48.39
C^{12}O^{16}	1.12832	1.93128	2.778	47.19
Li^{7}H^1	1.5957	7.5131	10.810	35.89
H^{1}Br^{18}	1.41444	8.46488	12.179	34.90
H^{1}F^{19}	0.91681	20.9557	30.149	27.37

^{a}Data taken from K. P. Huber and G. Herzberg, *Constants of Diatomic Molecules*, Van Nostrand Reinhold, New York, 1979; bond distances are in angstroms.
$^b\Theta_r = hcB/k_B = 1.4387\,B$ if B is in cm^{-1}.
^{c}Equation (7.3-5), in units of J mol^{-1} K^{-1}.

Table 7.3. Complete Calculation of the Standard Molar Entropy and Heat Capacity at 298.15 K of Four Gases

Molecule	$\tilde{\nu}_0^a$	Θ_v(K)	B^a	Θ_r(K)	$\langle \tilde{S}^{\circ} \rangle^b$	$\tilde{S}^{\circ\,c}$	$\langle \tilde{C}_p^{\circ} \rangle^{b,d}$	$\tilde{C}_p^{\circ\,c}$
HBr	2648.98	3811.19	8.4649	12.179	tr = 163.53	198.61	tr = 12.472	29.16
					rot = 34.90		rot = 8.314	
					vib = 0.00		vib = 0.004	
					$\langle \tilde{S}^{\circ} \rangle = 198.43$		$\langle \tilde{C}_p^{\circ} \rangle = 29.11$	
CO	2169.81	3121.79	1.9313	2.778	tr = 150.30	197.90	tr = 12.472	29.16
					rot = 47.19		rot = 8.314	
					vib = 0.00		vib = 0.026	
					$\langle \tilde{S}^{\circ} \rangle = 197.49$		$\langle \tilde{C}_p^{\circ} \rangle = 29.13$	
ICl	384.29	552.89	0.11416	0.1642	tr = 172.22	247.44	tr = 12.472	35.56
					rot = 70.71		rot = 8.314	
					vib = 4.28		vib = 6.292	
					$\langle \tilde{S}^{\circ} \rangle = 247.21$		$\langle \tilde{C}_p^{\circ} \rangle = 35.39$	
NO	1904.20	2739.65	1.67195	2.4055	tr = 151.16	210.65	tr = 12.472	29.84
					rot = 48.39		rot = 8.314	
					vib = 0.01		vib = 0.072	
					$\langle \tilde{S}^{\circ} \rangle = 199.56$		$\langle \tilde{C}_p^{\circ} \rangle = 29.17$	

^{a}Data taken from K. P. Huber and G. Herzberg, *Constants of Diatomic Molecules*, Van Nostrand Reinhold, New York, 1979, and are in cm^{-1}
^{b}Calculated values, in J mol^{-1} K^{-1}.
^{c}Experimental values taken from J. A. Dean (ed.), *Lange's Handbook of Chemistry*, 12th ed., McGraw-Hill, New York, 1979, pp. 9-16 to 9-39.
^{d}Calculated values obtained from $\tilde{C}_p^{\circ} = R + (\tilde{C}_{v,\text{tr}}^{\circ} + \tilde{C}_{v,\text{rot}}^{\circ} + \tilde{C}_{v,\text{vib}}^{\circ})$.

low. As in the case of rubidium, this suggests that we have failed to account for some additional factor.

Nitric oxide contains one unpaired electron. Therefore, the electronic molecular partition function for NO is approximately[3]

$$q_{\text{elec}} \simeq 2 + 2e^{-\Delta\varepsilon/k_B T}$$

where $\Delta\varepsilon = \varepsilon_2 - \varepsilon_1 = 119.8$ cm^{-1}. Carrying through the calculation, we find the electronic contributions to $\langle \tilde{S}^\circ \rangle$ and $\langle \tilde{C}_v^\circ \rangle$ are

$$\langle \tilde{S}_{\text{elec}}^\circ \rangle = R \ln q_{\text{elec}} + RT \left(\frac{\partial \ln q_{\text{elec}}}{\partial T} \right)_{N,V}$$

$$= 11.19 \text{ J mol}^{-1} \text{ K}^{-1} \tag{7.3-6a}$$

$$\langle \tilde{C}_{v,\text{elec}}^\circ \rangle = RT^2 \left(\frac{\partial^2 \ln q_{\text{elec}}}{\partial T^2} \right)_{N,V} + 2RT \left(\frac{\partial \ln q_{\text{elec}}}{\partial T} \right)_{N,V}$$

$$= 0.640 \text{ J mol}^{-1} \text{ K}^{-1} \tag{7.3-6b}$$

These contributions lead to a corrected molar entropy of $\langle \tilde{S}^\circ \rangle = 210.75$ J mol^{-1} K^{-1} and heat capacity of $\langle \tilde{C}_p^\circ \rangle = 29.81$ J mol^{-1} K^{-1}.

Observe that no *homonuclear* diatomic molecules are included in Tables 7.2 and 7.3. These present a complication that is now considered. Because homonuclear diatomic molecules are symmetric, their wave functions must also possess special symmetry. The nuclei of a homonuclear diatomic molecule are both either fermions or bosons (see Table 3.1), and thus a proper and complete wave function (translational + rotational + vibrational + electronic + nuclear spin) must be either antisymmetric or symmetric upon exchange of the two identical nuclei. Let us write

$$\Psi = \Psi_{\text{trans}}\Psi_{\text{rot}}\Psi_{\text{vib}}\Psi_{\text{elec}} \times \Psi_{\text{nuc}}$$

$$= \Psi' \times \Psi_{\text{nuc}}$$

The translational part of Ψ' is always symmetric upon exchange of the nuclei since Ψ_{trans} depends only upon the position of the center of mass. The vibrational part of Ψ' is symmetric upon exchange of the nuclei because Ψ_{vib} depends only upon the scalar distance R_{AB} between the two nuclei and this

---

[3]Angular momentum in molecules is an intricate subject (Herzberg, 1950; Levine, 1975; Landau and Lifshitz, 1977). In a diatomic molecule oriented along the z axis, the potential energy operator is independent of the confocal elliptical angle ϕ_i for each of the electrons i; in the kinetic energy operator the ϕ_i enter only in terms containing $\partial^2/\partial\phi_i^2$. It is seen, therefore, that the Hamiltonian for the molecule commutes with $\hat{L}_z$ but not with $\hat{L}_x$, $\hat{L}_y$, or $\hat{L}^2$, and thus only $\hat{L}_z$ is a constant of the motion. It can take values $M_L\hbar$, where $M_L = 0, \pm 1, \pm 2, \ldots$. The quantum number $\Lambda = |M_L|$ is used as the basis of the classification of electronic states: $\Lambda = 0, 1, 2, 3, \ldots$, are designated as Σ, Π, Δ, Φ, $\ldots$. States other than Σ are doubly degenerate since the energy would be expected to depend roughly on M_L^2 as is the case with H_2^+, and so both $+M_L$ and $-M_L$ correspond to the same value of Λ. For NO, the lowest level is $^2\Pi_{1/2}$; the next level, $^2\Pi_{3/2}$, lies 119.8 cm^{-1} above it.

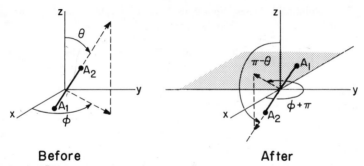

Before   **After**

FIGURE 7.2. Illustrating the changes in polar angles when the nuclei of a homonuclear diatomic molecule are exchanged.

coordinate is unaltered when the nuclei are exchanged. Finally, Ψ_{elec} is generally symmetric in the nuclei for molecules whose electronic ground state is a singlet. The symmetry of Ψ' thus depends for most molecules upon the symmetry of Ψ_{rot}. From Table 2.1, the rotational eigenfunctions of the (θ, ϕ) equation are

$$\Psi_{rot}(\theta, \phi) = \frac{1}{\sqrt{2\pi}} \left\{ \frac{(2J + 1)(J - |m|)!}{2(J + |m|)!} \right\}^{1/2} e^{im\phi} P_J^{|m|}(\cos \theta) \quad (7.3\text{-}7)$$

where $P_J^{|m|}$ is the associated Legendre polynomial (in the argument $\cos \theta$) of degree J and order $|m|$, with $|m| = 0, 1, 2, \ldots, J$. Figure 7.2 shows that when the nuclei of a diatomic molecule are exchanged, the original polar angles θ and ϕ become $\pi - \theta$ and $\phi + \pi$. This means that the factor $e^{im\phi}$ becomes multiplied by the quantity $e^{im\pi} = (-1)^m$. The function $P_J^{|m|}(\cos \theta)$ is the product of a power of $\sin \theta$ and a polynomial of $\cos \theta$ of degree $J - |m|$ containing only odd or only even powers of $\cos \theta$, depending on whether $J - |m|$ is an odd or even integer. Hence, the polynomial $P_J^{|m|}(\cos \theta)$ becomes multiplied by $[\cos(\pi - \theta)/\cos \theta]^{J-|m|} = (-1)^{J-m}$ since $\sin \theta = \sin(\pi - \theta)$. Overall, upon exchange of the nuclei, $\Psi_{rot}(\theta, \phi)$ becomes $\Psi_{rot}(\theta, \phi) \times (-1)^J$, and the wave function is then symmetric if J is an even integer (or zero).

We thus see that if the two nuclei in the homonuclear diatomic molecule are bosons, even-numbered J rotational wave functions must be multiplied by symmetric nuclear-spin wave functions, and odd-numbered J rotational wave functions must be multiplied by antisymmetric nuclear-spin wave functions. Opposite pairings apply to nuclei that are fermions (see Table 7.4).

The next task is to deduce the possible nuclear-spin states. A nucleus with spin I has a spin degeneracy[4] of $(2I + 1)$. For a pair of such nuclei there are then $(2I + 1)^2$ linearly independent nuclear-spin wave functions for the

---

[4] By analogy with the case for the electron, I for a nucleus means the maximum measurable component of the nuclear-spin angular momentum in units of $\hbar$. The quantum mechanically allowed values of the nuclear-spin angular momentum along some fixed direction in space are then $m_I \hbar$, where $m_I = I, I - 1, I - 2, \ldots, -I$.

Table 7.4.   Rotational Nuclear-Spin Wave Function Products for Homonuclear Diatomic Molecules

Rotatonal Function	Symmetry	Nuclear-Spin Function	Product Symmetry
Bosons			
J odd	A	A	S
J even	S	S	S
Fermions			
J odd	A	S	A
J even	S	A	A

molecule. The number of symmetric ones is the number $(2I + 1)$ of simple-product functions of form

$$\Psi_{m_I}(A_1) \cdot \Psi_{m_I}(A_2)$$

where the value of the spin quantum number m_I is the same for both nuclei, plus the number of symmetric combination functions of the form

$$\Psi_{m_I}(A_1) \cdot \Psi_{m_I'}(A_2) + \Psi_{m_I'}(A_1) \cdot \Psi_{m_I}(A_2)$$

where now m_I and m_I' are different for the two nuclei. The total number of symmetric nuclear-spin wave functions is therefore given by

$$g_{\text{nuc}}^{\text{sym}} = (2I + 1) + C(2I + 1, 2) \tag{7.3-8}$$

Upon computing $C(2I + 1, 2)$, the number of combinations of $(2I + 1)$ things taken two at a time, we find $g_{\text{nuc}}^{\text{sym}} = (I + 1)(2I + 1)$. Since there are $(2I + 1)^2$ total nuclear-spin wave functions, then the total number of antisymmetric nuclear-spin wave functions is $g_{\text{nuc}}^{\text{anti}} = (2I + 1)^2 - g_{\text{nuc}}^{\text{sym}} = I(2I + 1)$.

Let us now consider as an example the molecule Cl_2^{35}, the nuclei of which are fermions. The combined rotational nuclear-spin molecular partition function is

$$q_{\text{rot, nuc}} = (I + 1)(2I + 1) \sum_{\text{odd } J} (2J + 1)e^{-\Theta_r J(J+1)/T}$$

$$+ I(2I + 1) \sum_{\text{even } J} (2J + 1)e^{-\Theta_r J(J+1)/T} \tag{7.3-9}$$

The function $q_{\text{rot, nuc}}$ is evidently not factorable as a product $q_{\text{rot}} \cdot q_{\text{nuc}}$. However, in case $\Theta_r \ll T$, each summation can be replaced by an integration

**Table 7.5.   Calculation of the Standard Molar Entropy and Heat Capacity $\tilde{C}_p^\circ$
for Cl_2^{35} at 298.15 K**

$B(cm^{-1})$	$\Theta_r(K)$	$\tilde{\nu}_0(cm^{-1})$	$\Theta_v(K)$	$\langle \tilde{S}^\circ \rangle^a$	$\tilde{S}_{exp}^\circ$	$\langle \tilde{C}_p^\circ \rangle^a$	$\tilde{C}_{p,exp}^\circ$
0.24399	0.35104	559.72	805.29	tr = 161.88	229.96	tr = 12.47	33.95^b
				rot = 58.63^c		rot = 8.31	
				vib = 2.19		vib = 4.68	
				$\langle \tilde{S}^\circ \rangle = 222.70$		$\langle \tilde{C}_p^\circ \rangle = 33.78^d$	

^{a} Calculated values in units of J mol^{-1} K^{-1}.
^{b} The value for naturally occurring diatomic chlorine.
^{c} Calculated from the formula $\langle \tilde{S}_{rot}^\circ \rangle = R(1 + \ln\dfrac{T}{2\Theta_r})$.
$^d \langle \tilde{C}_p^\circ \rangle = R + [\langle \tilde{C}_{v,tr}^\circ \rangle + \langle \tilde{C}_{v,rot}^\circ \rangle + \langle \tilde{C}_{v,vib}^\circ \rangle]$.

and we have

$$q_{rot,nuc} = (I + 1)(2I + 1)\left(\tfrac{1}{2}\right)\int_0^\infty (2J + 1)e^{-\Theta_r J(J+1)/T}\, dJ$$

$$+I(2I + 1)\left(\tfrac{1}{2}\right)\int_0^\infty (2J + 1)e^{-\Theta_r J(J+1)/T}\, dJ$$

$$= (2I + 1)^2\left(\frac{T}{2\Theta_r}\right) \tag{7.3-10}$$

Therefore, at sufficiently high temperature, the rotational nuclear-spin part of
the molecular partition function does factor as $q_{rot} \cdot q_{nuc}$, where q_{rot} for a
homonuclear diatomic of two fermion nuclei is $\tfrac{1}{2}(T/\Theta_r)$, and $q_{nuc} = (2I + 1)^2$.
The factor of two in the denominator of eq. (7.3-10) is often referred to as a
symmetry number σ, for it can be interpreted as the number of indistinguishable ways that a diatomic chlorine molecule can be oriented. It owes its
presence to the inequality $\Theta_r \ll T$. As is customary, the nuclear portion of the
partition function is omitted when thermodynamic calculations are performed
[see eq. (7.1-3)]. For Cl_2, the relevant data and results are summarized in
Table 7.5.

Exercises

7.10. Show that the values of R_{AB}° and B are consistent with each other for
the first two entries in Table 7.2.

7.11. Assume that $\Theta_r \ll T$ for a diatomic gas, and treat J as a continuous
variable. Find an expression for the most probable value of J, and
calculate this most probable value for two selected diatomics. Fundamentally, why is the most populated vibrational level of a molecule at

room temperature the ground level, whereas the most populated rotational level is some excited one?

7.12. Calculate the standard molar entropy and molar heat capacity $\tilde{C}_p^{\circ}$ of $Br_{2(g)}$ at 298.15 K, and compare with the Handbook values of 245.35 and 36.02 J mol^{-1} K^{-1}. Use the following data for Br_2^{79}: B = 0.082107 cm^{-1}, $\tilde{\nu}_0$ = 325.321 cm^{-1}. Note the relation $\omega_0 = 2\pi\tilde{\nu}_0 c$.

7.13. Calculate the standard molar entropy of formation of HBr at 298.15 K, and compare with the Handbook value of 57.02 J mol^{-1} K^{-1}. Use the data for Br_2 obtained in Exercise 7.12, plus the following data for hydrogen: B = 60.853 cm^{-1}, $\tilde{\nu}_0$ = 4401.21 cm^{-1}. Also, you will need $\Delta\tilde{S}_{vap}^{\circ}(Br_2)$ = 93.06 J mol^{-1} K^{-1}.

7.14. As early as 1927 it had been predicted by Heisenberg that two forms of molecular hydrogen should exist. **Orthohydrogen** is that form in which the nuclear-spin wave function is symmetric with respect to exchange of the nuclei; **parahydrogen** is the form that is antisymmetric to exchange of the nuclei. Calculate the percentage of parahydrogen present at 75 K in equilibrium H_2. Giauque (1930) calculated this value as 51.78%, but he used a value for B that differs slightly from the current value.

7.15. Show explicitly that the five rotational wave functions corresponding to $J = 2$ are symmetric upon exchange of the nuclei in a homonuclear diatomic.

7.16. Is eq. (7.3-10) valid in the high-temperature limit for a diatomic molecule composed of two identical boson nuclei? At high temperatures what should be the percentage of paradeuterium (see Exercise 7.14) present in an equilibrium D_2 mixture? What should it be at 0 K?

7.4   INTERACTION OF VIBRATION AND ROTATION

In Sections 7.2 and 7.3 we have treated vibration and rotation as though they were separable. We now wish to carry out a more rigorous treatment that takes into account interaction terms. The validity of the Born–Oppenheimer approximation (see Section 3.4) is assumed, so that our starting point is the nuclear wave equation for a diatomic molecule AB.

$$\left\{-\frac{\hbar^2}{2m_A}\nabla_A^2 - \frac{\hbar^2}{2m_B}\nabla_B^2 + \mathscr{E}(Q)\right\}\phi(Q) = W\phi(Q)$$

The energy eigenvalues are symbolized by W, and the set of coordinates Q are the Cartesian coordinates x_A, y_A, z_A, x_B, y_B, z_B of the nuclei.

We first transform the wave equation to the center-of-mass frame (see Section 1.3). Define $M = m_A + m_B$ and

$$
\begin{cases}
X = \left(\dfrac{m_A}{M}\right)x_A + \left(\dfrac{m_B}{M}\right)x_B & u = x_B - x_A \\[2mm]
Y = \left(\dfrac{m_A}{M}\right)y_A + \left(\dfrac{m_B}{M}\right)y_B & v = y_B - y_A \\[2mm]
Z = \left(\dfrac{m_A}{M}\right)z_A + \left(\dfrac{m_B}{M}\right)z_B & w = z_B - z_A
\end{cases}
\tag{7.4-1}
$$

and let $\phi(x_A, y_A, z_A, x_B, y_B, z_B) = \Psi(X, Y, Z, u, v, w)$. Then from the partial differential calculus we have

$$
\frac{\partial}{\partial x_A} = \left(\frac{\partial X}{\partial x_A}\right)\frac{\partial}{\partial X} + \left(\frac{\partial Y}{\partial x_A}\right)\frac{\partial}{\partial Y} + \left(\frac{\partial Z}{\partial x_A}\right)\frac{\partial}{\partial Z}
$$

$$
+ \left(\frac{\partial u}{\partial x_A}\right)\frac{\partial}{\partial u} + \left(\frac{\partial v}{\partial x_A}\right)\frac{\partial}{\partial v} + \left(\frac{\partial w}{\partial x_A}\right)\frac{\partial}{\partial w}
$$

$$
= \left(\frac{m_A}{M}\right)\frac{\partial}{\partial X} - \frac{\partial}{\partial u}
\tag{7.4-2}
$$

and hence

$$
\frac{\partial^2}{\partial x_A^2} = \frac{m_A^2}{M^2}\frac{\partial^2}{\partial X^2} + \frac{\partial^2}{\partial u^2} - \frac{2m_A}{M}\frac{\partial^2}{\partial X \partial u}
\tag{7.4-3}
$$

Similar expressions for $\partial^2/\partial y_A^2$ and $\partial^2/\partial z_A^2$ lead to

$$
\frac{-\hbar^2}{2m_A}\nabla_A^2 = -\frac{m_A\hbar^2}{2M^2}\nabla_{XYZ}^2 - \frac{\hbar^2}{2m_A}\nabla_{uvw}^2
$$

$$
+ \frac{\hbar^2}{M}\left(\frac{\partial^2}{\partial X \partial u} + \frac{\partial^2}{\partial Y \partial v} + \frac{\partial^2}{\partial Z \partial w}\right)
\tag{7.4-4a}
$$

If this sequence of steps is carried through on the second Laplacian in the nuclear wave equation, then analogous to eq. (7.4-4a) one obtains

$$
\frac{-\hbar^2}{2m_B}\nabla_B^2 = -\frac{m_B\hbar^2}{2M^2}\nabla_{XYZ}^2 - \frac{\hbar^2}{2m_B}\nabla_{uvw}^2 - \frac{\hbar^2}{M}\left(\frac{\partial^2}{\partial X \partial u} + \frac{\partial^2}{\partial Y \partial v} + \frac{\partial^2}{\partial Z \partial w}\right)
\tag{7.4-4b}
$$

Insertion of eqs. (7.4-4) into the nuclear wave equation yields

$$\left[-\frac{\hbar^2}{2M}\nabla^2_{XYZ} - \frac{\hbar^2}{2\mu}\nabla^2_{uvw} + \mathcal{E}(Q) \right]\Psi = W\Psi \qquad (7.4\text{-}5)$$

where $\mu = m_A m_B/(m_A + m_B)$. Since the potential energy $\mathcal{E}(Q)$ is a function only of the internal coordinates u, v, w, the Hamiltonian in eq. (7.4-5) cleanly separates into center-of-mass and internal terms just as it does in classical mechanics (Section 1.3). From the discussion in Section 2.8, we can then write the eigenfunctions Ψ as a product of two functions, each involving separately the sets of coordinates $\{X, Y, Z\}$ and $\{u, v, w\}$.

Accordingly, let $\Psi = T(X, Y, Z)\psi(u, v, w)$ and substitute this into eq. (7.4-5). Then two equations result,

$$-\frac{\hbar^2}{2M}\nabla^2_{XYZ}T(X, Y, Z) = E_{tr}T(X, Y, Z) \qquad (7.4\text{-}6a)$$

$$-\frac{\hbar^2}{2\mu}\nabla^2_{uvw}\psi(u, v, w) + \mathcal{E}(u, v, w)\psi(u, v, w) = E_{int}\psi(u, v, w) \quad (7.4\text{-}6b)$$

with $W = E_{tr} + E_{int}$. Equation (7.4-6a) is the energy-eigenvalue equation for a freely translating particle of mass M; it gives the *translational* states of the diatomic as a whole. Equation (7.4-6b) is the energy-eigenvalue equation for the motion of a particle of mass μ under the influence of the potential $\mathcal{E}(u, v, w)$; it gives the internal states of the diatomic. Thus, the nuclear wave equation has been separated into translational and internal parts.

As in Section 7.2, we assume $\mathcal{E}(u, v, w)$ to depend only upon the *scalar* distance r_{AB}. Then taking u, v, w to be the polar coordinates r_{AB}, θ, ϕ, we see that eq. (7.4-6b) is of the same *general* form as the energy-eigenvalue equation for the hydrogen atom [see eq. (2.8-6)]. Hence, eq. (7.4-6b) is further separable into r_{AB}, θ, and ϕ equations:

$$\frac{d^2\Phi}{d\phi^2} = -m^2\Phi \qquad (7.4\text{-}7a)$$

$$\frac{1}{\Theta}\frac{d^2\Theta}{d\theta^2} + \frac{\cot\theta}{\Theta}\frac{d\Theta}{d\theta} - \frac{m^2}{\sin^2\theta} = -J(J+1) \qquad (7.4\text{-}7b)$$

$$\frac{1}{r_{AB}^2}\frac{d}{dr_{AB}}\left(r_{AB}^2\frac{dR}{dr_{AB}}\right) + \left[-\frac{J(J+1)}{r_{AB}^2} + \frac{2\mu}{\hbar^2}\{E_{int} - \mathcal{E}(r_{AB})\} \right]R = 0 \quad (7.4\text{-}7c)$$

The solutions of the angular equations are given in Table 2.1. Our next task is to solve the radial equation (7.4-7c).

Let $R(r_{AB}) = r_{AB}^{-1}T(r_{AB})$; substitution into eq. (7.4-7c) gives

$$\frac{d^2T}{dr_{AB}^2} + \left[-\frac{J(J+1)}{r_{AB}^2} + \frac{2\mu}{\hbar^2}\{E_{int} - \mathscr{E}(r_{AB})\} \right]T = 0 \qquad (7.4\text{-}8)$$

For small displacements $r_{AB} - r_{AB}^{\circ}$, the potential $\mathscr{E}(r_{AB})$ can be expanded in a Taylor series about the minimum r_{AB}°,

$$\mathscr{E}(r_{AB}) = \mathscr{E}(0) + (r_{AB} - r_{AB}^{\circ})\left(\frac{d\mathscr{E}}{dr_{AB}}\right)_0 + \tfrac{1}{2}(r_{AB} - r_{AB}^{\circ})^2\left(\frac{d^2\mathscr{E}}{dr_{AB}^2}\right)_0 + \cdots$$

$$\simeq \tfrac{1}{2}(r_{AB} - r_{AB}^{\circ})^2\left(\frac{d^2\mathscr{E}}{dr_{AB}^2}\right)_0$$

where the subscript zero denotes r_{AB}° and we arbitrarily set $\mathscr{E}(0) = 0$. Defining $x = r_{AB} - r_{AB}^{\circ}$, we can then rewrite eq. (7.4-8) approximately as

$$\frac{d^2T}{dx^2} + \left[-\frac{J(J+1)}{(x + r_{AB}^{\circ})^2} + \frac{2\mu E_{int}}{\hbar^2} - \left(\frac{\mu\omega_0 x}{\hbar}\right)^2 \right]T = 0 \qquad (7.4\text{-}9)$$

where by extension of eq. (7.2-1) we have defined $\mu\omega_0^2 = (d^2\mathscr{E}/dr_{AB}^2)_0$. For $x \ll r_{AB}^{\circ}$, we expand $(x + r_{AB}^{\circ})^{-2}$ by the binomial theorem,

$$(x + r_{AB}^{\circ})^{-2} \simeq r_{AB}^{\circ -2}\left\{ 1 - \frac{2x}{r_{AB}^{\circ}} + \frac{3x^2}{r_{AB}^{\circ 2}} \right\}$$

and substitute into eq. (7.4-9) to obtain

$$\frac{d^2T}{dx^2} + \left\{ \left(\frac{2\mu E_{int}}{\hbar^2} - J(J+1)r_{AB}^{\circ -2}\right) + 2J(J+1)r_{AB}^{\circ -3}x \right.$$

Constant   Linear

term   term

$$\left. - \left[\left(\frac{\mu\omega_0}{\hbar}\right)^2 + 3J(J+1)r_{AB}^{\circ -4}\right]x^2 \right\}T = 0 \qquad (7.4\text{-}10)$$

Quadratic

term

We can eliminate the linear term by making the change of variable

$$x = \rho - \tfrac{1}{2}\left(\frac{2J(J+1)r_{AB}^{\circ -3}}{(\mu\omega_0/\hbar)^2 + 3J(J+1)r_{AB}^{\circ -4}}\right)$$

which transforms eq. (7.4-10) into

$$\frac{d^2T}{d\rho^2} + \left\{ \left(\underbrace{\frac{2\mu E_{int}}{\hbar^2} - J(J+1)r_{AB}^{\circ -2} + \frac{J^2(J+1)^2 r_{AB}^{\circ -6}}{(\mu\omega_0/\hbar)^2 + 3J(J+1)r_{AB}^{\circ -4}}}_{\text{Constant term}} \right) \right.$$

$$\left. - \rho^2 \underbrace{\left[\left(\frac{\mu\omega_0}{\hbar}\right)^2 + 3J(J+1)r_{AB}^{\circ -4} \right]}_{\text{Quadratic term}} \right\} T = 0 \qquad (7.4\text{-}11)$$

Comparison with the harmonic-oscillator equation in Table 2.1 leads to the identifications

$$\left(n + \tfrac{1}{2}\right)\hbar\omega = E_{int} - \frac{J(J+1)\hbar^2 r_{AB}^{\circ -2}}{2\mu} + \frac{\hbar^2}{2\mu} \frac{J^2(J+1)^2 r_{AB}^{\circ -6}}{(\mu\omega_0/\hbar)^2 + 3J(J+1)r_{AB}^{\circ -4}}$$

$$(7.4\text{-}12\text{a})$$

$$\omega^2 = \omega_0^2 + \frac{3J(J+1)\hbar^2 r_{AB}^{\circ -4}}{\mu^2} \qquad (7.4\text{-}12\text{b})$$

Finally, rearranging eq. (7.4-12a) and introducing the definition of the rotational constant B [see below eq. (7.3-2b)], we find

$$E_{int} = \left(n + \tfrac{1}{2}\right)\hbar\omega + hcBJ(J+1) - \frac{2[hcBJ(J+1)]^2}{\mu r_{AB}^{\circ 2}\omega^2} \qquad (7.4\text{-}13)$$

Thus, the separation of eq. (7.4-6b) has automatically yielded internal energy eigenvalues that are the sum of a harmonic-oscillator contribution [first term in eq. (7.4-13)], a rigid-rotor contribution [second term in eq. (7.4-13)], and a correction term that may be interpreted as arising because the molecule actually stretches slightly as it rotates. Note that if $J = 0$, then eq. (7.4-13) reduces to the pure vibrational case, eq. (7.2-2), analogous to eq. (1.3-8) when $\ell = 0$. Provided J is not too large, then usually $\omega^2 \simeq \omega_0^2$ in eq. (7.4-12b), and eq. (7.4-13) may be rearranged with little loss of accuracy to

$$\frac{E_{int}}{hc} = \tilde{\nu}_0\left(n + \tfrac{1}{2}\right) + BJ(J+1) - DJ^2(J+1)^2 \qquad (7.4\text{-}14)$$

where $\tilde{\nu}_0 = \omega_0/2\pi c$, and the **centrifugal distortion constant** D is

$$D = \frac{\hbar^3}{16\pi^3 c^3 \tilde{\nu}_0^2 \mu^3 r_{AB}^{\circ 6}}$$

Experimental values of D can be inferred from spectra, and may not agree exactly with the values calculated theoretically.

7.5   COMPUTER HIGHLIGHT: DEDUCTION OF $\tilde{\nu}_0$, B, D FROM EXPERIMENT

The deduction of "best values" of the fundamental frequency $\tilde{\nu}_0$, the rotational constant B, and the centrifugal distortion constant D from spectroscopic data is a simple but natural problem for the computer. Let us consider for simplicity transitions in heteronuclear diatomic molecules in which $n = 0 \rightarrow n = 1$, and the rotational quantum number changes from J to $J \pm 1$, in accordance with spectroscopic selection rules for the harmonic oscillator and the rigid rotor. Then from eq. (7.4-14) we have

$$\Delta\left(\frac{E_{int}}{hc}\right) = \begin{cases} \tilde{\nu}_0 + 2B(J + 1) - 4D(J + 1)^3 & J \rightarrow J + 1 \\ \tilde{\nu}_0 - 2BJ + 4DJ^3 & J \rightarrow J - 1 \end{cases} \quad (7.5\text{-}1)$$

To find the optimum values of $\tilde{\nu}_0$, B, D from a set of experimental values of $\{J, \Delta(E_{int}/hc)\}$ pairs, we carry out a least-squares treatment of the data. Let X_i stand for a particular J (initial rotational state) and Y_i stand for the corresponding value of $\Delta(E_{int}/hc)$. The sum of the squares of the errors (SQ) is, for the case $J \rightarrow J + 1$,

$$SQ = \sum_{i=1}^{N} \left[Y_i - \left\{\tilde{\nu}_0 + 2B(X_i + 1) - 4D(X_i + 1)^3\right\}\right]^2$$

and this is to be minimized with respect to variations in the independent variables $\tilde{\nu}_0$, B, and D. This leads to the following set of three equations in three unknowns:

$$\begin{cases} \left(\dfrac{\partial(SQ)}{\partial\tilde{\nu}_0}\right)_{B,D} = \sum_{i=1}^{N}\left[-2Y_i + 2\tilde{\nu}_0 + 4B(X_i + 1) - 8D(X_i + 1)^3\right] = 0 \\[2mm] \left(\dfrac{\partial(SQ)}{\partial B}\right)_{\tilde{\nu}_0,D} = \sum_{i=1}^{N} 4(X_i + 1)\left[(\tilde{\nu}_0 - Y_i) + 2B(X_i + 1)\right. \\[2mm] \qquad\qquad\qquad\qquad\qquad\qquad\qquad\left. - 4D(X_i + 1)^3\right] = 0 \\[2mm] \left(\dfrac{\partial(SQ)}{\partial D}\right)_{\tilde{\nu}_0,B} = \sum_{i=1}^{N} 8(X_i + 1)^3\left[(Y_i - \tilde{\nu}_0) - 2B(X_i + 1)\right. \\[2mm] \qquad\qquad\qquad\qquad\qquad\qquad\qquad\left. + 4D(X_i + 1)^3\right] = 0 \end{cases} \quad (7.5\text{-}2)$$

Because the number of equations is small, the most direct method of solution is by elimination of any two of $\tilde{\nu}_0$, B, D from eq. (7.5-2). Alternatively, eq.

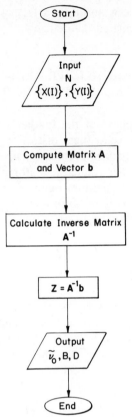

FIGURE 7.3. Flowchart for least-squares calculation of $\tilde{\nu}_0$, B, D.

(7.5-2) can be rewritten (with suitable change in notation) in matrix form as

$$
\begin{bmatrix}
N & 2\sum_{i=1}^{N}(X_i+1) & -4\sum_{i=1}^{N}(X_i+1)^3 \\
\sum_{i=1}^{N}(X_i+1) & 2\sum_{i=1}^{N}(X_i+1)^2 & -4\sum_{i=1}^{N}(X_i+1)^4 \\
-\sum_{i=1}^{N}(X_i+1)^3 & -2\sum_{i=1}^{N}(X_i+1)^4 & 4\sum_{i=1}^{N}(X_i+1)^6
\end{bmatrix}
\begin{bmatrix}
\tilde{\nu}_0 \\
B \\
D
\end{bmatrix}
=
\begin{bmatrix}
\sum_{i=1}^{N} Y_i \\
\sum_{i=1}^{N} Y_i(X_i+1) \\
-\sum_{i=1}^{N} Y_i(X_i+1)^3
\end{bmatrix}
$$

$$\mathbf{A} \qquad\qquad \mathbf{z} \qquad\qquad \mathbf{b}$$

and this can then be solved by standard matrix inversion using the formula in eq. (4.4-11) to give $\mathbf{z} = \mathbf{A}^{-1}\mathbf{b}$. A flowchart is shown in Fig. 7.3. The only input needed is the number (N) of data pairs, the set of X_i, and the set of Y_i. Equations analogous to eq. (7.5-2) can be developed for the second case of eq. (7.5-1).

7.6   A MORE REALISTIC POTENTIAL-ENERGY FUNCTION— THE MORSE POTENTIAL

Only for the first couple of energy levels is vibratory motion well described by Hooke's law (i.e., a quadratic potential energy). Spectral analysis and accurate ab initio calculations (see Section 4.6) show that the potential–energy curve has the general shape like that shown for H_2^+ in Fig. 3.6. This form can be approached by adding terms cubic and higher in ($r_{AB} - r_{AB}^o$) to the harmonic potential function. Alternatively, a concise three-parameter function that does a fair job of representing the potentials of diatomic molecules is the **Morse potential**, first proposed by P. M. Morse in 1929.

$$\mathscr{E}(r_{AB}) = D_e \left\{ 1 - e^{-a(r_{AB} - r_{AB}^o)} \right\}^2 \qquad (7.6\text{-}1)$$

In this form the zero of energy is at the minimum, in contrast to Exercise 1.19. The energy eigenvalues of the Morse potential are given by (Pauling and Wilson, 1935)

$$\left\{ \begin{aligned} \frac{E_{int}}{hc} &= \tilde{\nu}_0 \left(n + \tfrac{1}{2} \right) - x_0 \tilde{\nu}_0 \left(n + \tfrac{1}{2} \right)^2 + BJ(J+1) \\ &\quad - DJ^2(J+1)^2 - \alpha \left(n + \tfrac{1}{2} \right) J(J+1) \\ \tilde{\nu}_0 &= \frac{a}{\pi c} \left(\frac{D_e}{2\mu} \right)^{1/2} \\ x_0 &= \frac{hc}{4D_e} \tilde{\nu}_0 \\ \alpha &= \frac{3\hbar^2 \tilde{\nu}_0}{4a\mu D_e r_{AB}^{o\,3}} \left(1 - \frac{1}{a r_{AB}^o} \right) \end{aligned} \right. \qquad (7.6\text{-}2)$$

The quantities B and D have the same significance as in eq. (7.4-14); D_e is the electronic dissociation energy of the diatomic molecule (see Section 5.3), and α is the **vibration–rotation coupling constant**.

Values of $\tilde{\nu}_0$, x_0, r_{AB}^o, and D_e (or rather, D_0, the spectroscopic dissociation energy) may be inferred from experiment. Consequently, the Morse potential is overdetermined, and if it is fitted to the experimental values of $\tilde{\nu}_0$, x_0, and r_{AB}^o, the predicted value of the dissociation energy D_e is often in disagreement with the true value. Nevertheless, the Morse potential is important in showing the form of the correction factors to the rigid rotor–harmonic oscillator model. The parameters x_0, D, and α are generally very small.

All of the various corrections mentioned up to now have a small effect on the molecular partition function. The details of this may be found elsewhere (Pitzer, 1953). The result is that the internal molecular partition function can be expressed in terms of that for the idealized rigid rotor–harmonic oscillator (rr–ho) model

$$q_{int} \simeq q_{rr-ho}\left(1 + \frac{2Dk_BT}{hc\left(B - \frac{1}{2}\alpha\right)^2} + \frac{\alpha}{\left(B - \frac{1}{2}\alpha\right)(e^u - 1)} + \frac{2h\tilde{\nu}_0 x_0 c}{k_BT(e^u - 1)^2}\right)$$

$$q_{rr-ho} = \frac{1}{\sigma\beta} \frac{\exp\left(-\frac{1}{2}u + \frac{1}{4}y\right)}{1 - e^{-u}} \tag{7.6-3}$$

where $u = hc\tilde{\nu}_0/k_BT$, $y = hcx_0\tilde{\nu}_0/k_BT$, $\beta = (B - \frac{1}{2}\alpha)hc/k_BT$, and σ is the symmetry factor (either 1 or 2).

The correction terms in eq. (7.6-3) are important at high temperature where high-energy rotations are common and where vibrations may have large amplitudes. A different kind of correction is important at low temperature, for there the approximation of replacing the summation in eq. (7.3-4) by an integration may not be accurate. The **Euler–Maclaurin summation formula**, eq. (7.6-4), allows one to replace a summation by an integration plus a few terms (Dence, 1975).[5]

$$\sum_{r=0}^{n} g(r) = \int_0^n g(x)\,dx + \frac{1}{2}[g(n) + g(0)]$$

$$+ \sum_{r=1}^{} \frac{B_{2r}}{(2r)!}\left[g^{(2r-1)}(n) - g^{(2r-1)}(0)\right] \tag{7.6-4}$$

Here, the B_{2r}'s are **Bernoulli numbers**, the first several of which are given in Table 7.6. If $g(r)$ is a transcendental function, then the series on the right is an infinite series and may not converge. However, often the absolute error made in discarding all terms in the series on the right after the nth is less than the magnitude of the $(n + 1)$st. In the present case, $g(r) \equiv g(J) =$

[5] The Euler–Maclaurin formula was discovered by Leonhard Euler (1707–1783) in 1732 and independently by the Scottish mathematician Colin Maclaurin (1698–1746).

Table 7.6.   The First Few Bernoulli Numbers[a, b]

$B_0 = 1$	$B_4 = -\frac{1}{30}$
$B_1 = -\frac{1}{2}$	$B_5 = 0$
$B_2 = \frac{1}{6}$	$B_6 = \frac{1}{42}$
$B_3 = 0$	$B_{2n+1} = 0$

[a] Values for these numbers may disagree with those in other sources because various writers define them differently.
[b] Introduced by the Swiss mathematician James Bernoulli (1655–1705).

$(2J + 1)e^{-\beta J(J+1)}$, and eq. (7.6-4) becomes explicitly

$$\sum_{J=0}^{\infty} (2J + 1)e^{-\beta J(J+1)}$$

$$= \int_0^{\infty} (2J + 1)e^{-\beta J(J+1)}\, dJ + \frac{1}{2} + \frac{1}{12}[0 - 1 \cdot (2 - \beta)]$$

$$- \frac{1}{720}(\beta^3 - 12\beta^2 + 12\beta) + \cdots$$

$$= \frac{1}{\beta}\left(1 + \frac{\beta}{3} + \frac{\beta^2}{15} + \cdots + \right) \qquad (7.6\text{-}5)$$

The partition function given by eq. (7.6-3) can now be put in the final form

$$q_{int} = q_{rr-ho}\left[1 + \frac{2Dk_BT}{hc\left(B - \frac{1}{2}\alpha\right)^2} + \frac{\alpha}{\left(B - \frac{1}{2}\alpha\right)(e^u - 1)} + \frac{2h\tilde{\nu}_0 x_0 c}{k_BT(e^u - 1)^2}\right.$$

$$\left. + \frac{\left(B - \frac{1}{2}\alpha\right)hc}{3k_BT} + \frac{1}{15}\left(\frac{\left(B - \frac{1}{2}\alpha\right)hc}{k_BT}\right)^2\right] \qquad (7.6\text{-}6)$$

This equation has the form $q = q_{rr-ho} \cdot q_{corr}$, and is therefore convenient for calculations since the thermodynamic functions depend on $\ln q_{int} = \ln q_{rr-ho} + \ln q_{corr}$. The quantity $\ln q_{corr}$ is of the form $\ln(1 + x)$, with x small. To first order, $\ln(1 + x) \simeq x$, where now x means all of the terms in the brackets of eq. (7.6-6) except the unity. From $\ln q_{corr}$ we derive $\langle \tilde{A}_{corr} \rangle = -RT \ln q_{corr}$, the correction to the Helmholtz (and to the Gibbs) free energies. Successive differentiations of $\ln q_{corr}$ yield $\langle \tilde{S}_{corr} \rangle$ and $\langle \tilde{C}_{v, corr} \rangle$. The final results are

$$\langle \tilde{A}_{corr} \rangle = -RT\left(\frac{2\delta}{\beta} + \frac{\alpha\delta}{D(e^u - 1)} + \frac{2y}{(e^u - 1)^2} + \frac{\beta}{3} + \frac{\beta^2}{15} \right) \quad (7.6\text{-}7a)$$

$$\langle \tilde{S}_{corr} \rangle = R\left(\frac{4\delta}{\beta} + \frac{\alpha\delta}{D(e^u - 1)} + \frac{2y + \alpha u\delta e^u/D}{(e^u - 1)^2} - \frac{\beta^2}{15} \right.$$

$$\left. + \frac{2y(2ue^u - e^u + 1)}{(e^u - 1)^3} \right) \quad (7.6\text{-}7b)$$

$$\langle \tilde{C}_{v, corr} \rangle = R\left(\frac{4\delta}{\beta} + \frac{\delta u^2 e^u(e^u + 1)}{(e^u - 1)^3} + \frac{2\beta^2}{15} \right.$$

$$\left. + 4x_0 e^u u^2 \frac{u(2e^u + 1) - 2e^u + 2}{(e^u - 1)^4} \right) \quad (7.6\text{-}7c)$$

$$\delta = \frac{D}{B - \frac{1}{2}\alpha}$$

These expressions are most conveniently handled on a computer.

Exercises

7.17.   Carry through the three separations involved in eqs. (7.4-6) and (7.4-7). In eq. (7.4-9), why does a term linear in x not appear?

7.18.*   Plot qualitatively the first few harmonic eigenfunctions $T(\rho)$ for eq. (7.4-11). Note that there is a problem with the boundary conditions here because the true eigenfunctions of eq. (7.4-11) must vanish at

$$\rho = -r_{AB}^{\circ} + \frac{1}{2}\left(\frac{2J(J + 1)r_{AB}^{\circ -3}}{(\mu\omega_0/\hbar)^2 + 3J(J + 1)r_{AB}^{\circ -4}} \right)$$

Explain why. In general, the harmonic eigenfunctions of eq. (7.4-11) do not vanish at ρ in the preceding equation. It is argued, however, that the harmonic $T(\rho)$ decrease rapidly enough for sufficiently large negative ρ that the eigenvalues corresponding to the true and harmonic $T(\rho)$ functions may be taken to be identical.

7.19.   Calculate at what hypothetical value of J the centrifugal distortion energy for HBr (see Tables 7.1 and 7.2) would equal the ground vibrational energy. Also calculate the value of the rotational energy (rigid rotor) for this value of J and compare with the centrifugal distortion energy. What is the decrease in centrifugal distortion energy if J is now halved?

7.20.^{C} The following data on HCl pertain to the $J \rightarrow J + 1$ case of eq. (7.5-1) (Herzberg, 1950, p. 56).

$\Delta(E_{int}/hc)$ (cm^{-1})	J
2925.78	1
2963.24	3
2997.78	5
3029.96	7
3059.07	9

Determine optimum values of $\tilde{\nu}_0$, B, D after writing a suitable computer program to solve the eqs. (7.5-2).

7.21.^{C} Suppose the following data have been obtained from a spectroscopic study of the transient species $C^{12}H^1$.

$\Delta(E_{int}/hc)$ (cm^{-1})	J
2803.82	2
2746.32	4
2689.39	6
2633.32	8
2578.38	10

These numbers are for the case $J \rightarrow J - 1$ of eq. (7.5-1). Deduce equations analogous to eq. (7.5-2), and proceed as in Exercise 7.20 to find the values of $\tilde{\nu}_0$, B, and D.

7.22.^{C} For Na_2^{23} the values of $\tilde{\nu}_0$, x_0, and r_{AB}^o are 159.23 cm^{-1}, 0.004559, and 3.0786 Å (Herzberg, 1950, p. 554). Write a computer program to fit a Morse curve to these data and then plot the curve from 2 to 8 Å. Determine the value of the spectroscopic dissociation energy D_0 in units of kJ mol^{-1}.

7.23.^{C**} The best theoretical calculation (Wind, 1965) of the total energy for H_2^+ agrees completely with experiment. The following are values of $\mathscr{E}$ for six different internuclear distances.

r_{AB} (au)	$\mathscr{E}(r_{AB})$ (au)
1.0	−0.45179
1.5	−0.58232
1.99718 (r_{AB}^o)	−0.60263
2.5	−0.59382
3.0	−0.57756
4.0	−0.54608

A Morse potential is to be fitted to these data by requiring the potential to agree exactly at the equilibrium geometry and the sum of the squares of the deviations at the five *other* data points to be minimal. Deduce the value of the Morse parameter a that accomplishes this, and then calculate $\tilde{\nu}_0$ for H_2^+. (*Hint:* You may find it convenient to use the Newton–Raphson method of solution of transcendental equations. Recall also from Section 4.4 the definitions of atomic units.)

7.24. Verify that eq. (7.6-5) follows from eq. (7.6-4). For HBr, the vibration–rotation coupling constant α is 0.226 cm^{-1}. Evaluate the magnitude of the $\beta/3$ and $\beta^2/15$ corrections for HBr in eq. (7.6-5) at 50 and 250 K.

7.25. Consider the summation $\sum_{r=0}^{N}(r^2 + 1)^{-1}$. Show that it is given approximately by the expression

$$\sum \simeq \tan^{-1}N + \tfrac{1}{2}\left(\frac{N^2 + 2}{N^2 + 1}\right) - \tfrac{1}{6}\left(\frac{N}{(N^2 + 1)^2}\right)$$

and test the expression for $N = 1, 2, 5$.

7.26. Verify eq. (7.6-7b).

7.27.^{C} Assemble a computer program to calculate the standard molar entropy $\langle \tilde{S}^\circ \rangle$, Helmholtz free energy $\langle \tilde{A}^\circ \rangle$, and heat capacity $\langle \tilde{C}_p^\circ \rangle$ for a diatomic molecule at a variety of temperatures. Employ the corrections embodied in eqs. (7.6-7). Input for the program consists of values of the atomic weights for A and B, σ, $\tilde{\nu}_0$, r_{AB}°, B, α, x_0, and specifications of the temperatures (D is to be computed theoretically). Run your program on a molecule chosen from the following list.

	r_{AB}° (Å)	$\tilde{\nu}_0$ (cm^{-1})	B (cm^{-1})	α (cm^{-1})	x_0
Na_2^{23}	3.0786	159.23	0.1547	0.00079	0.00456
$C^{12}O^{16}$	1.1282	2169.81	1.9313	0.01749	0.00620
$I^{127}Cl^{35}$	2.3209	384.29	0.1142	0.00054	0.00381
Li^7H^1	1.5957	1405.65	7.5131	0.2132	0.01650
Cl_2^{35}	1.9880	564.9	0.2438	0.0017	0.00708

7.28. From the discussion surrounding eq. (7.6-5), we infer that for $\Theta_r \ll T$, one has for an ideal gas of rigid rotors

$$q_{\text{int}} = \frac{T}{\sigma\Theta_r}\left(1 + \frac{1}{3}\frac{\Theta_r}{T} + \frac{1}{15}\frac{\Theta_r^2}{T^2} + \cdots + \right)$$

Show that although the high-temperature limit of $\langle \tilde{C}_{v,\mathrm{rot}} \rangle$ is R, the high-temperature limit of $\langle \tilde{E}_{\mathrm{rot}} \rangle$ is *not RT*, but rather $R(T - \frac{1}{3}\Theta_r)$. The term $\Theta_r/3$ is significant for some hydrides; check the molecule HF in this regard.

7.7 INTERNAL ROTATION

The treatment of Sections 7.4 and 7.5 can be generalized to polyatomic molecules. Molecular vibrations are decomposed into a number of **normal modes** [$(3N - 6)$ for a nonlinear molecule of N atoms] instead of a single mode, as in the case of diatomics. The molecular vibrational partition function is

$$q_{\mathrm{vib}} = \prod_{i=1}^{3N-6} \frac{e^{-\Theta_{v,i}/2T}}{1 - e^{-\Theta_{v,i}/T}} \qquad (7.7\text{-}1)$$

where the characteristic vibrational temperature $\Theta_{v,i}$ now refers to the ith normal mode.

Within the rigid-rotor approximation (Davidson, 1962), the molecular rotational partition function takes the form

$$q_{\mathrm{rot}} = \frac{\sqrt{\pi}}{\sigma} \left(\frac{T}{\Theta_{r,x}} \right)^{1/2} \left(\frac{T}{\Theta_{r,y}} \right)^{1/2} \left(\frac{T}{\Theta_{r,z}} \right)^{1/2} \qquad (7.7\text{-}2)$$

where the characteristic rotational temperatures $\Theta_{r,x}$, $\Theta_{r,y}$, $\Theta_{r,z}$ are defined as $h^2/8\pi^2 k_B I_x$, and so on, and I_x, I_y, I_z are the three **principal moments of inertia**.[6] Polyatomic molecules may have all three moments of inertia equal (CH_4), two moments equal (CH_3Cl), or all three moments different (H_2O) (Fig. 7.4). The symmetry number σ is the number of indistinguishable ways that the molecule can be oriented in a fixed coordinate system; for example, $\sigma = 2$ for H_2O and $\sigma = 3$ for CH_3Cl. Now consider a molecule such as ethane. In addition to vibrations and *overall* rotations, the molecule can also undergo *internal* rotational motion of one methyl group with respect to the other. The following simplified analysis is appropriate for molecules like ethane that consist of two symmetric coaxial groups.

Imagine two bodies of mass m_1 and m_2 rotating about a common axis as in Fig. 7.5. If no external torques are being applied to the system, then the

---

[6]Like polarizability (see Section 5.4), inertia I is a second-rank tensor. For a body composed of N discrete masses, the ijth element of $\mathbf{I}$ is given by $\sum_{\alpha=1}^{N} m_\alpha (\delta_{ij} \sum_{k=1}^{3} x_{\alpha,k}^2 - x_{\alpha,i} \cdot x_{\alpha,j})$, where $x_{\alpha,k}$ is the k coordinate of particle α. The diagonal elements of $\mathbf{I}$ are the **moments of inertia**; Euler called the off-diagonal elements the **products of inertia**. In general, a coordinate system can be found in which the products of inertia are all zero. This is the **principal coordinate system**, and for a molecule possessing symmetry it coincides with the molecular symmetry axes.

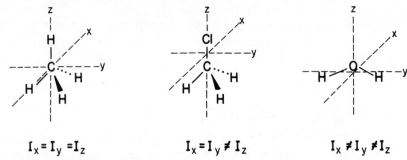

$$I_x = I_y = I_z \qquad\qquad I_x = I_y \neq I_z \qquad\qquad I_x \neq I_y \neq I_z$$

FIGURE 7.4. Illustrating the three principal moments of inertia of three molecules.

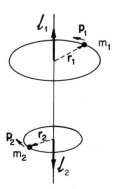

FIGURE 7.5. Illustrating internal rotation.

angular momentum vectors of the separate bodies are of equal magnitude and point in opposite directions since the total angular momentum is conserved. In terms of magnitudes, we have $|\ell_1| = |\ell_2|$, or

$$I_1\omega_1 = I_2\omega_2 \tag{7.7-3}$$

where I_1 and I_2 are the separate moments of inertia, and ω_1 and ω_2 are the separate angular velocities. Let ω be the relative angular velocity; then

$$\omega_1 + \omega_2 = \omega \tag{7.7-4}$$

and if ω_2 and ω_1 are eliminated between eqs. (7.7-3) and (7.7-4), we find

$$\omega_1 = \frac{I_2\omega}{I_1 + I_2} \qquad \omega_2 = \frac{I_1\omega}{I_1 + I_2} \tag{7.7-5}$$

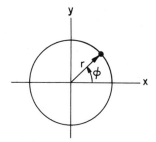

FIGURE 7.6. A particle moving on a ring.

The total kinetic energy of the system is, therefore,

$$T = \tfrac{1}{2}I_1\omega_1^2 + \tfrac{1}{2}I_2\omega_2^2$$

$$= \tfrac{1}{2}\left[I_1\left(\frac{I_2\omega}{I_1 + I_2}\right)^2 + I_2\left(\frac{I_1\omega}{I_1 + I_2}\right)^2\right]$$

$$= \tfrac{1}{2}I_r\omega^2 \tag{7.7-6}$$

where the **reduced moment of inertia** is defined as $I_r = I_1 I_2/(I_1 + I_2)$.

To put eq. (7.7-6) into quantum mechanical form, let us consider a simple model of a particle moving around a ring about the z-axis (Fig. 7.6). The kinetic energy of the particle is given by $\tfrac{1}{2}I_z\omega^2$, or equivalently in terms of Cartesian momenta,

$$T = \frac{p_x^2 + p_y^2}{2m} \tag{7.7-7}$$

If the origin of the coordinate system is at the center of the circle, then $x = r\cos\phi$ and $y = r\sin\phi$. Straightforward application of the chain rule with r fixed gives

$$\frac{\partial^2}{\partial x^2} = \frac{\sin\phi}{r}\left(\frac{\sin\phi}{r}\frac{\partial^2}{\partial\phi^2} + \frac{\cos\phi}{r}\frac{\partial}{\partial\phi}\right)$$

$$\frac{\partial^2}{\partial y^2} = \frac{\cos\phi}{r}\left(\frac{\cos\phi}{r}\frac{\partial^2}{\partial\phi^2} - \frac{\sin\phi}{r}\frac{\partial}{\partial\phi}\right) \tag{7.7-8}$$

Following the usual prescription for transforming a classical Hamiltonian into a quantum mechanical one (QM Postulate 3), we find that eq. (7.7-7) becomes

$$\hat{T} = \frac{-\hbar^2}{2mr^2}\frac{\partial^2}{\partial\phi^2} = \frac{-\hbar^2}{2I_z}\frac{\partial^2}{\partial\phi^2}$$

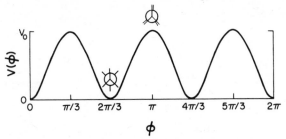

FIGURE 7.7. Qualitative sketch of a threefold barrier to internal rotation in a molecule such as ethane; the zero of the horizontal axis corresponds to a staggered conformation of the molecule.

Since this is the operator corresponding to the classical quantity $\frac{1}{2}I_z\omega^2$, we infer by analogy that the operator for T in eq. (7.7-6) must be $\hat{T} = (-\hbar^2/2I_r)(\partial^2/\partial\phi^2)$. The angle ϕ in this case represents the dihedral angle made by the two planes through the axis in Fig. 7.5 and containing the separate bodies m_1 and m_2.

Generally, as a molecule rotates internally, the potential energy changes with the angle ϕ. The potential $V(\phi)$ should be a complicated function of bond angles and lengths because these are expected to depend upon ϕ. However, this dependency is often slight, and for symmetric molecules such as ethane, a simple potential energy function that works remarkably well is

$$V(\phi) = \tfrac{1}{2}V_0(1 - \cos 3\phi) \tag{7.7-9}$$

The 3 enters because by symmetry the potential must possess a periodicity of 120° (Fig. 7.7). The energy-eigenvalue equation for internal rotation, considered independently of vibration and of overall molecular rotation, is then

$$\left(\frac{-\hbar^2}{2I_r}\frac{d^2}{d\phi^2} + \tfrac{1}{2}V_0(1 - \cos 3\phi)\right)\Psi(\phi) = E\Psi \tag{7.7-10}$$

where the quantity I_r for ethane is $[I(\text{methyl})]^2/2I(\text{methyl}) = \tfrac{1}{2}I(\text{methyl})$.

In order to put eq. (7.7-10) into a standard form, we make the substitutions $z = 3\phi/2$, $y(z) = \Psi(\phi)$, $\theta = 2I_rV_0/9\hbar^2$, and $a = 8I_r(E - \tfrac{1}{2}V_0)/9\hbar^2$.

$$\frac{d^2y}{dz^2} + (a + 2\theta\cos 2z)y = 0 \tag{7.7-11}$$

This differential equation is known as **Mathieu's equation** (Mathews and Walker, 1970; Whittaker and Watson, 1963). We require the solutions $y(z)$ to be periodic, $y(z + \pi) = y(z)$, and it is found that this cannot be satisfied for arbitrary a and θ. Equation (7.7-11), of course, can be solved numerically; however, it is instructive to return to eq. (7.7-10) and look at two extreme cases.

In the first case, the barrier is supposed very low, $V_0 \ll k_B T$. The potential $V(\phi)$ then never deviates much from its mean value of $\frac{1}{2}V_0$. The energy-eigenvalue equation is approximated by

$$\left(\frac{-\hbar^2}{2I_r} \frac{d^2}{d\phi^2} + \frac{1}{2}V_0 \right) \Psi(\phi) = E\Psi \tag{7.7-12}$$

the solutions of which are $\Psi(\phi) = \exp(\pm ib\phi)$, where $b = \hbar^{-1}[I_r(2E - V_0)]^{1/2}$. The boundary condition is $\Psi(\phi) = \Psi(\phi + 2\pi/3)$, and this leads to $b = 3j$, $j = 0, \pm 1, \pm 2, \ldots$. Substitution into the expression for b gives for the eigenvalues $E_j = \frac{1}{2}(9\hbar^2 I_r^{-1} j^2 + V_0)$. Finally, the molecular partition function for internal rotation, expressed as an integral, is just

$$q_{ir} = e^{-V_0/2k_B T} \int_{-\infty}^{\infty} e^{-9\hbar^2 j^2/2I_r k_B T} \, dj \simeq \frac{1}{3\hbar}(2\pi I_r k_B T)^{1/2} \tag{7.7-13}$$

and this contributes $\frac{1}{2}R$ to the molar heat capacity [compare eq. (7.3-5)].

In the second case, the barrier is considered very high, $V_0 \gg k_B/T$. Then in eq. (7.7-10) the variable ϕ will never be far from 0, so that $\cos 3\phi \simeq 1$. Expand the cosine in a Taylor series and retain only the first two terms: $\cos 3\phi \simeq 1 - \frac{9}{2}\phi^2$. The energy-eigenvalue equation now becomes approximately

$$\left[\frac{-\hbar^2}{2I_r} \frac{d^2}{d\phi^2} + \frac{1}{2}\left(\frac{9V_0}{2} \right)\phi^2 \right] \Psi(\phi) = E\Psi \tag{7.7-14}$$

and this equation is of the same form as that for the harmonic oscillator. The eigenvalues must therefore be

$$E_n = \left(n + \frac{1}{2} \right)\hbar\omega \qquad \omega = 3\left(\frac{V_0}{2I_r} \right)^{1/2} \quad (n = 0, 1, 2, \ldots,) \tag{7.7-15}$$

Consequently, the contribution to the heat capacity due to highly restricted rotation is given by eq. (7.2-5b). Similar conclusions apply to the other thermodynamic functions.

Thus, for extreme values of the barrier, the molecule behaves as if it has either an extra vibrational or extra rotational degree of freedom. For intermediate cases, the partition function and hence expressions for the thermodynamic quantities are too complicated to give here. Most molecules of interest are of the intermediate type. In these cases, Mathieu's equation (7.7-10) must be solved numerically (Pitzer, 1953). Figure 7.8 shows the graphical dependence of $\langle \tilde{C}_{p,ir}^{\circ} \rangle$ and of $\langle \tilde{S}_{ir}^{\circ} \rangle$ on barrier height as a function of temperature for ethane.

Various measurements on ethane indicate that the barrier to rotation is around 12 kJ mol^{-1}. For example, when $\tilde{C}_p^{\circ}$ was measured at 94 and 100 K

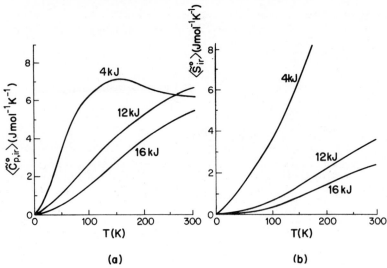

FIGURE 7.8. The contribution due to internal rotation to (a) the standard molar heat capacity, and (b) the standard molar entropy of an ethane-like molecule with reduced moment of inertia $I_r = 2.76 \times 10^{-47}$ kg m²; the curves are based on the numerical solution of eq. (7.7-11).

(Kistiakowsky et al., 1938), and the calculated contributions due to translation, vibration (this was negligible), and overall rotation were subtracted from the measured value, there remained unexplained 2.46 and 2.95 J mol⁻¹ K⁻¹. Figure 7.8(a) indicates that these values of $\langle \tilde{C}^\circ_{p,\text{ir}} \rangle$ correspond roughly to $V_0 = 12$ kJ mol⁻¹.

Other thermodynamic properties should also reflect the presence of barriers to internal rotation. The equilibrium constant or the enthalpy change for a reaction such as

$$
\begin{array}{ccc}
\text{H} & \text{H} & \text{H} \quad \text{H} \\
\diagdown & \diagup & | \quad | \\
\text{C} = \text{C} \quad + \text{H}_2 \rightarrow \text{H} - \text{C} - \text{C} - \text{H} \\
\diagup & \diagdown & | \quad | \\
\text{H} & \text{H} & \text{H} \quad \text{H}
\end{array}
$$

should be a probe of barrier height since internal rotation is plausible only in the product. Still other methods besides thermodynamic ones can be used to measure barrier heights. Various forms of spectroscopy, including nuclear magnetic resonance, infrared, and microwave spectroscopy, have been employed to great advantage. Table 7.7 lists the threefold barriers for a sampling of ethane-like molecules.

Quite another matter from the magnitude of some barrier to rotation is its origin, a question that has been pursued vigorously in the literature. The

**Table 7.7.   Selected Threefold Barriers to
Internal Rotation**

Molecule	V_0 (kJ mol^{-1})[a,b]
CH_3-CHO	4.85
CH_3-AsF_2	5.52
CH_3-SiH_3	7.11
CH_3-CH_3	12.03
CH_3-CF_3	13.6
CH_3-CH_2F	13.9
CH_3-CH_2Br	15.40
CH_3-CF_2Cl	15.48
CF_3-CF_2Cl	23.70

[a] Data taken from J. P. Lowe, "Barriers to Internal Rotation about Single Bonds," in A. Streitwieser, Jr., and R. W. Taft (Eds.), *Progress in Physical Organic Chemistry*, Vol. 6, Wiley, New York, 1968, p. 1.
[b] These data generally refer to gas phase measurements.

reader may enjoy looking at a paper (Scott and Scheraga, 1965) that attempts to account semiquantitatively for rotational barriers in terms of just two factors: nonbonded (van der Waals) interactions and exchange interactions of the electrons in bonds adjacent to the bond about which rotation occurs. The origin of internal rotational energy barriers, however, is much too complex an issue to be summarized here, and the interested reader is directed to the primary literature where reviews occur frequently.

7.8   SIMPLE CRYSTALLINE SOLIDS

We have so far restricted the discussion to ideal gases, since the canonical partition function then factors and we can evaluate the molecular partition function in closed form. Another model system, which allows a closed-form evaluation of the canonical partition function, is the **harmonic solid**, that is, an idealized solid in which the vibrations of the atoms are harmonic.

The classical Hamiltonian for a monatomic crystal containing N atoms is (Yourgrau et al., 1982)

$$H = \sum_{i=1}^{3N} \frac{p_i^2}{2m} + V(x_1, x_2, \ldots, x_{3N}) \qquad (7.8\text{-}1)$$

where the potential energy V is a function only of the Cartesian displacements

x_i of the atoms, and the momenta p_i are given by $p_i = m\dot{x}_i$. Since the zero of V is arbitrary, we take it to be when the vibrating atoms are at their equilibrium locations. By definition, the crystal is at mechanical equilibrium when each component f_i of force acting on a given atom is zero.

$$f_i = -\left(\frac{\partial V}{\partial x_i}\right)_{eq} = 0 \qquad (i = 1, 2, \ldots, 3N)$$

Hence, if $V(x_1, x_2, \ldots, x_{3N})$ is expanded in a Taylor series about the equilibrium configuration, the first nonzero terms are those quadratic in the x_i. The x_i are expected to be small compared to interatomic dimensions in the crystal, so that terms cubic, quartic, and of higher degree in the x_i are small compared to the quadratic terms. As a result, eq. (7.8-1) may be approximated by

$$H \simeq \sum_{i=1}^{3N} \frac{p_i^2}{2m} + \frac{1}{2} \sum_{j}^{3N} \sum_{i}^{3N} \left(\frac{\partial^2 V}{\partial x_i \partial x_j}\right)_{eq} x_i x_j \qquad (7.8\text{-}2)$$

from which one sees that the potential energy terms mix coordinates of different atoms. Individual atoms of the crystal may not be treated as independently vibrating bodies.

However, it is possible to define new coordinates that do permit a complete separation of the Hamiltonian. From Newton's Second Law we have

$$m\ddot{x}_i = -\left(\frac{\partial V}{\partial x_i}\right)$$

$$= -\sum_{j=1}^{3N} \left(\frac{\partial^2 V}{\partial x_i \partial x_j}\right)_{eq} x_j \qquad (7.8\text{-}3)$$

The $\frac{1}{2}$ of eq. (7.8-2) does not appear in eq. (7.8-3) because of the equality of the two mixed second-order partial derivatives of V with respect to x_i and x_j. If we let each atom pass through its equilibrium position at $t = 0$, then the substitution $x_i = A_i \sin \omega t$ transforms eq. (7.8-3) into

$$m\omega^2 A_i = \sum_{j=1}^{3N} \left(\frac{\partial^2 V}{\partial x_i \partial x_j}\right)_{eq} A_j \qquad (7.8\text{-}4)$$

where A_j is the amplitude of vibration of the jth atom.

Equation (7.8-4) can be thought of as a matrix equation, $m\omega^2 \mathbf{A} = \mathbf{D}\mathbf{A}$, where $\mathbf{A}$ is a column matrix of the amplitudes and $\mathbf{D}$ is a square matrix of the second-order partial derivatives of V. The matrix $\mathbf{D}$ is real; it is also symmetric. Therefore, $\mathbf{D}$ is Hermitian and can be diagonalized by an orthogonal transformation to yield the eigenvalues and eigenvectors. The procedure is analogous to that employed in treating the Roothaan equations [eqs. (4.2-11)]: We first set the following $3N \times 3N$ secular determinant equal to zero.

$$
\begin{bmatrix}
\left(\dfrac{\partial^2 V}{\partial x_1^2}\right)_{eq} - m\omega^2 & \left(\dfrac{\partial^2 V}{\partial x_1 \partial x_2}\right)_{eq} & \left(\dfrac{\partial^2 V}{\partial x_1 \partial x_3}\right)_{eq} & \cdots & \left(\dfrac{\partial^2 V}{\partial x_1 \partial x_{3N}}\right)_{eq} \\[2mm]
\left(\dfrac{\partial^2 V}{\partial x_2 \partial x_1}\right)_{eq} & \left(\dfrac{\partial^2 V}{\partial x_2^2}\right)_{eq} - m\omega^2 & \left(\dfrac{\partial^2 V}{\partial x_2 \partial x_3}\right)_{eq} & \cdots & \left(\dfrac{\partial^2 V}{\partial x_2 \partial x_{3N}}\right)_{eq} \\[2mm]
\vdots & \vdots & \vdots & & \vdots \\[2mm]
\left(\dfrac{\partial^2 V}{\partial x_{3N} \partial x_1}\right)_{eq} & \left(\dfrac{\partial^2 V}{\partial x_{3N} \partial x_2}\right)_{eq} & \left(\dfrac{\partial^2 V}{\partial x_{3N} \partial x_3}\right)_{eq} & \cdots & \left(\dfrac{\partial^2 V}{\partial x_{3N}^2}\right)_{eq} - m\omega^2
\end{bmatrix}
$$

Expansion of the secular determinant leads to a polynomial equation of degree $3N$ in ω^2, the solution of which yields $3N$ values of $\omega_k > 0$, the **eigenfrequencies** of the crystal. The corresponding **eigenvectors** each involves collective motions of all of the atoms of the crystal. For each value of ω_k, the eqs. (7.8-4) yield a set of ratios of $A_i^{(k)}$'s; this provides $3N(3N - 1)$ equations. Hence, $3N$ additional conditions can be imposed upon the $A_i^{(k)}$'s. We take these to be conditions of normalization.

$$
\sum_{i=1}^{3N} A_i^{(k)} A_i^{(k)} = 1 \qquad (k = 1, 2, \ldots, 3N)
$$

However, since **D** is Hermitian, its eigenvectors corresponding to distinct eigenvalues (i.e., distinct values of ω) are orthogonal (recall Property 2 in Section 2.3), or can be made so in the case of degenerate eigenvalues. Thus, this equation can be generalized to

$$
\sum_{i=1}^{3N} A_i^{(k)} A_i^{(\ell)} = \delta_{k\ell} \quad (k, \ell = 1, 2, \ldots, 3N) \tag{7.8-5a}
$$

Now let the symbol **A** stand for the square matrix of all of the eigenvectors of **D** arranged in columns. We see immediately that eq. (7.8-5a) is equivalent to the matrix equation $\mathbf{A}^T \mathbf{A} = \mathbf{1}$, where $\mathbf{A}^T$ is the transpose of **A**. It follows that **A** and $\mathbf{A}^T$ are orthogonal matrices [recall eq. (5.7-2)]. An orthogonal matrix **M** has the demonstrable property that for any two columns i, j,

$$
\sum_n M_{ni} M_{nj} = \delta_{ij}
$$

If we apply this property to **A**, we obtain eq. (7.8-5a), and if we apply it to $\mathbf{A}^T$, then we find

$$\sum_{k=1}^{3N} A_i^{(k)} A_j^{(k)} = \delta_{ij} \tag{7.8-5b}$$

Let us now define new coordinates, referred to as **normal coordinates**, that will replace the Cartesian coordinates.

$$q_k = \sum_{i=1}^{3N} A_i^{(k)} x_i \quad (k = 1, 2, \ldots, 3N) \tag{7.8-6}$$

We can then show that the inverse transformation is given by

$$x_i = \sum_{k=1}^{3N} A_i^{(k)} q_k \quad (i = 1, 2, \ldots, 3N) \tag{7.8-7}$$

and combination of eqs. (7.8-4), (7.8-5), and (7.8-7) transforms the Hamiltonian of eq. (7.8-2) into

$$H(q, p) = \tfrac{1}{2} \sum_{k=1}^{3N} \left(\frac{p_k^2}{m} + m\omega_k^2 q_k^2 \right) \tag{7.8-8}$$

where $p_k = m\dot{q}_k$. In this Hamiltonian there is no mixing of coordinates in the potential energy terms. Thus, even though each frequency ω_k entails the motion of all atoms, the different frequencies correspond to *independent* vibrations. The Hamiltonian associated with each such independent vibration, $\tfrac{1}{2}(p_k^2/m + m\omega_k^2 q_k^2)$, is of the same form as that of a harmonic oscillator of mass m and frequency ω_k [recall eq. (1.4-8)].

We conclude that the crystal can be treated as a collection of harmonic oscillators, where each oscillator is a normal mode involving the concerted movement of all atoms. Reference to eq. (7.7-1) then gives for the canonical vibrational partition function

$$Q_{\text{vib}} = \prod_{k=1}^{3N} e^{-\hbar\omega_k/2k_B T} \left(1 - e^{-\hbar\omega_k/k_B T} \right)^{-1} \tag{7.8-9}$$

To evaluate Q_{vib}, we need the eigenfrequencies ω_k. Solution of the secular equation for the general monatomic crystal is very difficult; although the presence of symmetry usually permits enormous simplifications. In any case, the vibrational levels of a crystal containing on the order of 10^{23} atoms are expected to be very closely spaced. This suggests that we approximate the crystal as an elastic continuum (Montroll, 1967). Following the Dutch physical chemist Peter J. W. Debye (1884–1966), we denote by $g(\omega)\,d\omega$ the number of modes with frequencies lying between ω and $\omega + d\omega$. Thus, the number of modes with frequencies less than ω can be expressed as

$$N(\omega) = \int_0^\omega g(\omega)\, d\omega \qquad (7.8\text{-}10)$$

Since the real crystal contains only $3N$ atoms, and consequently a total of $3N$ modes, there must exist a maximum frequency, called the **Debye frequency** ω_D, for which

$$\int_0^{\omega_D} g(\omega)\, d\omega = 3N \qquad (7.8\text{-}11)$$

A wave in an *isotropic* continuum satisfies the classical wave equation (Feynman et al., 1963)

$$\frac{1}{v^2}\frac{\partial^2 \Psi}{\partial t^2} = \nabla^2 \Psi \qquad (7.8\text{-}12)$$

where $\Psi(x, y, z, t)$ is the displacement at any point and v is the velocity of propagation of the wave. Let the continuum have the shape of the crystal, and suppose this to be a rectangular box with sides ℓ_1, ℓ_2, ℓ_3. Assuming the amplitudes of the waves in the crystal to vanish at the boundaries, and carrying out a separation of variables in eq. (7.8-11), we find for the solutions

$$\Psi = Be^{i\omega t}\sin\frac{\pi n_1 x}{\ell_1}\sin\frac{\pi n_2 y}{\ell_2}\sin\frac{\pi n_3 z}{\ell_3} \qquad (7.8\text{-}13)$$

where n_1, n_2, n_3 are nonnegative integers. Substitution of eq. (7.8-13) into eq. (7.8-12) then shows that

$$\omega^2 = \pi^2 v^2\left[\left(\frac{n_1}{\ell_1}\right)^2 + \left(\frac{n_2}{\ell_2}\right)^2 + \left(\frac{n_3}{\ell_3}\right)^2\right] \qquad (7.8\text{-}14)$$

Equation (7.8-14) defines an ellipsoid in (n_1, n_2, n_3) space. From the multivariate calculus (Courant and John, 1974), one can establish that the volume of the general ellipsoid (see Fig. 7.9)

$$\left(\frac{x}{a}\right)^2 + \left(\frac{y}{b}\right)^2 + \left(\frac{z}{c}\right)^2 = 1$$

is $4\pi abc/3$. The volume of the ellipsoid defined by eq. (7.8-14) is therefore $4V(\omega/v)^3/3\pi^2$, where V is the volume of the crystal. For large n_1, n_2, n_3, the number of frequencies $N(\omega)$ less than a given ω should be proportional to the volume of the ellipsoid, and since the n's are nonnegative, only one octant of

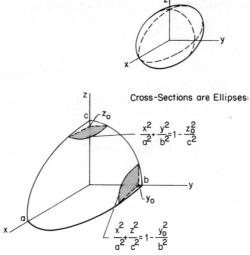

FIGURE 7.9. The positive octant of the ellipsoid $(x/a)^2 + (y/b)^2 + (z/c)^2 = 1$.

the ellipsoid need be considered.

$$N(\omega) = \tfrac{1}{8} \times \frac{4}{3\pi^2} V \left(\frac{\omega}{v}\right)^3$$

Two kinds of waves are propagated in a continuum: **longitudinal waves** (displacement is parallel to direction of propagation) with velocity v_ℓ, and **transverse waves** (displacement is perpendicular to direction of propagation) with velocity v_t. There are two transverse waves for every longitudinal wave since there are two independent directions in space perpendicular to the direction of propagation. The formula for the number of frequencies less than a given ω should be amended to read

$$N(\omega) = \frac{1}{6\pi^2} V\omega^3 \left(\frac{2}{v_t^3} + \frac{1}{v_\ell^3}\right) \tag{7.8-15}$$

From eqs. (7.8-10) and (7.8-15) we have

$$g(\omega) = \frac{dN}{d\omega}$$

$$= \frac{V\omega^2}{2\pi^2} \left(\frac{2}{v_t^3} + \frac{1}{v_\ell^3}\right) \tag{7.8-16}$$

Substituting expression (7.8-16) into eq. (7.8-11) and performing the integra-

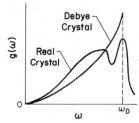

FIGURE 7.10. Comparison of frequency spectra for a Debye crystal and a real crystal (after P. A. Egelstaff, *Thermal Neutron Scattering*, Academic, New York, 1965, p. 211, with permission).

tion, we obtain

$$3N = \frac{V\omega_D^3}{6\pi^2}\left(\frac{2}{v_t^3} + \frac{1}{v_\ell^3}\right) \tag{7.8-17}$$

Finally, combination of eqs. (7.8-16) and (7.8-17) yields

$$g(\omega) = \begin{cases} \dfrac{9N}{\omega_D^3}\omega^2 & \omega < \omega_D \\[2mm] 0 & \omega > \omega_D \end{cases} \tag{7.8-18}$$

Thus, $g(\omega)$ is a parabola that opens upward and cuts off abruptly at $\omega = \omega_D$; more realistic frequency spectra of anisotropic solids as revealed by neutron scattering show secondary structure (Fig. 7.10).

If the frequencies were distributed discretely, the Helmholtz free energy would be given from eq. (7.8-9) as

$$\langle A_{\text{vib}} \rangle = -k_B T \ln Q_{\text{vib}}$$
$$= \sum_{i=1}^{3N} \tfrac{1}{2}\hbar\omega_i + k_B T \sum_{i=1}^{3N} \ln(1 - e^{-\hbar\omega_i/k_B T})$$

where the first summation represents the zero-point energy of the crystal. But since the frequencies of the Debye solid are continuously distributed, the summations are replaced by integrations,

$$\langle A_{\text{vib}} \rangle = \frac{9N}{\omega_L^3}\int_0^{\omega_D}\left[\frac{\hbar\omega}{2} + k_B T \ln(1 - e^{-\hbar\omega/k_B T})\right]\omega^2\, d\omega \tag{7.8-19}$$

after making use of eq. (7.8-18). The internal energy of the crystal is given by

Table 7.8.   The Characteristic Debye Temperature of Some Monatomic Solids

Solid	Θ_D (K)[a]	Solid	Θ_D (K)[a]
Pb	88	Zn	250
K	100	Cu	315
Cd	172	Co	385
Pt	225	Fe	420

[a] These values are obtained from different sources; different procedures yield different values of Θ_D.

$E = A - T(\partial A/\partial T)_{V,N}$; combination of this with eq. (7.8-19) yields

$$\langle E_{\text{vib}} \rangle = \tfrac{9}{8}N\hbar\omega_D + 3Nk_BTD\left(\frac{\Theta_D}{T}\right)$$

$$D(u) = \frac{3}{u^3}\int_0^u \frac{x^3\,dx}{e^x - 1}$$

(7.8-20)

where the **characteristic Debye temperature** $\Theta_D = \hbar\omega_D/k_B$. This quantity is a parameter that is to be determined for each particular substance by the fitting of experimental data. Such fitting can be achieved by examination of heat capacity data; the calculated heat capacity at constant volume would be given by the derivative with respect to temperature of eq. (7.8-20). Employing the Leibniz rule for differentiation of an integral (Dence, 1975), we obtain

$$\langle C_v \rangle = 3Nk_B\left[4D\left(\frac{\Theta_D}{T}\right) - \frac{3\Theta_D/T}{e^{\Theta_D/T} - 1}\right] \qquad (7.8\text{-}21)$$

Table 7.8 lists the characteristic Debye temperature for a selection of monatomic metals.

The formula in eq. (7.8-21) as well as analogous formulas for the other thermodynamic functions are all functions of the variable Θ_D/T. The Debye model of a solid is therefore consistent with a **Law of Corresponding States**, much as the van der Waals equation of state is for gases. Two different solids, at generally different temperatures that are identical multiples of the respective characteristic Debye temperatures, should possess identical values of any of their thermodynamic functions.

7.9   COMPUTER HIGHLIGHT: EVALUATION OF THE DEBYE FUNCTION $D(u)$

The integral in eq. (7.8-20) defined by the Debye function, $D(u)$, cannot be evaluated analytically, although tables of it obtained by numerical integration

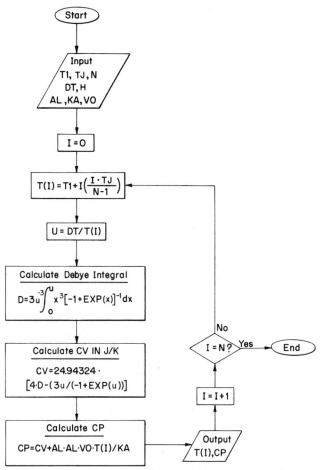

FIGURE 7.11. Flowchart for calculation of $\langle \tilde{C}_p \rangle$ for a Debye crystal at various temperatures.

are available (Abramowitz, Stegun, 1965). We may, however, take advantage of the power of a computer and perform our own integration.

Simpson's rule[7] is a simple but often effective numerical method for estimating integrals over finite ranges. Let the interval of integration for some function $f(x)$ be $a \le x \le b$; divide this interval into an even number n of subintervals of width $h = (b - a)/n$. Then the area A is given approximately by (Dence, 1975)

$$A = \int_a^b f(x)\, dx \simeq \frac{h}{3}(y_1 + 4y_2 + 2y_3 + 4y_4 + \cdots + y_n) \qquad (7.9\text{-}1)$$

[7]After Thomas Simpson (1710–1761), a British silk weaver by trade.

Table 7.9. **Debye Heat Capacity Calculations for Solid Lead**

T (K)	Θ_D/T	$\langle \tilde{C}_v \rangle$ (calculated)[a]	$\langle \tilde{C}_p \rangle$ (calculated)[a, b]	$\tilde{C}_p$ (experimental)[a, c]
10	8.8	2.69	2.69	2.76[d]
20	4.4	11.11	11.11	11.01
40	2.2	19.81	19.82	19.57
60	1.466	22.45	22.47	22.43
80	1.1	23.50	23.53	23.69
100	0.88	24.00	24.03	24.43
150	0.587	24.52	24.57	25.27
200	0.44	24.70	24.77	25.87
250	0.352	24.79	24.87	26.36
298.2	0.295	24.83	24.93	26.82

[a]All heat capacities in units of J mol^{-1} K^{-1}.
[b]The values of α and κ are not very sensitive to temperature; mean values of $\alpha = 24 \times 10^{-6}$ K^{-1} and $\kappa = 2.0 \times 10^{-6}$ atm^{-1} were employed for all temperatures (Landolt-Börnstein, *Zahlenwerte und Funktionen*, Vol. II, Part 1, 7th ed., Springer-Verlag, Berlin, 1971, pp. 445, 446). A mean value of $\tilde{V} = 11.34$ cm^3 was also used.
[c]Landolt-Börnstein, *Zahlenwerte und Funktionen*, Vol. II, Part 4, 6th ed., Springer-Verlag, Berlin, 1961, p. 479, except where noted.
[d]D. E. Gray (Ed.), *American Institute of Physics Handbook*, 3rd ed., McGraw-Hill, New York, 1972, p. 4-106.

where $y_1 = f(a)$, $y_2 = f(a + h)$, $y_3 = f(a + 2h), \ldots, y_n = f(b)$. The approximation improves with increasing n.

Figure 7.11 gives a flowchart of program logic for a calculation of $\langle \tilde{C}_p \rangle$ based on Simpson's Rule. The input consists of the lowest temperature of the interval ($T1$), the number (N) of other temperatures besides $T1$ at which $\tilde{C}_p$ is to be computed, the size of the temperature jumps (TJ), the characteristic Debye temperature (DT), and the size of the integration subinterval (H). For purposes of comparison with experiment, one desires $\tilde{C}_p$ instead of $\tilde{C}_v$. These are related according to the expression

$$\tilde{C}_p = \tilde{C}_v + \frac{\alpha^2 \tilde{V} T}{\kappa} \qquad (7.9\text{-}2)$$

where $\alpha = V^{-1}(\partial V/\partial T)_P$ is the **coefficient of thermal expansion**, and $\kappa = -V^{-1}(\partial V/\partial P)_T$ is the **coefficient of isothermal compressibility**. Since $\tilde{C}_p - \tilde{C}_v$ is generally small for solids and since α, κ, $\tilde{V}$ are not very sensitive to temperature, mean values of these quantities (AL, KA, VO) may also be inputted to the program.

The results of some sample calculations on metallic lead are shown in Table 7.9 and Fig. 7.12. The trend of the data is fairly well reproduced, although at the higher temperatures there is increasing deviation from the true values. Other solids display somewhat better agreement between theory and experi-

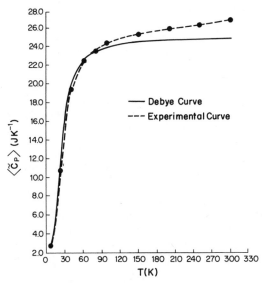

FIGURE 7.12. Comparison of the heat capacity $\tilde{C}_p$ of lead at various temperatures according to experiment and to the Debye expression.

ment; this example is therefore useful in pointing out that any theory that neglects details of the particle interactions is necessarily oversimplified. In fact, lead is an example of a monatomic solid whose experimental $\tilde{C}_v$ [as determined from eq. (7.9-2)] at temperatures above a certain value (roughly 150 K) exceeds the classical Dulong and Petit value of $3R$. One possible explanation is that the vibrations of the crystal may have considerable anharmonic character that presumably could be accounted for by suitable modification of eq. (7.8-9). Improved treatments of the vibrations of crystals have appeared since Debye's early paper in 1912, but these are outside the scope of the present book.

Exercises

7.29. In an interesting paper, Blinder (1975) discussed additivity rules in molecular thermodynamics from the statistical standpoint. For example, in the ideal gas limit, an N-atom nonlinear molecule has a heat capacity given closely by

$$\langle \tilde{C}_p^{\circ} \rangle = 4R + R \sum_{i=1}^{3N-6} \frac{x_i^2 e^{x_i}}{\left(e^{x_i} - 1\right)^2} \qquad \left(x_i = hc\tilde{\nu}_i / k_B T \right)$$

Explain this formula. We now represent the translational, rotational, and PV contributions to $\tilde{C}_p^\circ$ as six *fictitious* vibrational modes of some appropriate frequency, and thus rewrite this formula as

$$\langle \tilde{C}_p^\circ \rangle = \sum_{i=1}^{3N} C(\tilde{\nu}_i)$$

If, however, the heat capacity is to be approximately represented as a sum of atomic contributions, then

$$\langle \tilde{C}_p^\circ \rangle = \sum_{i=1}^{N} C_{p,i}$$

thus making each atomic contribution correspond to a set of three vibrational modes. For simplicity, we take the three modes in any such set to be of equal frequency. Hence, we have

$$C_{p,i} = \frac{3Rx_i^2 e^{x_i}}{\left(e^{x_i} - 1\right)^2}$$

From an examination of several compounds, the following set of atomic frequencies ($\tilde{\nu}_i$) at 25° C is obtained.

Atom	$\tilde{\nu}_i$ (cm^{-1})	Atom	$\tilde{\nu}_i$ (cm^{-1})
H	1096	F	715
C	499	Cl	508
N	553	Br	433
O	553	S	354

Test this set of effective atomic frequencies on the compounds $C_2H_5OH_{(g)}$, NH_3, CH_3SH, SO_2, CH_3Br, and any others you may choose.

7.30. Consult Blinder's paper and read the sections on entropy. Test his scheme for additivity of entropy on a set of molecules.

7.31. The oxygen difluoride molecule has a bond angle of 103°11′ and bond lengths of 1.41 Å. Let the molecule be oriented in the yz plane with the center of mass at the origin as is the H_2O molecule in Fig. 7.4. Verify that the three principal moments of inertia are $I_x = 91.4 \times 10^{-47}$ kg m^2, $I_y = 14.3 \times 10^{-47}$ kg m^2, $I_z = 77.0 \times 10^{-47}$ kg m^2.

7.32. A theorem in classical mechanics known as **Steiner's theorem** (after the Swiss geometrician Jakob Steiner, 1796–1863) says that if **I** is the inertia tensor in a coordinate system where the center of mass is at the origin, then in a new system whose origin is at $\mathbf{R}$ and whose axes are *parallel to those of the old coordinate system*, the elements of the new inertia tensor $\mathbf{I}'$ are given by

$$I'_{ij} = I_{ij} + \sum_{\alpha} m_{\alpha}\left(R^2 \delta_{ij} - R_i R_j\right)$$

Apply Steiner's theorem and compute the elements of $\mathbf{I}'$ if the coordinate system in Exercise 7.31 is translated so that the oxygen is located along the y' axis of the new system at $y' = 1$ Å. Check some of your numbers I'_{ij} by calculating them directly from the definition given in footnote 6 in Section 7.7. Is the trace of $\mathbf{I}'$ equal to the trace of $\mathbf{I}$? Should the two be equal?

7.33. * Herman (1938) determined the normal coordinate vibrational frequencies of $TiCl_4$.

$\tilde{\nu}_1 \ (g = 1)$ (cm^{-1})	$\tilde{\nu}_2 \ (g = 2)$ (cm^{-1})	$\tilde{\nu}_3 \ (g = 3)$ (cm^{-1})	$\tilde{\nu}_4 \ (g = 3)$ (cm^{-1})
386	119	491	139

The degeneracies are indicated in parentheses; the Ti-Cl bond distance is 2.185 Å. Calculate $\langle \tilde{C}_p^{\circ} \rangle$ and $\langle \tilde{S}^{\circ} \rangle$ at 298.15 K, and compare with the Handbook values of 95.40 J mol^{-1} K^{-1} and 353.1 J mol^{-1} K^{-1}, respectively. Make no corrections for vibrational anharmonicity or centrifugal stretching.

7.34. Verify eqs. (7.7-8) and the subsequent expression for the operator $\hat{T}$ for the rotational kinetic energy.

7.35. Verify the steps leading to eq. (7.7-13). What should be the contribution to the standard molar entropy of an ethane-like molecule if internal rotation is free?

7.36. From Table 7.7, we see that replacement of three hydrogens in ethane by fluorines raises the internal rotational barrier by only 1.6 kJ mol^{-1}. However, if the three hydrogens in 1-chloro-1,1-difluoroethane are replaced by fluorines, the barrier rises by more than 8 kJ mol^{-1}. Suggest a reason for the difference in results of these two replacements.

7.37. The linear molecule dimethylcadmium has been studied (Li, 1956). Its third-law entropy at 298.15 K and 1 atm is 302.9 J mol^{-1} K^{-1}. The vibrational contribution to $\langle \tilde{S}^{\circ} \rangle$ is calculated to be 36.6 J mol^{-1} K^{-1}, while translation and overall rotation together contribute another 253.8 J mol^{-1} K^{-1}. The reduced moment of inertia may be taken to be the same as that employed in Fig. 7.8, namely, 2.76×10^{-47} kg m^2. In view of Exercise 7.35, show that Li is justified in saying that internal rotation in $(CH_3)_2Cd$ is free.

7.38. The calculated molecular internal rotational partition function for ethane at 500 K is 1.910. Calculate the effect on the equilibrium constant of the ethylene hydrogenation reaction relative to the hypothetical case where the ethane molecule is in a frozen conformation. Calculate the effect relative to the hypothetical case where internal rotation is completely free.

7.39.* Consult the paper by Scott and Scheraga (see References) and try your hand at the estimation of V_0 for some molecule not mentioned in the article. Look up or estimate any structural data you need. A possible molecule is CF_3CF_2Br, for which the experimental V_0 is 26.8 kJ mol^{-1} (Risgin and Taylor, 1959).

7.40. If the actual rotational barrier in ethane were considered to be high, what would be the frequency of the $n = 0 \rightarrow n = 1$ torsional transition? The experimental value is 289 cm^{-1}.

7.41.* The normal vibrational frequencies of oxygen dichloride are as indicated (Shimanouchi, 1972).

$\tilde{\nu}_1$ (cm^{-1})	$\tilde{\nu}_2$ (cm^{-1})	$\tilde{\nu}_3$ (cm^{-1})
639	296	686

After looking up any needed structural data, calculate the standard molar Gibbs free energy of formation of OCl_2 at 298.15 K. What zero point for the energy scale have you used in this calculation?

7.42. In order to arrive at the secular determinant for the eigenfrequencies of a harmonic crystal, why is it necessary to postulate that the motions of the atoms are synchronous?

7.43. Work through the steps in the derivation of eq. (7.8-8). The upper limit of the summation should be, more correctly, $3N - 6$; explain why. Why then is it permissible to use $3N$ as the upper limit?

7.44. In this and the next two exercises we explore briefly some properties of orthogonal matrices. Let **M** be an orthogonal matrix whose elements are m_{ij}. Why is it true that $\det|\mathbf{M}| = \pm 1$? Confirm that the following is an orthogonal matrix. For it, what is $\det|\mathbf{M}|$?

$$\mathbf{M} = \begin{bmatrix} 0 & \cos\theta & \sin\theta \\ \cos\phi & -\sin\theta\sin\phi & \cos\theta\sin\phi \\ \sin\phi & \sin\theta\cos\phi & -\cos\theta\cos\phi \end{bmatrix}$$

7.45. Suppose for an orthogonal matrix $\det|\mathbf{M}| = +1$. Let $\{M_{ij}\}$ be the set of all cofactors of $\det|\mathbf{M}|$. Prove that $m_{ij} = M_{ij}$. (*Hint*: consider the inverse of **M**.)

7.46. Using the result of Exercise 7.45, expand det|**M**| along elements of the kth column? On the other hand, formulate the sum of the products of corresponding elements in columns k and ℓ. Show by illustration with a 4×4 matrix that this sum is identical to the expansion along elements of the kth column of the determinant of the matrix **M'** formed by replacing the elements of any column n by the elements of column k. Since **M'** now contains two identical columns, what does this imply about $\Sigma_i m_{ik} m_{i\ell}$?

7.47. Lattice waves are called **phonons**. If in the Debye treatment one supposes v_t and v_ℓ to be of comparable magnitude, calculate the velocity of propagation of phonons in lead. Lead is a face-centered cubic crystal (4 atoms per unit cell) of dimension 4.9505 Å (Weast and Astle, 1980, p. B-208).

7.48. Work through the steps in the derivation of eq. (7.8-21) from the definition of the Helmholtz free energy. Then show $\lim_{T \to \infty}\langle C_v \rangle = 3Nk_B$.

7.49. Select one solid each from the two columns of Table 7.8, and for the pair look up entropy data at a variety of temperatures and check how well a Law of Corresponding States is obeyed.

7.50. In eq. (7.8-20), $9N\hbar\omega_D/8$ can be interpreted as the zero-point energy of the crystal if it can be shown that

$$\lim_{T \to 0} 3Nk_B T D\left(\frac{\Theta_D}{T}\right) = 0$$

To do this, it is sufficient to show that the integral $\int_0^\infty x^3(e^x - 1)^{-1}\,dx$ converges; explain. Show that the integral is, indeed, convergent. Deduce the molar zero-point energy of lead.

7.51.[C] Write the program outlined in Fig. 7.11 and apply it to one of the solids in Table 7.8. Look up mean values of α, κ, and $\tilde{V}$ for the solid, and compare your calculated $\langle \tilde{C}_p \rangle$ values with experiment.

7.52. In eq. (7.8-20), one has $D(u) \simeq D(\infty)$ at very low temperatures. The integrand of $D(u)$ is identical to the convergent series $x^3\Sigma_{n=1}^\infty e^{-nx}$. This can be integrated termwise. Make use of the known result

$$\sum_{n=1}^\infty n^{-4} = \frac{\pi^4}{90}$$

and show that at low temperature one has **Debye's T-cubed Law**:

$$\langle \tilde{C}_v \rangle = \frac{12}{5}\pi^4 R \left(\frac{T}{\Theta_D}\right)^3 \qquad T \ll \Theta_D$$

The law has been verified approximately for many substances.

7.53.* Show that the entropy of a Debye crystal is given by

$$\langle S \rangle = 3Nk_B \left[\frac{4}{3} D\left(\frac{\Theta_D}{T} \right) - \ln(1 - e^{-\Theta_D/T}) \right]$$

and that in the high-temperature limit this goes as

$$\langle S \rangle \simeq 3Nk_B \left[\frac{4}{3} - \ln\left(\frac{\Theta_D}{T} \right) + \frac{1}{40}\left(\frac{\Theta_D}{T} \right)^2 - \cdots + \right]$$

[*Hints*: Expand each term in the first equation carefully out to second order in Θ_D/T; for the logarithmic term, use a Taylor series expansion whose leading term is $\ln(\Theta_D/T)$.] An experimental value of Θ_D for potassium is 91 K. Estimate the molar entropy of potassium at 298.15 K and compare with the experimental value of 63.6 J mol^{-1} K^{-1}.

7.54. Some experimental $\tilde{C}_p$ data for gold at very low temperatures are given here (Weast and Astle, 1980, p. D-183)

T (K)	$\tilde{C}_p$ (J mol^{-1} K^{-1})	T (K)	$\tilde{C}_p$ (J mol^{-1} K^{-1})
2.0	0.004924	6.0	0.09849
3.0	0.01379	8.0	0.23636
4.0	0.03152	10.0	0.43333

Theory suggests that at these low temperatures (where we may ignore any difference between $\tilde{C}_p$ and $\tilde{C}_v$) the heat capacity should be representable by the function

$$\langle \tilde{C}_p \rangle = \frac{12}{5} \pi^4 R \left(\frac{T}{\Theta_D} \right)^3 + \gamma T$$

Write a computer program to perform a least-squares fitting of the data to the two-term function, and ascertain optimum values of the parameters Θ_D and γ.

7.55. At low temperatures the lattice vibrations in graphite contribute a term to $\tilde{C}_v$ that is proportional to T^2 instead of the more usual T^3 in the Debye Law. Suggest a structural reason for a T^2 law, and then carry out a derivation of this law. Look up some heat capacity data for graphite (Wostenholm and Yates, 1973) and deduce a value for Θ_D.

7.56. In Exercise 7.51 Simpson's rule was used as the means of computing $D(u)$. Many other methods of quadrature are known (Norris, 1981). One, referred to as the **Gaussian quadrature**, is often more accurate for the same number of subdivisions of the interval of integration. Briefly, the integral

$$I = \int_a^b f(x)\, dx$$

is given approximately by the formula

$$I = \frac{b-a}{2} \sum_{i=1}^{n} A_i f(x_i)$$

where $x_i = (b+a)/2 + t_i(b-a)/2$, the t_i are the zeros of the nth Legendre polynomial $P_n(t)$, the A_i are coefficients that depend on the choice of n, and n is the number of points of subdivision. For example, a four-point ($n = 4$) Gaussian quadrature has

$$t_1 = -0.8611363$$
$$t_2 = -0.3399810 \qquad A_1 = A_4 = 0.3478548$$
$$t_3 = +0.3399810 \qquad A_2 = A_3 = 0.6521452$$
$$t_4 = +0.8611363$$

Apply Gaussian quadrature to the calculation of $D(1.4)$, the tabulated value for which is 0.570793 (Abramowitz and Stegun, 1965). Compare with the result from Simpson's rule with a division of the integration interval into six subintervals.

7.57.[C] A formula for $\langle C_v \rangle$ that is simpler than Debye's and which was derived from the Maxwell distribution law for velocities as its only statistical basis is one due to Compton (1915). The pertinent formula is eq. (5) in that paper. Using the data in Table 7.9 for lead, write a program that can test how well Compton's formula works.

REFERENCES

Abramowitz, M. and Stegun, I. A., *Handbook of Mathematical Functions*, Dover, New York, 1965, pp. 998–1000. Short tables of Einstein functions for $0 \le x \le 6$ and of Debye-type functions, $D_n(u)$, for $0 \le u \le 10$.

Blinder, S. M., "On Additivity Rules in Molecular Thermodynamics," *J. Am. Chem. Soc.*, **97**, 978 (1975).

Compton, A. H., "The Variation of the Specific Heat of Solids with Temperature," *Phys. Rev.*, **6**, 377 (1915).

Courant, R. and John, F., *Introduction to Calculus and Analysis*, Vol. 2, Wiley, New York, 1974, p. 417. Derivation of the formula for the volume of an ellipsoid.

Davidson, N., *Statistical Mechanics*, McGraw-Hill, New York, 1962, pp. 177–178. The partition function for overall rotation of a polyatomic molecule is derived here by classical statistical mechanics.

Dence, J. B., *Mathematical Techniques in Chemistry*, Wiley, New York, 1975, pp. 73, 124–127, 189–195. The first citation gives the result for the differentiation of an integral in which the variable of differentiation occurs both in the limits and in the integrand; the second citation provides a derivation of Simpson's rule, and the third citation discusses the Euler–Maclaurin summation formula.

Feynman, R. P., Leighton, R. B., and Sands, M., *The Feynman Lectures on Physics*, Vol. 1, Addison-Wesley, Reading, 1963, Chap. 47. Gives a derivation of the wave equation for sound.

Giauque, W. F., "The Entropy of Hydrogen and the Third Law of Thermodynamics; The Free Energy and Dissociation of Hydrogen," *J. Am. Chem. Soc.*, **52**, 4816 (1930).

Herman, R. C., "Note on the Heat Capacities and Energies of $SiCl_4$, $TiCl_4$ and $SnCl_4$," *J. Chem. Phys.*, **6**, 406 (1938).

Herzberg, G., *Spectra of Diatomic Molecules*, 2nd ed., Van Nostrand Reinhold, New York, 1950, pp. 212–214. A brief summary of angular momentum in molecules.

Kistiakowsky, G. B., Lacher, J. R., and Stitt, F., "Hindered Internal Rotation of Ethane," *J. Chem. Phys.*, **6**, 407 (1938).

Landau, L. D. and Lifshitz, E. M., *Quantum Mechanics-Nonrelativistic Theory*, 3rd ed., Pergamon Press, Oxford, 1977. See Chapters 4 and 11 for discussions of angular momentum and diatomic molecules.

Landolt-Börnstein, *Zahlenwerte und Funktionen*, 6th ed., Vol. 2, part 4, Springer-Verlag, Berlin, 1961, p. 399.

Levine, I. N., *Molecular Spectroscopy*, Wiley-Interscience, New York, 1975, pp. 56–62, 188–191.

Li, J. C. M., "The Thermodynamic Properties of Cadmium Dimethyl," *J. Am. Chem. Soc.*, **78**, 1081 (1956).

Mathews, J. and Walker, R. L., *Mathematical Methods of Physics*, 2nd ed., Benjamin, New York, 1970, pp. 198–204. A brief and not too deep discussion of solutions to the Mathieu equation; it is seen that Mathieu functions are infinite series.

Montroll, E. W., "Vibrations of Crystal Lattices and Thermodynamic Properties of Solids," in E. U. Condon and H. Odishaw (Eds.), *Handbook of Physics*, 2nd ed., McGraw-Hill, New York, 1967, pp. 5-147 to 5-156. A short authoritative review article on the thermal physics of solids; develops the models of Debye and of Born and von Kármán.

Moore, C. E., *Atomic Energy Levels*, Vol. I, National Bureau of Standards, Circular 467, Washington, D.C., 1949, p. 145.

Norris, A. C., *Computational Chemistry*, Wiley, New York, 1981, pp. 170–174. A brief discussion of Gaussian quadrature.

Pauling, L. C., *The Nature of the Chemical Bond*, 3rd ed., Cornell University Press, Ithaca, 1960, pp. 41–53. These pages give a clear and elementary discussion of term symbols for Russell–Saunders states of atoms.

Pauling, L. C. and Wilson, E. B., Jr., *Introduction to Quantum Mechanics*, McGraw-Hill, New York, 1935, pp. 264–274. Our discussion of the quantum mechanical separation of vibration and rotation is based on the approach here.

Pitzer, K. S., *Quantum Chemistry*, Prentice-Hall, Englewood Cliffs, 1953, Appendix 14. The author, a pioneer in the study of internal rotation, discusses anharmonicity and rotational stretching effects on energy levels. The tables in Appendix 18 show how thermodynamic properties are influenced by the internal rotation barrier height.

Risgin, O. and Taylor, R. C., "Infrared and Raman Spectra of the Pentafluoroethyl Halides," *Spectrochim. Acta*, **15**, 1036 (1959).

Scott, R. A. and Scheraga, H. A., "Method for Calculating Internal Rotation Barriers," *J. Chem. Phys.*, **42**, 2209 (1965). A simple method, based on some ideas of Pauling, to estimate the threefold barrier to internal rotation in ethanelike molecules.

Shimanouchi, T., *Tables of Molecular Vibrational Frequencies: Consolidated Volume I*, NSRDS-NBS 39, National Bureau of Standards, Washington, D.C., 1972.

Weast, R. C. and Astle, M. J. (Eds.), *CRC Handbook of Chemistry and Physics*, 61st ed., CRC Press, Boca Raton, 1980.

Whittaker, E. T. and Watson, G. N., *A Course of Modern Analysis*, 4th ed., Cambridge University Press, London, 1963, Chap. XIX. A deeper discussion of the Mathieu equation than that given in the Mathews and Walker reference.

Wind, H., "Electron Energy for H_2^+ in the Ground State," *J. Chem. Phys.*, **42**, 2371 (1965).

Wostenholm, G. H. and Yates, B., "The Influence of Heat Treatment upon the Low Temperature Heat Capacity of Pyrolytic Graphite," *Phil. Mag.*, **27**, 185 (1973).

Yourgrau, W., van der Merwe, A. and Raw, G., *Treatise on Irreversible and Statistical Thermophysics*, Dover, New York, 1982, pp. 203–217. Our discussion of lattice eigenfrequencies and of the Debye model is based in part on these pages.

CHAPTER 8

Fluids

Chapter 7 dealt with the statistical thermodynamics of isothermal systems of noninteracting particles—the ideal gas and the harmonic solid. Such systems are especially easy to treat because the partition function factors into single-particle partition functions. If the particles interact with one another, the partition function cannot be so factored and an entirely new approach is necessary. In this final chapter we introduce some of the more well-established methods of handling such systems. Our considerations are limited to gases and liquids, that is, **fluids**.

8.1 EMPIRICAL EQUATIONS OF STATE

We approach the subject of systems of interacting particles from a macroscopic perspective. The equation of state of a gas, insofar as it deviates from that of an ideal gas, reflects the statistically averaged effects of molecular interactions. Since the derivation of the equation of state *a priori* from such interactions (i.e., the Hamiltonian) is a difficult task, we begin at a more primitive level by considering a simple physical model.

We view the gas as made up of impenetrable spheres that are free to move in a volume $(V - nb)$, where V is the volume of the containing vessel and b is the **excluded molar volume**, that is, the volume per mole unavailable to the spheres. The equation of state of the ideal gas should then be modified to read

$$P_{ideal}(V - nb) = nRT \qquad (8.1\text{-}1)$$

At distances of separation larger than a few molecular diameters, real atoms and neutral molecules experience a net attraction for one another (Maitland et al., 1981). Hence, if a particle is brought from the interior of the gas to the vessel wall (regarded as structureless), work must be done against the attractions of the surrounding molecules. In comparing the ideal gas and real gas at identical T and V, we then expect the pressure at the wall, P_{real}, to be less for the real gas. We take the pressure ratio to be the Boltzmann factor,

$$\frac{P_{\text{real}}}{P_{\text{ideal}}} = e^{-\Delta \tilde{E}/RT} \tag{8.1-2}$$

where $\Delta \tilde{E}$ is the excess molar potential energy at the wall of molecules of the real gas. Intuitively, $\Delta \tilde{E}$ is expected to increase with the density, so we write $\Delta \tilde{E} = nA/V$, where A is a constant and n/V is the number of moles per unit volume.

Combining eqs. (8.1-2) and (8.1-1) and dropping the subscript "real," we have

$$P(V - nb) = nRTe^{-nA/RTV} \tag{8.1-3a}$$

where P now means the true measured pressure of the gas and V means the true measured volume of the vessel. This is **Dieterici's equation** (Dieterici, 1899). Examination of a number of gases shows that if b is assumed temperature independent, then A varies with temperature as $A = aT^{-1/2}$. The Dieterici equation is then modified to read

$$P(V - nb) = nRTe^{-na/RT^{3/2}V} \tag{8.1-3b}$$

where b and a are now constants depending only on the nature of the gas.

Figure 8.1 displays isotherms for a gas obeying Dieterici's equation of state (8.1-3b). These curves are similar to isotherms measured for real gases. For example, the very steep portion (for $V \leq 3$ mL), which actually corresponds to the liquid phase in the case of a real substance, suggests that Dieterici's equation can "predict" a **critical isotherm**. To see this, we observe from Fig. 8.1 that an isotherm (e.g., $T = 256$ K) of a Dieterici fluid may possess a relative maximum and a relative minimum. The positions of these two extrema can be found from the relation $(\partial P/\partial V)_T = 0$, which for eq. (8.1-3b) takes the form

$$\tilde{V}^2 - \frac{a\tilde{V}}{RT^{3/2}} + \frac{ab}{RT^{3/2}} = 0 \tag{8.1-4}$$

As the temperature increases, the extrema approach each other, that is, the two roots of eq. (8.1-4) approach equality. At some unique temperature, the **critical temperature** T_{cr}, the extrema coalesce and the roots of eq. (8.1-4) become identical. We can therefore determine the condition that the second and third

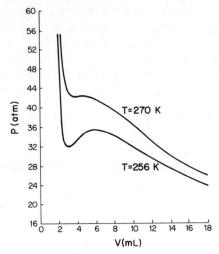

FIGURE 8.1. Isotherms according to the modified Dieterici equation for 0.03 mole of a gas with $a = 100$ L^2 atm K$^{1/2}$ mol^{-2} and $b = 0.0020$ L mol^{-1}.

terms of eq. (8.1-4) must fulfill in order to insure this. The result is

$$T_{cr} = \left(\frac{a}{4bR}\right)^{2/3} \qquad P_{cr} = \frac{R}{be^2}\left(\frac{a}{4bR}\right)^{2/3}$$

$$\tilde{V}_{cr} = 2b \qquad Z_{cr} = \frac{P_{cr}\tilde{V}_{cr}}{RT_{cr}} = \frac{2}{e^2} = 0.271 \qquad (8.1\text{-}5)$$

In Table 8.1 we give the **compressibility factor** Z_{cr} at the critical point for a collection of real gases. The average is in good agreement with the value of 0.271 given by eq. (8.1-5).

Using Dieterici's equation of state, we can estimate the **Boyle temperature**, T_B, defined by

$$\lim_{P \to 0}\left(\frac{\partial Z}{\partial P}\right)_{T_B} = 0 \qquad (8.1\text{-}6)$$

that is, the temperature at which the partial derivative vanishes in the limit of zero pressure. Let us expand Z as a power series in the pressure (this procedure is justified in Section 8.4).

$$Z = 1 + A_1(T)P + A_2(T)P^2 + A_3(T)P^3 + \cdots \qquad (8.1\text{-}7)$$

Table 8.1.   Compressibility Factor at the Critical Point for Various Gases

Gas	P_{cr} (atm)	$\tilde{V}_{cr}$ (L mol^{-1})	T_{cr} (K)	Z_{cr}^{a}
Ammonia	111.3	0.0724$_4$	405.6	0.242
Benzene	48.34	0.258$_4$	562.1	0.271
Carbon dioxide	72.85	0.0940$_2$	304.2	0.274
Chlorine	76.1	0.123$_8$	417.2	0.275
Diethyl ether	35.9	0.279$_7$	466.7	0.262
Methane	45.44	0.0989$_6$	190.6	0.288
Neon	26.86	0.0417$_4$	44.5	0.307
Sulfur trioxide	83.8	0.126$_4$	491.4	0.263
Water	218.3	0.0554$_3$	647.4	0.228
				Av = 0.268

^{a} Calculated from critical data in J. A. Dean (Ed.), *Lange's Handbook of Chemistry*, 12th ed., McGraw-Hill, New York, 1979, pp. 9–176 to 9-184.

It follows from eqs. (8.1-6) and (8.1-7) that

$$A_1(T_b) = 0 \tag{8.1-8}$$

Now replacing P in eq. (8.1-7) by $ZRT/\tilde{V}$, we obtain

$$Z = 1 + A_1(T)\left(\frac{ZRT}{\tilde{V}}\right) + A_2(T)\left(\frac{ZRT}{\tilde{V}}\right)^2 + \cdots \tag{8.1-9}$$

On the other hand, Z is given by eq. (8.1-3b) as

$$Z = \frac{\tilde{V}}{\tilde{V} - b}e^{-a/RT^{3/2}\tilde{V}}$$

which may be expanded in powers of $1/\tilde{V}$. Term by term comparison with eq. (8.1-9) shows that

$$A_1(T) = \frac{1}{ZRT}(b - aR^{-1}T^{-3/2}) \tag{8.1-10}$$

From eqs (8.1-8) and (8.1-10) we deduce that

$$T_B = \left(\frac{a}{bR}\right)^{2/3} = (16)^{1/3}T_{cr} \tag{8.1-11}$$

where the second equality follows from eq. (8.1-5). Table 8.2 compares T_B calculated on the basis of Dieterici's equation of state [i.e., from eq. (8.1-11)]

Table 8.2. Boyle Temperatures for Some Gases

Gas	H$_2$	N$_2$	CO	O$_2$	Ar	CH$_4$
T_B (calculated, K)	84	318	335	390	380	480
T_B (experimental, K)^a	110	325	343	402	408	506

^{a}Interpolated from tables in Landolt-Börnstein, *Zahlenwerte und Funktionen*, 6th ed., Vol. II, Part 1, Springer-Verlag, Berlin, 1971, pp. 249–257.

with experimental results for various gases; the agreement is within 6% for all gases except H$_2$.

Finally, let us use Dieterici's equation of state to calculate the internal energy. If the general thermodynamic relation

$$\left(\frac{\partial E}{\partial V}\right)_T = T\left(\frac{\partial P}{\partial T}\right)_V - P$$

is applied to one mole of a Dieterici fluid, one obtains

$$\left(\frac{\partial \tilde{E}}{\partial \tilde{V}}\right)_T = \frac{3a}{2T^{1/2}}\frac{e^{-a/RT^{3/2}\tilde{V}}}{\tilde{V}(\tilde{V}-b)} \tag{8.1-12}$$

Now make the substitution $\tilde{x} = a(1 - b/\tilde{V})/(bRT^{3/2})$ in eq. (8.1-12) and

Table 8.3. Calculation of the Dependence of Internal Energy on Volume for CH$_4$ at 300 K Using the Dieterici Equation

i	ρ^a (mol dm^{-3})	$\tilde{V}$ (dm^3)	P (calc) (atm)	$\tilde{x}_i$	$[Ei(\tilde{x}_i) - Ei(\tilde{x}_1)]^b$	$\tilde{E}(\exp)^c$ (J mol^{-1})	$\Delta\tilde{E}(\exp)$ (J mol^{-1})	$\Delta\tilde{E}$ (calc)^d (J mol^{-1})
1	0.01	100	0.246	2.0246	0	−2431	0	0
2	0.10	10	2.45	2.0156	−0.02613	−2455	−24	−26
3	1.00	1	23.4	1.9254	−0.36197	−2700	−269	−362
4	2.00	0.50	44.7	1.8252	−0.71078	−2968	−537	−711

^{a}Data from S. Angus, B. Armstrong, and K. M. deReuck, *International Thermodynamic Tables of the Fluid State—Methane*, International Union of Pure and Applied Chemistry, Pergamon Press, Oxford, 1978, pp. 191–193.
^{b}Calculated using Simpson's rule with six evenly spaced intervals.
^{c}The absolute values in this column are referred to a particular reference state (see footnote a); only the differences are of relevance.
^{d}It is convenient to use $R = 0.08206$ L atm mol^{-1} K^{-1} throughout, and then to convert the final answer to SI units by multiplying by the factor 101.325 J L^{-1} atm^{-1}.

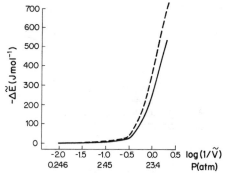

FIGURE 8.2. Variation of the internal energy of methane at 300 K relative to its energy at $\tilde{V} = 100$ L as a function of its molar volume; experiment (—), calculated (---).

integrate,

$$\Delta \tilde{E} = \frac{3}{2} ab^{-1} T^{-1/2} e^{-a/RT^{3/2}b} \int_{\tilde{x}_1}^{\tilde{x}_2} \frac{e^x}{x} \, dx$$

$$= \frac{3}{2} ab^{-1} T^{-1/2} e^{-a/RT^{3/2}b} [Ei(\tilde{x}_2) - Ei(\tilde{x}_1)] \qquad (8.1\text{-}13)$$

where $Ei(x)$ is a standard integral known as the **exponential integral**. Even though $Ei(x)$ is tabulated (Abramowitz and Stegun, 1965), let us evaluate the integral in eq. (8.1-13) numerically, as in Section 7.9. For the case of methane, we calculate from the critical data of Table 8.1

$$\begin{cases} b = \dfrac{1}{2} \tilde{V}_{cr} \quad = 0.04948 \text{ L mol}^{-1} \\[2mm] a = 4bRT_{cr}^{3/2} = 42.7372 \text{ L}^2 \text{ atm K}^{1/2} \text{ mol}^{-2} \end{cases}$$

Additional data and calculations are summarized in Table 8.3, and the final results are presented graphically in Fig. 8.2. At low pressures (less than about 7 atm) agreement between experiment and calculation is good. At molar volumes less than 1 L (or pressures greater than 20 atm) the calculated internal energies begin to diverge widely from the experimental values.

In summary, a simple physical model has led to an **empirical equation of state** (i.e., an equation whose parameters are obtained by fitting the equation to experimental data) from which we are able to calculate with moderate success a number of experimental quantities. Other equations of state can be constructed that are more accurate than the Dieterici equation, especially if they contain more than two parameters. An interesting summary of work on various empirical equations of state is contained in a recent monograph

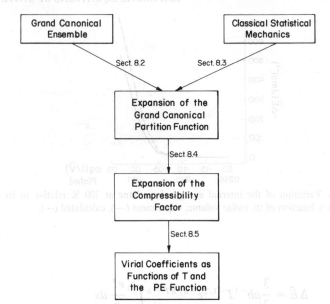

FIGURE 8.3. Steps leading to the virial expansion.

(Kreglewski, 1984). However, an empirical equation of state provides only a limited understanding of a fluid; it lacks a fundamental basis. In the next few sections our goal is to derive from fairly rigorous considerations a general equation of state in the form of a series expansion that is applicable to real gases. This expansion is known as the **virial equation**. To orient the reader, a flowchart of the steps is provided in Fig. 8.3.

8.2. THE GRAND CANONICAL ENSEMBLE

We consider a system at fixed temperature T, fixed volume V, and fixed chemical potential μ. Both the energy and the number of particles in the system are allowed to fluctuate. An ensemble is then constructed by immersing a collection of systems of volume V in a heat bath at temperature T and in contact with a large reservoir of particles. The walls of the containers of the systems permit exchange of energy and particles. When equilibrium is reached (as determined by equality of the temperatures and chemical potentials throughout), the entire ensemble is removed from the bath and reservoir and isolated. Any particular system chosen "at random" from the ensemble contains n particles and is in quantum state i (see Fig. 8.4). A physical analogy to a grand canonical ensemble is a one-liter container full of water at $25°C$ that is subdivided by hypothetical boundaries into compartments that are each of dimensions 10^{-6} m $\times$ 10^{-6} m $\times$ 10^{-6} m (Davidson, 1962).

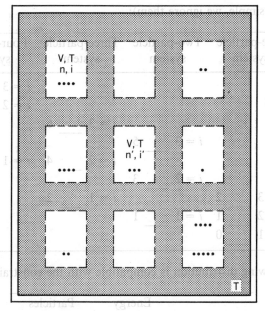

FIGURE 8.4. Illustrating the grand canonical ensemble.

Let the ensemble consist of $\mathcal{N}$ systems and let N_{ni} stand for the number of systems, each of which contains n particles and is in quantum state i. An allowed distribution is a set of numbers $\{N_{ni}\}$ consistent with the three constraints

$$\sum_n \sum_i N_{ni} = \mathcal{N} \tag{8.2-1a}$$

$$\sum_n \sum_i N_{ni}E_{ni} = E_0 \tag{8.2-1b}$$

$$\sum_n \sum_i nN_{ni} = N_0 \tag{8.2-1c}$$

We proceed as in Section 6.4 (Yourgrau et al., 1982). The number of ways, W, that a given distribution $\{N_{ni}\}$ can be realized is

$$W = \frac{\mathcal{N}!}{\prod_n \prod_i (N_{ni}!)} \tag{8.2-2}$$

To illustrate, we imagine an ensemble consisting of $\mathcal{N} = 6$ systems, a total of $N_0 = 18$ particles, and having a total energy $E_0 = 30$. Suppose the following sets of energy levels are available to systems that contain 1, 2, 3, or 4 particles (systems containing more particles are possible in this example, but to keep the discussion simple, we ignore them):

One-particle system	Two-particle system	Three-particle system	Four-particle system
			$i = 3$ ___ 10
			$i = 2$ ___ 7
		$i = 3$ ___ 6	
	$i = 3$ ___ 5		
		$i = 2$ ___ 4	$i = 1$ ___ 4
	$i = 2$ ___ 3		
$i = 3$ ___ 2		$i = 1$ ___ 2	
$i = 2$ ___ 1	$i = 1$ ___ 1		
$i = 1$ ___ 0			

(E increases upward, indicated by the arrow at the left.)

Then the following distribution is compatible with the constraints (8.2-1).

	Energy	Particles
$N_{23} = 2$	10	4
$N_{33} = 2$	12	6
$N_{41} = 2$	8	8
6	30	18

The ensemble corresponding to this distribution is pictured in Fig. 8.5(a). The number of ways the distribution can be realized is, according to eq. (8.2-2)

$$W = \frac{6!}{2!2!2!} = 90$$

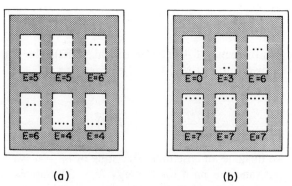

(a) (b)

FIGURE 8.5. Two different distributions in an ensemble of six systems where the total number of particles is 18 and the total energy is 30 units.

Still another distribution that is consistent with the constraints is the following:

	Energy	Particles
$N_{11} = 1$	0	1
$N_{22} = 1$	3	2
$N_{33} = 1$	6	3
$N_{42} = 3$	21	12
$\overline{6}$	$\overline{30}$	$\overline{18}$

This distribution can be realized in 120 ways. The corresponding ensemble is shown in Fig. 8.5(b).

The total number of physically conceivable *ensemble states* for a grand canonical ensemble is

$$\Omega(\mathcal{N}, E_0, N_0) = \sum_{\{N_{ni}\}} W(\{N_{ni}\})$$

where the sum is over all distributions that are compatible with eqs. (8.2-1). We emphasize that Ω is a function of the three parameters $\mathcal{N}$, E_0, N_0 [see eq. (6.4-4)]. According to the Prinicple of Equal *a Priori* Probability (Section 6.1), the $\Omega(\mathcal{N}, E_0, N_0)$ ensemble states are equally probable, and the probability that a given distribution is realized is

$$p(\{N_{ni}\}) = \frac{W(\{N_{ni}\})}{\Omega(\mathcal{N}, E_0, N_0)} \tag{8.2-3}$$

We now define new quantities $\mathcal{N} - 1 = M$, $N_{ni} - 1 = M_{ni}$, $N_{mj} = M_{mj}$ (for either $m \neq n$ or $j \neq i$). The analog of eq. (6.4-9) then becomes

$$\langle N_{ni} \rangle \Omega(\mathcal{N}, E_0, N_0) = \mathcal{N} \sum \frac{(\mathcal{N} - 1)!}{N_{11}!N_{12}!N_{21}!N_{22}! \cdots (N_{ni} - 1)! \cdots N_{kl}!}$$

$$= \mathcal{N} \sum \frac{M!}{M_{11}!M_{12}!M_{21}!M_{22}! \cdots M_{ni}! \cdots M_{kl}!}$$

$$= \mathcal{N} \Omega(\mathcal{N} - 1, E_0 - E_{ni}, N_0 - n) \tag{8.2-4}$$

where the summation is now subject to the new constraints

$$\sum_m \sum_j M_{mj} = \mathcal{N} - 1 \qquad \sum_m \sum_j M_{mj} E_{mj} = E_0 - E_{ni}$$

$$\sum_m \sum_j m M_{mj} = N_0 - n \tag{8.2-5}$$

We now write an equation analogous to eq. (6.4-12), except that an additional term related to the third of the constraints in eq. (8.2-5) must be added.

$$\ln \Omega(\mathcal{N} - 1, E_0 - E_{ni}, N_0 - n)$$

$$= \ln \Omega(\mathcal{N}, E_0, N_0) - \left(\frac{\partial \ln \Omega(\mathcal{N}, E_0, N_0)}{\partial \mathcal{N}} \right)$$

$$- E_{ni} \left(\frac{\partial \ln \Omega(\mathcal{N}, E_0, N_0)}{\partial E_0} \right) - n \left(\frac{\partial \ln \Omega(\mathcal{N}, E_0, N_0)}{\partial N_0} \right)$$

$$= \ln \Omega(\mathcal{N}, E_0, N_0) - \alpha - \beta E_{ni} - \gamma n \qquad (8.2\text{-}6)$$

Taking antilogs and invoking condition (8.2-1a), we obtain the probability p_{ni} of finding the system to be in state i and to contain n particles:

$$p_{ni} = \frac{e^{-\beta E_{ni} - \gamma n}}{\sum_n \sum_i e^{-\beta E_{ni} - \gamma n}}$$

$$= \frac{e^{-\beta E_{ni} - \gamma n}}{\Xi(\beta, \gamma, V)} \qquad (8.2\text{-}7)$$

The denominator Ξ of eq. (8.2-7) is called the **grand canonical partition function**, and eq. (8.2-7) is the fundamental distribution law for the grand canonical ensemble.[1] Note that the parameters of the grand canonical ensemble are β, γ and V and that the β and γ are related to thermodynamic variables characterizing an open macroscopic system. Thus, Ξ is to be regarded as a function of β, γ, and V. We proceed next to determine the relation of γ and β to the thermodynamic state variables using SM Postulate 2 (Section 6.3).

The average internal energy of the system may be expressed as

$$\langle E \rangle = \sum_n \sum_i p_{ni} E_{ni}$$

the differential of which is

$$d\langle E \rangle = \sum_n \sum_i (p_{ni} \, dE_{ni} + E_{ni} \, dp_{ni})$$

For E_{ni} we may substitute from eq. (8.2-7) to obtain

$$d\langle E \rangle = \sum_n \sum_i \left(p_{ni} \, dE_{ni} - \frac{1}{\beta} (\gamma n + \ln p_{ni} + \ln \Xi) \, dp_{ni} \right) \qquad (8.2\text{-}8)$$

---

[1] The symbol Ξ is tough to write but easy to pronounce ("zi"). It appears to be the most widely employed symbol for the grand canonical partition function.

The eigenvalues E_{ni} are functions only of n and V, but since the first term in eq. (8.2-8) is summed over all values of n, the result is a function only of V. Using this fact as well as the relation

$$d\langle n \rangle = \sum_n \sum_i n \, dp_{ni}$$

we can rewrite eq. (8.2-8) as

$$d\langle E \rangle = \sum_n \sum_i \left[p_{ni} \left(\frac{\partial E_{ni}}{\partial V} \right) dV - \left(\frac{1}{\beta} \ln p_{ni} + \frac{1}{\beta} \ln \Xi \right) dp_{ni} \right] - \frac{\gamma}{\beta} d\langle n \rangle$$

$$= -\langle P \rangle \, dV - \frac{1}{\beta} \sum_n \sum_i (\ln p_{ni} + \ln \Xi) \, dp_{ni} - \frac{\gamma}{\beta} d\langle n \rangle \qquad (8.2\text{-}9a)$$

In writing the second line of eq. (8.2-9a) we have used the relation $P_{ni} = -\partial E_{ni}/\partial V$, where P_{ni} is the pressure of the system containing n particles in state i. This relation is based on the thermodynamic result that for a reversible expansion in which the system does mechanical work only, the decrease in the internal energy is given by

$$dE_{ni} = \left(\frac{\partial E_{ni}}{\partial V} \right) dV = -P_{ni} \, dV$$

The quantity $\ln \Xi$ in eq. (8.2-9a) is just a number, that is, it does not depend on the indices n and i. Thus

$$\sum_n \sum_i \ln \Xi \, dp_{ni} = \ln \Xi \sum_n \sum_i dp_{ni}$$

$$= 0$$

since $\sum_n \sum_i p_{ni} = 1$. Equation (8.2-9a) may therefore be rewritten as

$$d\langle E \rangle = -\langle P \rangle \, dV - \frac{1}{\beta} d\left(\sum_n \sum_i (p_{ni} \ln p_{ni}) \right) - \frac{\gamma}{\beta} d\langle n \rangle \qquad (8.2\text{-}9b)$$

But from thermodynamics, for a reversible process involving only mechanical work in an open system

$$dE = -P \, dV + T \, dS + \mu \, dn \qquad (8.2\text{-}10)$$

where μ is the **chemical potential**. Comparing eqs. (8.2-9b) and (8.2-10) and invoking SM Postulate 2, we conclude

$$\beta = \frac{1}{k_B T}$$

$$\langle S \rangle = -k_B \sum_n \sum_i (p_{ni} \ln p_{ni})$$

$$\mu = \frac{-\gamma}{\beta} = -k_B T \gamma$$

The parameters β and γ have now been identified, and it is thus possible to derive explicit formulas for the thermodynamic functions in terms of Ξ. These are

$$\langle n \rangle = k_B T \left(\frac{\partial \ln \Xi}{\partial \mu} \right)_{V,T} \tag{8.2-11a}$$

$$\langle S \rangle = k_B \ln \Xi + k_B T \left(\frac{\partial \ln \Xi}{\partial T} \right)_{V,\mu} \tag{8.2-11b}$$

$$\langle P \rangle = k_B T \frac{\ln \Xi}{V} \tag{8.2-11c}$$

$$\langle A \rangle = k_B T \mu \left(\frac{\partial \ln \Xi}{\partial \mu} \right)_{V,T} - k_B T \ln \Xi \tag{8.2-11d}$$

$$\langle E \rangle = k_B T \left[T \left(\frac{\partial \ln \Xi}{\partial T} \right)_{V,\mu} + \mu \left(\frac{\partial \ln \Xi}{\partial \mu} \right)_{V,T} \right] \tag{8.2-11e}$$

$$\langle H \rangle = k_B T \left[T \left(\frac{\partial \ln \Xi}{\partial T} \right)_{V,\mu} + \mu \left(\frac{\partial \ln \Xi}{\partial \mu} \right)_{V,T} + \ln \Xi \right] \tag{8.2-11f}$$

$$\langle G \rangle = k_B T \mu \left(\frac{\partial \ln \Xi}{\partial \mu} \right)_{V,T} \tag{8.2-11g}$$

To summarize, the grand canonical ensemble pertains to an open, isothermal, constant-volume system, all of whose thermodynamic properties can be computed from the grand canonical partition function [eqs. (8.2-11)]. Thus, the role of $\Xi(T, V, \mu)$ for an open, isothermal system is similar to that of $Q(T, V, N)$ for a closed, isothermal system. A further connection between the canonical and the grand canonical ensembles is seen by freezing (mentally) the composition of the systems in the grand canonical ensemble by inserting between the systems "membranes" that are permeable to heat but not to particles. One then has a collection of canonical ensembles. This can be seen by summing over i first and n second in the definition of Ξ.

$$
\Xi = \sum_n \sum_i \exp\left(\frac{-E_{ni}}{k_B T} + \frac{n\mu}{k_B T} \right)
$$
$$
= \sum_n e^{n\mu/k_B T} \sum_i e^{-E_{ni}/k_B T}
$$
$$
= Q(T, V, 0) + e^{\mu/k_B T} Q(T, V, 1) + e^{2\mu/k_B T} Q(T, V, 2) + \cdots \quad (8.2\text{-}12)
$$

This expansion allows one to view the many-particle system as a one-particle system plus a two-particle system, and so on, and proves very useful in the development of the virial equation for a real gas (Section 8.4).

Exercises

8.1.   Derive the relations given in eq. (8.1-5). Deduce values of the Dieterici parameters for isobutane, for which $T_{cr} = 134.98\,°\text{C}$ and $\rho_{cr} = 0.221$ g mL^{-1}.

8.2.[C]   Write a computer program to calculate and plot the isotherms for a Dieterici fluid. Input consists of values of ρ_{cr}, T_{cr}, T and a range of volumes. Plot out three different isotherms for isobutane.

8.3.   Sketch qualitatively plots of Z versus P at several temperatures for a real gas and explain graphically the meaning of the Boyle temperature. Why does a gas behave very nearly ideally at T_B over a range of pressures?

8.4.   A model equation of state which has sometimes been employed is the **Redlich–Kwong equation**, introduced in 1949,

$$
P = \frac{nRT}{V - nb} - \frac{n^2 a}{T^{1/2} V (V + nb)}
$$

where a, b are independent of T, V. For such a gas, deduce the Boyle temperature and then calculate T_B for CH_4 (see Tables 8.1 and 8.2).

8.5.[C]   Write a computer program to calculate $\Delta \tilde{E}$ given by eq. (8.1-13); use 12 evenly spaced intervals in the numerical integration. Then consult the reference in footnote a of Table 8.3 to obtain data for methane at 400 K. Calculate some values of $\Delta \tilde{E}$ relative to $\tilde{V} = 100$ L for densities out to 5.0 mol dm^{-3}.

8.6.   For the simple, hypothetical ensemble in Section 8.2, find two other distributions that are a $(2 + 2 + 2)$ partition of the six systems. Find two other distributions that are a $(1 + 1 + 1 + 3)$ partition of the

systems. Find five other distributions of any form that are consistent
with the original constraints.

8.7.   From eq. (8.2-9b), derive eq. (8.2-11a). From $\langle S \rangle = -k_B \Sigma_n \Sigma_i (p_{ni} \ln p_{ni})$
and eq. (8.2-7), derive eq. (8.2-11c).

8.3   CLASSICAL STATISTICAL MECHANICS

In applying the grand canonical ensemble, it is convenient to introduce the
classical limit where possible. According to the **Bohr Correspondence Principle**
(Bohr, 1928), the conclusions of quantum mechanics must pass over into those
of classical mechanics when the variables characterizing a system are large
compared with atomic units (e.g., extension in space much larger than a_0, and
angular momentum much larger than $\hbar$). We therefore need to examine how
the quantum statistical mechanical relations are modified in the limit of large
quantum numbers (i.e., the classical limit).

We recall (Section 1.2) that in classical mechanics the state of a system of n
particles in f dimensions is completely specified by giving the fn generalized
coordinates and fn generalized momenta at any instant. For a single particle,
we may imagine motion to occur in a $2f$-dimensional space, known as **phase
space**, f of whose axes are for the generalized coordinates and f for the
momenta. Suppose the coordinates are limited to a range q_i to $(q_i + \delta q_i)$, and
the momenta to a range p_i to $(p_i + \delta p_i)$. Then these ranges delimit a region in
phase space whose **hypervolume** δV (SI units of $J^f s^f$) is

$$\delta V = \int_{\{q_i\}}^{\{q_i + \delta q_i\}} \cdots \int \prod_{i=1}^{f} dq_i \int_{\{p_i\}}^{\{p_i + \delta p_i\}} \cdots \int \prod_{i=1}^{f} dp_i \tag{8.3-1}$$

Now consider a point particle confined to a one-dimensional region of
length L. According to quantum mechanics, its energy is that of a particle in a
box in quantum state n_x (see Table 2.1).

$$E = \frac{h^2}{8mL^2} n_x^2$$

The number of quantum states (g) with energy between 0 and E can be
expressed as

$$g = n_x = \frac{2L}{h}(2mE)^{1/2} = \frac{1}{h}\int_0^L dx \int_{-(2mE)^{1/2}}^{(2mE)^{1/2}} dp_x.$$

The limits on the momentum integration follow from the classical relation $E = p_x^2/2m$ for a free particle. Making use of eq. (8.3-1), we have

$$\frac{\text{phase space hypervolume}}{\text{number of states}} = \frac{\delta V}{g}$$

$$= \frac{\int_0^L dx \int_{-(2mE)^{1/2}}^{(2mE)^{1/2}} dp_x}{2L(2mE)^{1/2}/h} = h \qquad (8.3\text{-}2)$$

If the particle has two degrees of freedom and is restricted to moving in a square region $L \times L$, the energy

$$E = \frac{h^2}{8mL^2}\left(n_x^2 + n_y^2\right)$$

may be represented by points in a Cartesian plane whose axes are n_x and n_y [Fig. 8.6(a)]. For an arbitrary value of E, the quantum numbers n_x and n_y must lie within a circle of radius $R = 2L(2mE)^{1/2}/h$. For sufficiently large n_x and n_y, the number of eigenstates with energy between 0 and E is essentially equal to the area of the (upper right) quadrant of the circle.

$$g = \frac{1}{4}\pi R^2 = \frac{\pi L^2}{h^2}(2mE)$$

The corresponding hyperplane in phase space is obtained by making use of a polar coordinate system [Fig. 8.6(b)], where

$$dp_x\, dp_y = (m\, dv_x)(m\, dv_y)$$

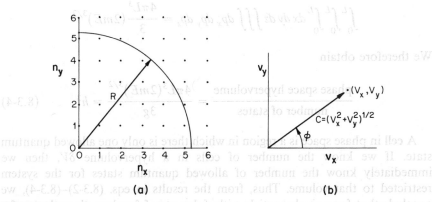

FIGURE 8.6. (a) Representation of the eigenstates of a particle in a square, (b) transformation of Cartesian velocity components into polar form.

and

$$v_x = c \cos \phi \qquad v_y = c \sin \phi \qquad c \geq 0$$

Then we require $c^2 = v_x^2 + v_y^2 \leq 2E/m$, and hence

$$\int_0^L \int_0^L dx \, dy \iint dp_x \, dp_y = L^2 \iint (m \, dv_x)(m \, dv_y)$$

$$= L^2 m^2 \int_0^{2\pi} d\phi \int_0^{(2E/m)^{1/2}} c \, dc$$

$$= \pi L^2 (2mE)$$

Thus, we find

$$\frac{\text{phase space hypervolume}}{\text{number of states}} = \frac{\pi L^2 (2mE)}{\pi L^2 (2mE)/h^2} = h^2 \qquad (8.3\text{-}3)$$

Finally, in three dimensions the energy of a point particle confined to a cubic region $L \times L \times L$ is

$$E = \frac{h^2}{8mL^2} \left(n_x^2 + n_y^2 + n_z^2 \right)$$

and for an arbitrary E, the quantum numbers are restricted to one octant of a sphere of radius $R = 2L(2mE)^{1/2}/h$. Analysis similar to the previous leads to

$$g = \frac{4\pi L^3}{3h^3} (2mE)^{3/2}$$

while the corresponding hypervolume in phase space is

$$\int_0^L \int_0^L \int_0^L dx \, dy \, dz \iiint dp_x \, dp_y \, dp_z = \frac{4\pi L^3}{3} (2mE)^{3/2}$$

We therefore obtain

$$\frac{\text{phase space hypervolume}}{\text{number of states}} = \frac{4\pi L^3 (2mE)^{3/2}}{3g} = h^3 \qquad (8.3\text{-}4)$$

A **cell** in phase space is a region in which there is only one allowed quantum state. If we know the number of cells in a hypervolume δV, then we immediately know the number of allowed quantum states for the system restricted to that volume. Thus, from the results in eqs. (8.3-2)–(8.3-4), we conclude that for a single particle with f degrees of freedom, the volume of a cell is h^f.

From the quantal viewpoint, the molecular translational partition function is given by

$$q_{\text{trans}} = \sum_i e^{-E_i/k_B T}$$

where the index i stands for a discrete collection of numbers in general. On the other hand, in the classical view energy is a continuous quantity, which for a freely translating particle is given by the classical Hamiltonian $H(p) = (1/2m)\mathbf{p} \cdot \mathbf{p}$. In the classical limit the summation ought, therefore, to be replaced by an integration over the phase-space volume and the integral multiplied by the factor $1/h^f$, which gives the number of states per unit volume of phase space. We have then the correspondence

$$\sum_i e^{-E_i/k_B T} \rightarrow \frac{1}{h^f} \int \cdots \int e^{-H(p)/k_B T} \, d^f q \, d^f p$$

For a system of n *independent*, identical point particles in three dimensions, the Hamiltonian is

$$H(p) = \frac{1}{2m} \sum_{j=1}^{n} \mathbf{p}_j \cdot \mathbf{p}_j$$

and we obtain

$$Q_{\text{trans}} = \frac{q_{\text{trans}}^n}{n!} \tag{8.3-5a}$$

where

$$q_{\text{trans}} \equiv \frac{1}{h^3} \int \cdots \int e^{-p_i^2/2mk_B T} \, d^3 q_i \, d^3 p_i \tag{8.3-5b}$$

and eq. (8.3-5a) follows from the separability of $H(p)$. The factor $n!$ corrects approximately for the indistinguishability of the particles (see Section 6.9); the integral in eq. (8.3-5b) is now a sixfold integral. Equation (8.3-5b) is the classical limit of eq. (6.9-4).

If the particles interact with one another this argument fails. However, it can be shown that expressions analogous to eqs. (8.3-5) hold also for systems of interacting particles (Hill, 1960). For example, the limiting partition function for a system comprising n indistinguishable, yet interacting, molecules is

$$Q_{\text{class}} = \frac{1}{h^{nf} n!} \int \cdots \int e^{-H(p,\,q)/k_B T} \prod_{i=1}^{n} d^f q_i \, d^f p_i \tag{8.3-6}$$

where f now includes *internal* degrees of freedom, that is, rotational, vibrational and electronic, as well as translational degrees of freedom.

In some cases the classical limit may not be a good approximation for all degrees of freedom of a system. For example, while translation and rotation can usually be treated classically even at quite low temperatures, vibrations cannot be so treated. However, if the Hamiltonian can be written as a sum of terms involving "quantal" and "classical" degrees of freedom separately, then the partition function factors as

$$Q = Q_{quant} Q_{class} \qquad (8.3\text{-}7)$$

In the case of n indistinguishable interacting molecules, each having t degrees of translational freedom, r degrees of rotational freedom, and $(f - t - r)$ degrees of vibrational freedom

$$Q_{class} = \frac{1}{h^{(t+r)n} n!} \int \cdots \int e^{-H_{class}/k_B T} \prod_{i=1}^{n} d^{t+r} q_i \, d^{t+r} p_i$$

$$Q_{quant} = q_{vib}^n$$

where Q_{quant} is the canonical vibrational partition function (7.2-3).

8.4   THE VIRIAL EQUATION

Equation (8.2-12) can be rewritten as

$$\Xi = \sum_{n=0}^{\infty} Q_n(T, V) \lambda^n \qquad (8.4\text{-}1)$$

where $\lambda \equiv \exp(\mu/k_B T)$, and the reservoir of particles used in the construction of the ensemble is taken to be limitless. To simplify notation, we define $Q_n(T, V) \equiv Q(T, V, n)$. Thus, $Q_1(T, V)$ becomes identical to $q(T, V)$. We also define $Q_0(T, V) = 1$, since for no particles there are no energy states to sum over and $e^{-0} = 1$.

Alternatively, we have from eq. (8.2-11c)

$$\Xi = e^{\langle P \rangle V/k_B T} \qquad (8.4\text{-}2)$$

But $\langle P \rangle/k_B T = V^{-1} \ln \Xi$ can be regarded as a function of λ and expanded in a power series as

$$\langle P \rangle/k_B T = V^{-1} \ln \Xi = \sum_{i=1}^{\infty} b_i \lambda^i$$

The series starts with the first power of λ, since $\ln \Xi = 0$ in the limit $\lambda = 0$. Now substituting the expansion for $\langle P \rangle / k_B T$ into eq. (8.4-2), we obtain

$$\Xi = \exp\left(V \sum_{i=1}^{\infty} b_i \lambda^i \right)$$

$$= \prod_{i=1}^{\infty} \exp\left(V b_i \lambda^i \right)$$

$$= \prod_{i=1}^{\infty} \left(\sum_{m_i=0}^{\infty} \frac{\left(V b_i \lambda^i \right)^{m_i}}{(m_i)!} \right) \tag{8.4-3}$$

where the last line follows from the Maclaurin series $e^x = \sum_m (x^m/m!)$. The coefficient of any small power of λ can be most easily picked out by simply writing out the first few factors of the extended product:

$$\Xi = \left[1 + V b_1 \lambda + \frac{1}{2}(V b_1 \lambda)^2 + \frac{1}{6}(V b_1 \lambda)^3 + \cdots \right]$$

$$\times \left[1 + V b_2 \lambda^2 + \frac{1}{2}(V b_2 \lambda^2)^2 + \frac{1}{6}(V b_2 \lambda^2)^3 + \cdots \right] \times \cdots \times$$

Equations (8.4-1) and (8.4-3) are alternative power series for Ξ. If they are both to be valid, the coefficients of like powers of λ must be equal. Comparing the two expressions, we therefore require

$$Q_1 = V b_1$$

$$Q_2 = V b_2 + \frac{1}{2}\left(V^2 b_1^2 \right)$$

$$Q_3 = V b_3 + (V b_1)(V b_2) + \frac{1}{6}(V b_1)^3 \tag{8.4-4}$$

These equations can be solved sequentially from the top down to yield

$$b_1 = V^{-1} Q_1$$

$$b_2 = V^{-1}\left(Q_2 - \frac{1}{2} Q_1^2 \right)$$

$$b_3 = V^{-1}\left(Q_3 - Q_1 Q_2 + \frac{1}{3} Q_1^3 \right) \tag{8.4-5}$$

Our goal is to obtain an expansion in powers of the density (an observable). From eq. (8.2-11a) we have

$$\langle n \rangle = k_B T \left(\frac{\partial \ln \Xi}{\partial \mu} \right)_{T,V} = \lambda \left(\frac{\partial \ln \Xi}{\partial \lambda} \right)_{T,V}$$

$$= \lambda \frac{\partial}{\partial \lambda} \left(V \sum_{i=1}^{\infty} b_i \lambda^i \right)$$

$$= V \sum_{i=1}^{\infty} i b_i \lambda^i \qquad (8.4\text{-}6)$$

where the second line follows from the original definition of the λ-series expansion of $\ln \Xi$. Therefore, the density can be expressed as

$$\rho = \frac{\langle n \rangle}{V} = \sum_{i=1}^{\infty} i b_i \lambda^i \qquad (8.4\text{-}7)$$

We now invert this series and express λ as an expansion in powers of ρ.[2] Note from eq. (8.4-7) that as $\lambda \to 0$, $\rho \to b_1 \lambda$. Thus, as $\rho \to 0$, $\lambda \to b_1^{-1}\rho = (V/Q_1)\rho$. Hence, we let

$$\lambda = V\rho/Q_1 + \sum_{i=2}^{\infty} A_i \left(\frac{V\rho}{Q_1} \right)^i$$

and then substitute this expansion into eq. (8.4-7) to obtain

$$\rho = b_1 \left[\frac{V\rho}{Q_1} + \sum_{i=2}^{\infty} A_i \left(\frac{V\rho}{Q_1} \right)^i \right] + 2b_2 \left[\frac{V\rho}{Q_1} + \sum_{i=2}^{\infty} A_i \left(\frac{V\rho}{Q_1} \right)^i \right]^2 + \cdots \qquad (8.4\text{-}8)$$

Equating coefficients of like powers of ρ on both sides of eq. (8.4-8), we find

$$0 = b_1 A_2 \left(\frac{V}{Q_1} \right)^2 + 2b_2 \left(\frac{V}{Q_1} \right)^2$$

$$0 = b_1 A_3 \left(\frac{V}{Q_1} \right)^3 + 2b_2(2A_2) \left(\frac{V}{Q_1} \right)^3 + 3b_3 \left(\frac{V}{Q_1} \right)^3$$

which can be solved sequentially for the A_i's:

$$A_2 = -2b_2 \left(\frac{V}{Q_1} \right)$$

$$A_3 = 8b_2^2 \left(\frac{V}{Q_1} \right)^2 - 3b_3 \left(\frac{V}{Q_1} \right) \qquad (8.4\text{-}9)$$

---

[2] The inversion of an infinite series requires justification (Fulks, 1969).

Replacing λ by its expansion in powers of ρ gives

$$
\frac{PV}{k_BT} = V\left\{ b_1\left[\frac{V\rho}{Q_1} + \sum_{i=2}^{\infty} A_i\left(\frac{V\rho}{Q_1}\right)^i\right] + b_2\left[\frac{V\rho}{Q_1} + \sum_{i=2}^{\infty} A_i\left(\frac{V\rho}{Q_1}\right)^i\right]^2 + \cdots \right\}
$$

$$
= V\left\{ \rho - b_2\left(\frac{V}{Q_1}\right)^2\rho^2 + \left[4b_2^2\left(\frac{V}{Q_1}\right) - 2b_3\right]\left(\frac{V}{Q_1}\right)^3\rho^3 - \cdots \right\} \quad (8.4\text{-}10)
$$

where we have used eqs. (8.4-5) and (8.4-9). Finally, division of both sides of eq. (8.4-10) by the number of particles $\langle n \rangle$ converts the left-hand side into the compressibility factor.

$$
Z = \frac{PV}{\langle n \rangle k_BT} = 1 - b_2\left(\frac{V}{Q_1}\right)^2\rho + \left[4b_2^2\left(\frac{V}{Q_1}\right) - 2b_3\right]\left(\frac{V}{Q_1}\right)^3\rho^2 - \cdots
$$

$$
= 1 + B(T)\rho + C(T)\rho^2 + \cdots \quad (8.4\text{-}11)
$$

Equation (8.4-11) is the usual form of the **virial equation** for a real gas, and the temperature-dependent coefficients $B(T)$ and $C(T)$ are known as the **second** and **third virial coefficients**. Since $\rho = \langle n \rangle / V$, eq. (8.4-11) is also an expansion in inverse powers of the volume. Alternatively, the virial equation is sometimes expressed as a power series in the pressure.

$$
Z = 1 + B'(T)P + C'(T)P^2 + \cdots \quad (8.4\text{-}12)
$$

The two expressions (8.4-11) and (8.4-12) are, of course, equal and it is straightforward to show that $B'(T) = B(T)/k_BT$, $C'(T) = (k_BT)^{-2}\{C(T) - [B(T)]^2\}$.

In summary, the grand canonical ensemble has given us a rigorous expansion of the compressibility factor in terms of either the density or the pressure (Hill, 1960). Equations (8.4-11) and (8.4-12) both reduce to the ideal gas result in the limit as $V \to \infty$ or $P \to 0$. Further use of the virial equation requires explicit evaluation of the virial coefficients, and this can be done only if the interactions among the particles are known, that is, if the potential energy $U(q)$ is known.

8.5   THE VIRIAL COEFFICIENTS

From eqs. (8.4-11) and (8.4-5) it is clear that calculation of the virial coefficients requires evaluation of the canonical partition function $Q_n(V, T)$ for various orders n. We assume for the remainder of this Section that the classical limit is valid. Moreover, for simplicity we restrict our considerations to systems of identical atoms, where there are no internal degrees of freedom.

Then from eq. (8.3-6) we have

$$Q_n(V, T) = \frac{1}{h^{3n}n!} \int \cdots \int e^{-H/k_B T} \prod_{i=1}^{n} d^3q_i \, d^3p_i \qquad (8.5\text{-}1)$$

where

$$H = \sum_{i=1}^{n} \frac{p_i^2}{2m} + U(\mathbf{r}_1, \mathbf{r}_2, \ldots, \mathbf{r}_n)$$

Carrying out the integrations over momenta explicitly, we obtain

$$Q_n = \frac{Z}{n! \Lambda^{3n}} \qquad (8.5\text{-}2)$$

where $\Lambda = (h^2/2\pi m k_B T)^{1/2}$ and the **configuration integral** Z (not to be confused with the compressibility factor) is given by

$$Z = \int \cdots \int e^{-U(\mathbf{r}_1, \mathbf{r}_2, \ldots, \mathbf{r}_n)/k_B T} \prod_{i=1}^{n} d^3r_i \qquad (8.5\text{-}3)$$

For the one atom case $U = 0$ and

$$Q_1 = V/\Lambda^3 \qquad (8.5\text{-}4)$$

in agreement with eq. (6.9-5). In the first nontrivial case, $n = 2$, we expect U to be a function only of the scalar distance $r_{12} = |\mathbf{r}_2 - \mathbf{r}_1|$ and not to depend on the absolute locations of the volume elements d^3r_1 and d^3r_2 (the vessel is assumed large enough that surface effects can be ignored). Hence, if we change from variables (r_1, r_2) to (r_1, r_{12}), then Q_2 can be rewritten as

$$Q_2(T, V) = \frac{1}{2} \Lambda^{-6} \int d^3r_1 \iiint e^{-U(r_{12})/k_B T} \sin\theta \, r_{12}^2 \, dr_{12} \, d\theta \, d\phi$$

$$= \frac{2\pi V}{\Lambda^6} \int_0^{r_{12}(\text{max})} r_{12}^2 e^{-U(r_{12})/k_B T} \, dr_{12} \qquad (8.5\text{-}5)$$

Here, we have used spherical polar coordinates (r_{12}, θ, ϕ), with the position of atom 1 as the origin, for the integration over the position of atom 2. Combination of eqs. (8.4-5), (8.4-11), (8.5-4), and (8.5-5) yields finally for $B(T)$

$$B(T) = 2\pi \int_0^{r_{12}(\text{max})} (1 - e^{-U(r_{12})/k_B T}) r_{12}^2 \, dr_{12}$$

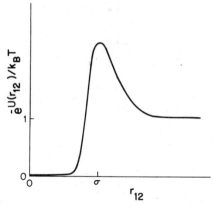

FIGURE 8.7. Typical behavior of the function $e^{-U(r_{12})/k_BT}$ for a pair of atoms.

Since the dimensions of the vessel $[r_{12}(\text{max})]$ are large compared to atomic distances and since $e^{-U(r_{12})/k_BT}$ should approach 1 for large r_{12}, no significant error is made in replacing $r_{12}(\text{max})$ by ∞ (Fig. 8.7). Dropping the subscript on r_{12}, we have finally

$$B(T) = 2\pi \int_0^{\infty} \left(1 - e^{-U(r)/k_BT}\right) r^2 \, dr \qquad (8.5\text{-}6)$$

We note that this formula is valid only for atoms with spherically symmetric interactions. If one wishes to treat diatomic molecules, for example, it is necessary to return to eq. (8.3-6) and include the internal degrees of freedom explicitly.

The third virial coefficient $C(T)$ is a bit trickier to work out, and we indicate it here only for atoms with spherically symmetric interactions. From eqs. (8.4-5) and (8.4-11), $C(T)$ is given by

$$
\begin{aligned}
C(T) &= \left[4b_2^2\left(\frac{V}{Q_1}\right) - 2b_3\right]\left(\frac{V}{Q_1}\right)^3 \\
&= \frac{2V^2}{Q_1^3}\left(\frac{2Q_2^2}{Q_1} - Q_1 Q_2 + \frac{1}{6}Q_1^3 - Q_3\right) \qquad (8.5\text{-}7)
\end{aligned}
$$

An explicit formula can be obtained by recasting each of the terms in brackets as integrals with the same volume element $d^3r_1 \, d^3r_2 \, d^3r_3$. This permits the integrands of the various integrals to be combined. Make the definitions

$x = e^{-U(r_{12})/k_BT}$, $y = e^{-U(r_{13})/k_BT}$, and $z = e^{-U(r_{23})/k_BT}$. Then we have

$$-Q_3 = -\frac{1}{6\Lambda^9} \iiint xyz \, d^3r_1 \, d^3r_2 \, d^3r_3 \tag{8.5-8a}$$

$$\frac{Q_1^3}{6} = \frac{V^3}{6\Lambda^9} = \frac{1}{6\Lambda^9} \iiint d^3r_1 \, d^3r_2 \, d^3r_3 \tag{8.5-8b}$$

$$-Q_1Q_2 = -\frac{V}{2\Lambda^9} \iint x \, d^3r_1 \, d^3r_2$$

$$= -\frac{1}{2\Lambda^9} \iiint x \, d^3r_1 \, d^3r_2 \, d^3r_3$$

$$= -\frac{1}{6\Lambda^9} \iiint (x + y + z) \, d^3r_1 \, d^3r_2 \, d^3r_3 \tag{8.5-8c}$$

$$\frac{2Q_2^2}{Q_1} = \frac{1}{2V\Lambda^9} \iint x \, d^3r_1 \, d^3r_2 \iint x \, d^3r_1 \, d^3r_2$$

$$= \frac{1}{6\Lambda^9} \iiint (xy + xz + yz) \, d^3r_1 \, d^3r_2 \, d^3r_3 \tag{8.5-8d}$$

Addition of eqs. (8.5-8) and substitution into eq. (8.5-7) yields

$$C(T) = \frac{-1}{3V} \iiint [xyz - (xy + xz + yz)$$

$$+ (x + y + z) - 1] \, d^3r_1 \, d^3r_2 \, d^3r_3 \tag{8.5-9}$$

Careful inspection reveals that the expression in brackets in eq. (8.5-9) is factorable. The final result is

$$C(T) = \frac{-1}{3V} \iiint (e^{-U(r_{12})/k_BT} - 1)(e^{-U(r_{13})/k_BT} - 1)$$

$$\times (e^{-U(r_{23})/k_BT} - 1) \, d^3r_1 \, d^3r_2 \, d^3r_3 \tag{8.5-10}$$

Although this expression looks formidable, it can be reduced to easily evaluable power series in T^{-1} for some common intermolecular potentials (Alder and Pople, 1957).

Let us now return to eq. (8.5-6) and evaluate the second virial coefficient for various model potentials. The simplest is the so-called **hard-sphere (HS) potential**,

$$U_{HS}(r_{12}) = \begin{cases} \infty & r_{12} < \sigma_D \\ 0 & r_{12} > \sigma_D \end{cases}$$

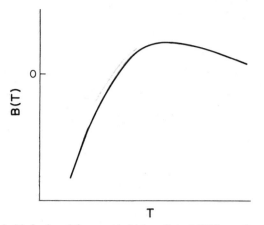

FIGURE 8.8. Typical behavior of the second virial coefficient $B(T)$ as a function of temperature.

where σ_D is the diameter of the sphere. Substitution of U_{HS} into eq. (8.5-6) yields

$$B(T) = 2\pi \int_0^{\sigma_D} r^2 \, dr$$

$$= \frac{2}{3}\pi\sigma_D^3 \tag{8.5-11}$$

With the hard-sphere potential, $B(T)$ is therefore temperature independent. Typically, however, a plot of $B(T)$ versus T for a real gas has the shape of the curve in Fig. 8.8. Since $B(T)$ flattens out at high temperature, we might expect eq. (8.5-11) to agree with the high-T limit of $B(T)$. Using $\sigma_D = 3.30$ Å for the hard-sphere diameter of N_2, as estimated by the relation $b = 16\pi\sigma_D^3 N_0/3 = \tilde{V}_{cr}/2$ from eq. (8.1-5), we calculate $B_{hi-T} = 45$ cm³ mol⁻¹; the experimental value at 700 K is 24 cm⁻³ mol⁻¹. Clearly, the agreement is poor.

The **square-well (SW) potential** is more realistic since it contains both attractive and repulsive regions (Fig. 8.9). It is expressed as

$$U_{SW}(r_{12}) = \begin{cases} \infty & r_{12} < \sigma_D \\ -\varepsilon & \sigma_D < r_{12} < d\sigma_D \\ 0 & r_{12} > d\sigma_D \end{cases} \tag{8.5-12}$$

Substitution of U_{SW} into eq. (8.5-6) now gives

$$B(T) = \frac{2\pi\sigma_D^3}{3}\left[e^{\varepsilon/k_B T}(1 - d^3) + d\right] \tag{8.5-13}$$

Table 8.4 shows a comparison of the $B(T)$ calculated from eq. (8.5-13) and

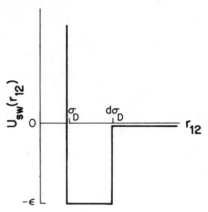

FIGURE 8.9. The square-well potential.

Table 8.4. Square-Well Second Virial Coefficients for Methane

T (K)	273	290	300	320	340	373	400	450	500
$B(T)_{\text{calc}}{}^{a}$ (cm³ mol⁻¹)	-53.6^{c}	-46.2	-42.3	-35.4	-29.4^{c}	-21.3	-15.7	-7.5	-1.10^{c}
$B(T)_{\text{exp}}{}^{b}$ (cm³ mol⁻¹)	-53.6	-46.3	-42.3	-35.4	-29.4	-21.1	-15.7	-7.4	-1.1

[a] Calculated from eq. (8.5-13) with σ_D = 3.378 Å, ε/k_B = 131 K, d = 1.6404.
[b] D. E. Gray (Ed.), *American Institute of Physics Handbook*, 3rd ed., McGraw-Hill, New York, 1972, p. 4-217.
[c] These temperatures were used for the calibration points.

experimental data for methane. The parameters σ_D, ε, and d were determined by requiring expression (8.5-13) to agree with experiment at three temperatures. The agreement at other temperatures is excellent, so we conclude that the square-well model is able to account partially for the role of the attractive forces. However, we note that eq. (8.5-13) cannot predict a maximum in $B(T)$ in any case (see Fig. 8.8) and so in this regard the square-well model is defective.

Lennard-Jones (LJ) suggested in 1924 that a still more realistic potential is given by the two-term expression (Jones, 1924)

$$U_{\text{LJ}}(r_{12}) = 4\varepsilon\left[\left(\frac{\sigma}{r_{12}}\right)^{12} - \left(\frac{\sigma}{r_{12}}\right)^{6}\right] \qquad (8.5\text{-}14)$$

The parameters ε, σ have the meaning shown in Fig. 8.10. It should be noted that, whereas σ_D in eq. (8.5-13) can clearly be interpreted as a hard-sphere diameter, the σ in eq. (8.5-14) cannot. In fact, we might expect values of σ to

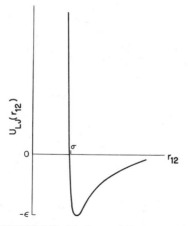

FIGURE 8.10. The Lennard-Jones potential.

**Table 8.5.   Comparison of Square-Well Hard-Sphere Diameters (σ_D)
and the Lennard-Jones σ Parameters**

Substance	σ_D (Å)[a]	σ (Å)[b]
Water	2.61	2.71[d]
Ammonia	2.90	3.22[d]
Argon	3.16	3.47
Nitrogen	3.30	3.74
Methane	3.45[c]	3.79
Butane	4.81	5.41
Carbon tetrachloride	4.59[c]	5.61
Neopentane	5.25[c]	5.76

[a] From analysis of second virial coefficient data using a square-well potential; from J. O. Hirshfelder, C. F. Curtiss, and R. B. Bird, *The Molecular Theory of Gases and Liquids*, Wiley, New York, 1954, p. 160, except where noted.
[b] F. M. Mourits and F. H. A. Rummens, *Can. J. Chem.*, **55**, 3007 (1977).
[c] From the best fit data of Table II in A. F. Collings and I. L. McLaughlin, *J. Chem. Phys.*, **73**, 3390 (1980).
[d] Stockmayer constants.

be a few tenths of an angstrom larger than corresponding values of σ_D. This is indicated clearly in Table 8.5.

$B(T)$ for the Lennard-Jones potential cannot be evaluated analytically, although it can be given as a power series expansion (Hirschfelder et al., 1954). Here we take a simpler, numerical approach. Suppose σ, ε have been determined in advance in some independent manner. For example, let us take for methane the tabulated values $\sigma = 3.783$ Å and $\varepsilon/k_B = 148.9$ K. Assume $T = 273$ K. Then the integral involved in $B(T)$ can be split into three contributions:

$$\int_0^\infty \left(1 - \exp\left\{-\frac{4\varepsilon}{k_BT}\left[\left(\frac{\sigma}{r}\right)^{12} - \left(\frac{\sigma}{r}\right)^6\right]\right\}\right) r^2\,dr$$

$$= \int_0^3 (\cdots)r^2\,dr + \int_3^7 (\cdots)r^2\,dr + \int_7^\infty (\cdots)r^2\,dr$$

$$\quad\quad\quad\quad \text{I} \quad\quad\quad\quad\quad\quad \text{II} \quad\quad\quad\quad\quad\quad \text{III}$$

For the parameters chosen the exponential is nearly zero in the range $0 < r < 3$. Accordingly, we approximate the first integral as

$$I = \int_0^3 r^2\,dr = 9.00 \text{ Å}^3$$

In integral III the 12th power term is much smaller than the 6th power term; because the 6th power term is sufficiently small, the exponential e^x is closely approximated by $1 + x$. Hence, we have

$$\text{III} = \int_7^\infty \frac{-4\varepsilon}{k_BT}\left(\frac{\sigma}{r}\right)^6 r^2\,dr = \frac{-4(148.9)}{273}(3.783)^6 \int_7^\infty \frac{dr}{r^4}$$

$$= -6.214 \text{ Å}^3$$

Integral II must be handled numerically, but this is straightforward since the range of integration is small. Choosing 12 evenly spaced intervals of length $\frac{1}{3}$ and applying Simpson's Rule (see Section 7.9) we obtain

$$\text{II} = \int_3^7 \left(1 - \exp\left\{\frac{-4\varepsilon}{k_BT}\left[\left(\frac{\sigma}{r}\right)^{12} - \left(\frac{\sigma}{r}\right)^6\right]\right\}\right) r^2\,dr$$

$$= -17.038 \text{ Å}^3$$

Summing the three contributions, we obtain finally

$$B(273) = 2\pi(-17.038 - 6.214 + 9.000) \text{ Å}^3 \times 10^{-24} \text{ cm}^3 \text{ Å}^{-3}$$

$$\times 6.022 \times 10^{23} \text{ mol}^{-1}$$

$$= -53.9 \text{ cm}^3 \text{ mol}^{-1}$$

in good agreement with experiment (see Table 8.4).

Actually, eq. (8.5-6), together with accurate experimental data for $B(T)$, can be used to work out "best values" of the Lennard-Jones parameters. Such values are quite sensitive to the accuracy of the data, and least-squares fitting is necessary in order not to place undue emphasis on a few arbitrarily chosen data points. Some results are given in Table 8.6. Alternatively, it is possible to arrive at "best values" of Lennard-Jones parameters by analysis of tempera-

Table 8.6.   Lennard-Jones Parameters for Some Nonpolar Gases

	Ar	N_2	CH_4	Kr	CO_2	CF_4	Neopentane	C_6H_6
σ (Å)[a]	3.504	3.745	3.783	3.827	4.328	4.744	7.445	8.569
ε/k_B (K)[a]	117.7	95.2	148.9	164.0	198.2	151.5	232.5	242.7

[a]A. E. Sherwood and J. M. Prausnitz, *J. Chem. Phys.*, **41**, 429 (1964); these parameters have been obtained from virial coefficient data.

ture-dependent gas viscosities. These values generally differ, sometimes considerably, from corresponding values obtained from second virial coefficient data (Mourits and Rummens, 1977). Such disparities indicate that the Lennard-Jones potential, while an improvement over the square well, is still an oversimplification of the true interaction between two nonpolar molecules. Many other potential-energy functions have been proposed; over two dozen are in current use in the literature (Fitts, 1966).

Exercises

8.8. Prepare carefully on graph paper a figure similar to Fig. 8.6(a), but with n_x, n_y extending to 20. Draw the quarter circle with $R = 20$. Compare the number of lattice points for which $n_x^2 + n_y^2 \le 400$, $n_x > 0$, $n_y > 0$, with the area of the circle.

8.9.[C] Write a computer program that accomplishes the same task as in Exercise 8.8, but now let $n_x\, n_y$ extend to 200. Let the circle have radius $R = 200$, and then determine the number of lattice points for which $n_x^2 + n_y^2 \le 40{,}000$, $n_x > 0$, $n_y > 0$.

8.10. Deduce the expression for the *classical* molecular partition function of a one-dimensional harmonic oscillator and show that this agrees with the quantum mechanical result in the limit of high temperature.

8.11.* Refer to eq. (1.3-4), which gives the Lagrangian for the relative motion of two particles, and to eq. (1.2-14), which gives the conjugate generalized momenta. For a rigid rotor show that the rotational Hamiltonian is $(p_\theta^2 + p_\phi^2 \csc^2\theta)/2I$. Then write out, as completely as possible, the canonical partition functions $Q_1(T, V)$ and $Q_2(T, V)$ for systems of 1 and 2 rigid rotors.

8.12. Verify the form of eq. (8.5-8d).

8.13. In the sequence of coefficients in eq. (8.4-5), find the next member, b_4. Then express the fourth virial coefficient, $D(T)$, in terms of the b_n's.

8.14. As stated in the text in connection with eq. (8.4-12), establish the relations between $B(T)$, $C(T)$, and $B'(T)$ and $C'(T)$.

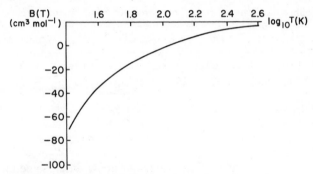

FIGURE 8.11. Plot of $B(T)$ versus $\log_{10} T$ for molecular hydrogen.

8.15. Calculate $B(T)$ for methane at various temperatures by using the Dieterici equation, and compare with the data in Table 8.4.

8.16. Compare the hard-sphere diameters deduced from the Dieterici equation for the molecules of Table 8.5 with the diameters obtained from $B(T)$ data analyzed with a square-well potential.

8.17. Figure 8.11 shows the plot of $B(T)$ versus $\log_{10} T$ for molecular hydrogen. Estimate the Boyle temperature for hydrogen.

8.18. A helium–helium interatomic potential (Slater and Kirkwood, 1931) has the analytical form

$$U(r) = \left\{ 7.70 e^{-2.43 r / a_0} - 0.680 r^{-6} a_0^6 \right\} \times 10^{-10} \text{ erg}$$

where a_0 is the Bohr radius. Compute the potential energy in units of kilojoules per mole of pairs of helium in the most favorable configuration.

8.19. Consider the infinite tail in the integral of eq. (8.5-6):

$$2\pi \int_R^\infty \left(1 - e^{-U(r)/k_B T} \right) r^2 \, dr$$

where R is large. Show that $U(r)$ must decay more rapidly than r^{-3} if $B(T)$ is to exist. Do the square-well and the Lennard-Jones potentials satisfy this restriction?

8.20. The **Sutherland potential** represents an extreme case of a Lennard-Jones potential in which the repulsive term is infinitely strong.

$$U(r) = \begin{cases} -A r^{-6} & r > \sigma \\ \infty & r < \sigma \end{cases}$$

Show that the second virial coefficient is given by

$$B(T) = \frac{-2\pi\sigma^3}{3} \sum_{n=0}^{\infty} \frac{1}{n!}\left(\frac{1}{2n-1}\right)\left(\frac{A}{k_B T\sigma^6}\right)^n$$

8.21. The following values are reported for methane (see footnote b in Table 8.4)

T (K)	$T(dB/dT)$ (cm^3 mol^{-1})	C(cm^6 mol^{-2})
273	129	2900
340	94	2200
400	76	1800
600	45	1300

Calculate the quantities in the second and third columns using the Dieterici equation of state.

8.22.[C] The following values are recommended for the second virial coefficient of nitrogen (Dymond and Smith, 1980):

T (K)	$B(T)$ (cm^3 mol^{-1})	T (K)	$B(T)$ (cm^3 mol^{-1})
75	-275	200	-35.2
80	-243	250	-16.2
90	-197	300	-4.2
100	-160	400	$+9.0$
110	-132	500	16.9
125	-104	600	21.3
150	-71.5	700	24.0

Write a computer program to calculate $B(T)$ from a Lennard-Jones potential, apply it to N_2 using the parameters in Table 8.6, and compare with the experimental data listed here.

8.23.[C] Write a computer program to evaluate $B(T)$ from the potential of Exercise 8.20. Limit the summation to just 10 terms, and apply the program to methane at various temperatures, using the same value of σ as for the Lennard-Jones potential. Let the A parameter be $376\ k_B\sigma^6$. Is the Sutherland potential defective in the same way as is the square-well potential? What does your answer suggest about the potential?

8.24. The nonideality of a gas should be manifest in its thermodynamic properties. A mole of a real gas and a mole of the corresponding hypothetical ideal gas are in separate containers of volume $\tilde{V}$ and at a temperature T. Prove that if only the second virial coefficient is significant, then $\tilde{S}_{real} - \tilde{S}_{id} = R[B + T(dB/dT)]/\tilde{V}$. Estimate this for methane at 273 K.

8.25. First prove the thermodynamic identity

$$\mu_{JT}\tilde{C}_p = T\left(\frac{\partial \tilde{V}}{\partial T}\right)_P - \tilde{V}$$

where μ_{JT} is the Joule–Thomson coefficient, $\mu_{JT} = (\partial T/\partial P)_H$. Then show that in the limit of zero pressure one has

$$\mu_{JT}^\circ = \lim_{P\to 0} \mu_{JT} = \frac{T\dfrac{d\tilde{B}}{dT} - \tilde{B}}{\tilde{C}_p^\circ}$$

where $\tilde{B}$ is the molar virial coefficient of eq. (8.4-11) and $\tilde{C}_p^\circ$ is the zero-pressure value of the molar heat capacity.

8.26.[C] Apply the result of Exercise 8.25 to the case of argon at 273 K, for which $\tilde{C}_p^\circ = 20.8$ J K^{-1} mol^{-1}. Calculate μ_{JT} and compare with the experimental value of 0.437 K atm^{-1} (Rybolt, 1981). Do this problem, first using the parameters in Table 8.6, and second using the Dieterici equation of state.

8.27.* Walter Stockmayer proposed that the interaction between two *polar* molecules be described by a potential function that consists of a Lennard-Jones potential plus a term representing the interaction of point dipoles embedded at the centers of the two molecules (Stockmayer, 1941). His potential has the form

$$U(r) = 4\varepsilon\left[\left(\frac{\sigma}{r}\right)^{12} - \left(\frac{\sigma}{r}\right)^6\right] - \frac{\mu^2}{4\pi\varepsilon_0 r^3}\xi(\theta_1, \theta_2, \phi)$$

where μ is the dipole moment of the molecule, ε_0 is the permittivity of vacuum, and ξ is defined as (see Fig. 8.12):

$$\xi(\theta_1, \theta_2, \phi) = 2\cos\theta_1\cos\theta_2 - \sin\theta_1\sin\theta_2\cos\phi$$

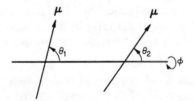

FIGURE 8.12. Definition of angles for use in the Stockmayer potential.

Table 8.7. Second Virial Coefficients for CH_3Cl as Computed from the Stockmayer Potential

T (K)	239	250	255	283	311	366	422	450
$B(T)_{calc}{}^a$ $(cm^3\ mol^{-1})$	-783	-681	-639	-471	-374	-256	-183	-159
$B(T)_{exp}{}^b$ $(cm^3\ mol^{-1})$	-764	-668	-637	-500	-401	-265	-184	-155

[a] From Table II-A in J. O. Hirschfelder, C. F. Curtiss, and R. B. Bird, *Molecular Theory of Gases and Liquids*, Wiley, New York, 1954, p. 1147. Interpolations between values of the parameters T^* were done graphically. Values of the potential parameters employed were $\mu = 6.24 \times 10^{-30}$ C m, $\sigma = 3.399 \times 10^{-10}$ m, $\varepsilon/k_B = 380$ K.
[b] Data taken from Landolt-Börnstein, *Zahlenwerte und Funktionen*, 6th ed., Vol. II, Part 1, Springer-Verlag, Berlin, 1971, p. 259.

From the classical theory of electromagnetic fields, the interaction between two point dipoles separated by a distance **r** is

$$E = \frac{3(\mu \cdot r)(\mu \cdot r)}{4\pi\varepsilon_0 r^5} - \frac{\mu \cdot \mu}{4\pi\varepsilon_0 r^3}$$

Show that this leads to the expression for the dipolar part of the Stockmayer potential. The **Stockmayer potential** may be inserted into eq. (8.3-6) and with some labor integrated over all r and all angular orientations. Table 8.7 illustrates how well the Stockmayer potential works for a given choice of σ and ε for the polar substance chloromethane (CH_3Cl).

8.28. The grand canonical partition function has many other applications besides the derivation of the virial equation. One such application is the deduction of the **Brunauer–Emmett–Teller isotherm** for the adsorption of vapors by solid surfaces. An interesting discussion is given in a short monograph (Guggenheim, 1966). Read the pertinent pages there and write out for your own clarification the derivation of the BET isotherm, which we may express as

$$\frac{P}{V}\left(\frac{P^0}{P^0 - P}\right) = \frac{P}{V_m} + \frac{P_h}{V_m}$$

Here, the parameter P_h is a pressure (independent of P) characteristic of the surface, gas and temperature, and the parameter V_m is a pressure- and temperature-independent volume characteristic of the surface and gas. Treat the following data for the adsorption of krypton on the mineral barium muscovite at 77 K ($P^0 = 2.63$ Torr), and deduce values of P_h and V_m.

P (Torr)	V (cm³ at STP)
1.098	0.571
1.034	0.546
0.970	0.519
0.880	0.492
0.798	0.462
0.702	0.435
0.550	0.405
0.368	0.372

8.6   DISTRIBUTION FUNCTIONS

The convergence of the virial series eq. (8.4-11) depends upon ρ not being too large. It is generally believed (no "proof" is known) that this series fails to converge at densities well below those of typical liquids. The concept of **distribution functions**, pioneered principally by the American physical chemist John G. Kirkwood (1907–1959), provides an alternative to the virial expansion that avoids the problem of convergence. In the modern theory of fluids distribution functions play a central role. We continue to assume that the classical limit provides an adequate description for our purposes. By analogy to eq. (6.4-15), we can then express the probability of observing a system of n molecules to be in a state where molecule 1 is in phase element $d^3p_1 d^3q_1$ at (p_1, q_1), molecule 2 is in phase element $d^3p_2 d^3q_2$ at (p_2, q_2), and so on, as

$$\frac{e^{-E_i/k_B T}}{Q} \rightarrow \frac{e^{-H(p,q)/k_B T} \prod_{i=1}^{n} d^3p_i \prod_{j=1}^{n} d^3q_j}{\int \cdots \int e^{-H(p,q)/k_B T} \prod_{i=1}^{n} d^3p_i \prod_{j=1}^{n} d^3q_j} \qquad (8.6\text{-}1)$$

irrespective of the internal quantum state (vibrational, electronic) of the system. If the momentum distribution is of no interest to us, we can integrate over the momenta and from eq. (8.5-2) write

$$f^{(n)}(\mathbf{r}_1, \mathbf{r}_2, \ldots, \mathbf{r}_n) = \frac{e^{-U(\mathbf{r}_1, \mathbf{r}_2, \ldots, \mathbf{r}_n)/k_B T} \prod_{i=1}^{n} d^3 r_i}{Z} \qquad (8.6\text{-}2)$$

for the probability of finding molecule 1 in $d^3 r_1$, simultaneously molecule 2 in $d^3 r_2$, and so on. We refer to $f^{(n)}$ as the **nth order specific configurational** distribution function ["specific," because particular molecules are assigned to the volume elements $d^3 r_i$ in eq. (8.6-2)].

Lower-order distribution functions can be obtained by integrating over the coordinates of all but a number k of molecules. The **kth-order specific configurational distribution function** for a system of n molecules is then defined to be $(k < n)$

$$f^{(k)}(\mathbf{r}_1, \mathbf{r}_2, \ldots, \mathbf{r}_k) = \frac{\int \cdots \int e^{-U(\mathbf{r}_1, \mathbf{r}_2, \ldots, \mathbf{r}_n)/k_B T} \prod_{i=k+1}^{n} d^3 r_i}{Z} \qquad (8.6\text{-}3)$$

Note that $f^{(k)}$ is a function of only k coordinates, but the integrand is a function of all n coordinates. The meaning of $f^{(k)}(\mathbf{r}_1, \mathbf{r}_2, \ldots, \mathbf{r}_k)$ is that $f^{(k)} \prod_{i=1}^{k} d^3 r_i$ is the joint probability of finding molecule 1 in volume element $d^3 r_1$ at $\mathbf{r}_1$, molecule 2 in volume element $d^3 r_2$ at $\mathbf{r}_2, \ldots$, and molecule k in volume element $d^3 r_k$ at $\mathbf{r}_k$, irrespective of the configuration of the remaining $(n - k)$ molecules.

This joint probability has no physical meaning, however, because no experiment can distinguish among identical molecules. We can define a distribution function $p^{(k)}(\mathbf{r}_1, \mathbf{r}_2, \ldots, \mathbf{r}_k)$, however, that gives the probability *density* for simultaneously finding *any* molecule in $d^3 r_1$, any of the remaining $n - 1$ molecules in $d^3 r_2, \ldots$, and any of the remaining $(n - k + 1)$ molecules in $d^3 r_k$.

$$p^{(k)}(\mathbf{r}_1, \mathbf{r}_2, \ldots, \mathbf{r}_k) = \frac{n!}{(n - k)!} \frac{\int \cdots \int e^{-U(\mathbf{r}_1, \mathbf{r}_2, \ldots, \mathbf{r}_n)/k_B T} \prod_{i=k+1}^{n} d^3 r_i}{Z} \qquad (8.6\text{-}4)$$

The combinatorial factor is the number of permutations of n molecules taken k at a time. The function $p^{(k)}$ is referred to as the **kth-order generic configurational distribution function**.

Except for points close to the wall of the vessel, the probability of finding a given molecule must be the same at all locations in a homogeneous fluid. Normalization of eq. (8.6-3) requires that

$$\int f^{(1)}(\mathbf{r}_1) \, d^3 r_1 = 1$$

and if $f^{(1)}(\mathbf{r}_1)$ is constant, then it follows that $f^{(1)}(\mathbf{r}_1) = 1/V$, and from eq. (8.6-4) then $p^{(1)}(\mathbf{r}_1) = n/V = \rho_0$, the bulk **number density** of the fluid.

Next, consider the second-order generic configurational distribution function. If intermolecular forces were negligible, for example, if the two volume

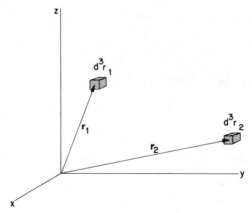

FIGURE 8.13. Two point particles in distantly separated volume elements show little correlation.

elements were very far apart (Fig. 8.13), then one would have

$$p^{(2)}(\mathbf{r}_1, \mathbf{r}_2) = \frac{n!}{(n-2)!} \frac{V^{n-2}}{V^n}$$

$$= \left(\frac{n}{V}\right) \frac{n-1}{V}$$

$$\simeq \rho_0^2$$

This is what one would expect since without forces the molecules are randomly distributed and the chances of finding two molecules somewhere is just the product of the one–particle distribution functions. However, intermolecular forces do operate and the positions of the molecules are correlated. It is useful then to define a dimensionless **k-body correlation function** $g^{(k)}(\mathbf{r}_1, \mathbf{r}_2, \ldots, \mathbf{r}_k)$ as

$$g^{(k)}(\mathbf{r}_1, \mathbf{r}_2, \ldots, \mathbf{r}_k) = \frac{1}{\rho_0^k} p^{(k)}(\mathbf{r}_1, \mathbf{r}_2, \ldots, \mathbf{r}_k) \tag{8.6-5}$$

This becomes unity for a purely random distribution. In the case of the two-body or pair correlation function, $g^{(2)}(\mathbf{r}_1, \mathbf{r}_2)$, we have explicitly

$$g^{(2)}(\mathbf{r}_1, \mathbf{r}_2) = \frac{n(n-1)}{\rho_0^2} \frac{\int \cdots \int e^{-U(\mathbf{r}_1, \mathbf{r}_2, \ldots, \mathbf{r}_n)/k_B T} \prod_{i=3}^{n} d^3 r_i}{Z} \tag{8.6-6}$$

from which it follows that

$$\iint g^{(2)}(\mathbf{r}_1, \mathbf{r}_2) \, d^3 r_1 \, d^3 r_2 = \frac{n(n-1)}{\rho_0^2}$$

$$\simeq V^2 \tag{8.6-7}$$

For those homogeneous liquids in which the potential energy function is spherically symmetric, $g^{(2)}(\mathbf{r}_1, \mathbf{r}_2)$ depends only on the scalar distance between the volume elements. Let us, therefore, rename $g^{(2)}(\mathbf{r}_1, \mathbf{r}_2)$ as $g(r_{12})$. Changing to new coordinates, as in the evaluation of $Q_2(T, V)$ in eq. (8.5-5) and writing the volume element as $d^3r_1\, d^3r_{12}$, we have

$$\left(\int d^3r_1\right) g(r_{12})\, d^3r_{12} = V g(r_{12})\, d^3r_{12}$$

and so from eq. (8.6-7) it follows that

$$\int g(r_{12})\, d^3r_{12} = V$$

Since molecule 1 is taken as the origin, r_{12} is the radius vector to molecule 2 and $d^3r_{12} = 4\pi r_{12}^2\, dr_{12}$. The function $4\pi r_{12}^2 g(r_{12})$ is then termed the **radial distribution function**. Accordingly, $4\pi\rho_0 r_{12}^2 g(r_{12})\, dr_{12}$ is the number of molecules lying in a thin spherical shell at a distance r_{12} from a given molecule, $\rho_0 g(r_{12})$ is the "local" density of molecules 2 at a distance r_{12} from reference molecule 1.

We now examine semiquantitatively the form of the pair correlation function. For a very dilute gas the influence of other particles on a given pair (say, 1 and 2) can be neglected. Then from eq. (8.6-6) we have, after approximating Z by that for an ideal gas (i.e., V^n),

$$g^{(2)}(\mathbf{r}_1, \mathbf{r}_2) \simeq \frac{n(n-1)}{\rho_0^2} \frac{e^{-U(\mathbf{r}_1, \mathbf{r}_2)/k_B T} \int \cdots \int \prod_{i=3}^{n} d^3r_i}{Z}$$

or

$$g(r_{12}) \simeq \frac{n(n-1)}{\rho_0^2} \frac{e^{-U(r_{12})/k_B T} V^{n-2}}{V^n}$$

$$= e^{-U(r_{12})/k_B T}$$

This is shown in Fig. 8.14(a) for the Lennard-Jones potential.

On the other hand, for an ideal solid composed of atoms "frozen" at the lattice points, one has

$$g(r_{12}) = \frac{\text{sum of neighbors in all shells of infinitesimal thickness}}{4\pi r_{12}^2 \rho_0}$$

$$= \sum_i z_i \frac{\delta(r_{12} - R_i)}{4\pi r_{12}^2 \rho_0} \tag{8.6-8}$$

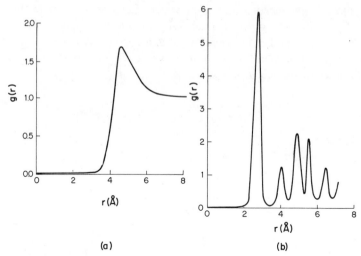

FIGURE 8.14. Schematic plots of the pair distribution function for (*a*) a very dilute gas at 298 K with the Lennard-Jones parameters of methane ($\sigma = 3.783$ Å, $\varepsilon/k_B = 148.9$ K), and (*b*) a face-centered cubic crystal with the size and density of platinum ($a = 3.932$ Å, $\rho_0 = 21.45$ g/cm³) and each delta function represented by the Gaussian function $(25/\pi)^{1/2}\exp[-25(r - R_i)^2]$.

where z_1 is the number of nearest neighbors at a distance R_1 from the reference atom, z_2 is the number of next nearest neighbors at a distance R_2, and so on, and $\delta(r_{12} - R_i)$ is a Dirac delta function. From a rigorous viewpoint, the expression (8.6-8) has meaning only when it is integrated over the argument r_{12} (see Section 2.5). However, eq. (8.6-8) can be pictured loosely as a sequence of unevenly spaced, infinitely high spikes of zero width. Since the atoms of a solid are not perfectly at rest, even at 0 K, we may obtain a truer picture of $g(r_{12})$ by replacing the delta functions in eq. (8.6-8) by Gaussian functions, as in eq. (2.5-9). The result is shown in Fig. 8.14(*b*).

By pictorial interpolation we guess that $g(r_{12})$ for the liquid has broader peaks than for the solid and that the intensities of the peaks fall off rapidly with increasing r_{12}. Note that $g(0) = 0$ because there is zero probability of two molecules occupying the same position. As $r_{12} \to \infty$, $g(r_{12}) \to 1$, since $U(r_{12}) \to 0$ and there is no longer any correlation between the molecules. In a reasonably dense liquid there is a first coordination shell of molecules about a given molecule and the first maximum in $g(r_{12})$ corresponds physically to this.

8.7   EXPERIMENTAL DETERMINATION OF $g(r)$

It is fortunate that the pair correlation function $g(r)$ can be determined indirectly by experimental means. The technique most commonly used is X-ray diffraction (Gingrich, 1943; Karnicky and Pings, 1976), and more recently neutron diffraction (Bosi et al., 1980). We discuss only X-ray diffraction (Green, 1969; Kohler, 1972).

For our purposes, a classical picture of X-ray scattering is sufficient, although quantum theory leads to modification of certain details (Compton and Allison, 1935). When a beam of X-rays passes through a substance, the electrons are set into vibration and then radiate X-rays in all directions. The radiation emitted by these electrons is called **scattered** or **secondary radiation**. If the incident X-rays are soft (i.e., if their wavelength is not too small), the secondary radiation is of essentially the same wavelength as the incident radiation and is said to be scattered **coherently**.

Consider a plane electromagnetic wave (beam of X-rays) propagating in the $+x$ direction and having an associated electric field **E** pointing in the y direction. Let the wave be incident upon an electron of charge $-e$ [Fig. 8.15(a)]. The wave can be represented by the function

$$\mathbf{E} = \mathbf{e}_y E_0 \cos(\omega t - kx) \tag{8.7-1}$$

where $\mathbf{e}_y$ is a unit vector in the $+y$ direction, E_0 is the **amplitude** of the wave, $\omega = 2\pi c/\lambda$ is the angular frequency of the wave, x is a plane relative to an arbitrary origin, and $k = 2\pi/\lambda$ is the wave number.

If the electron is located at the origin, it experiences a fluctuating electric field $E_y = E_0 \cos \omega t$. The acceleration (in the y direction) of the electron is given by Newton's Second Law (Section 1.1) as

$$a = \frac{-eE_0 \cos \omega t}{m}$$

and this fluctuates also. According to classical electromagnetic theory, an accelerated charge radiates an electromagnetic wave, the *amplitude* of whose associated electric field at a point of observation **R** is

$$E_\phi = \left| \frac{ea \sin \phi}{4\pi\varepsilon_0 c^2 R} \right|$$

$$= \frac{e^2 E_0 |\sin \phi|}{4\pi\varepsilon_0 mc^2 R} \tag{8.7-2}$$

Here, ϕ is the angle between the y axis and **R** [Fig. 8.15(a)].

The **intensity of radiation**, I, at a point where the electric field is **E** is defined to be the time average $I = c\varepsilon_0 \langle \mathbf{E} \cdot \mathbf{E} \rangle$. The scattered radiation is of the form $\mathbf{E}_\phi = \mathbf{1} E_\phi \cos(\omega t - \beta)$, where β is a constant, **1** is a unit vector as shown

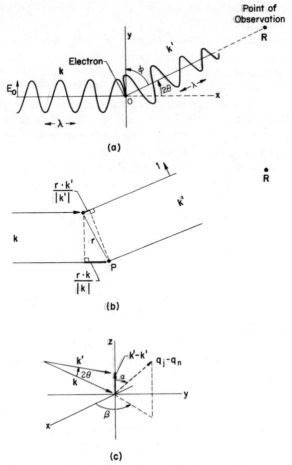

FIGURE 8.15. Coherent scattering of a plane electromagnetic wave (a) at one center, (b) at two centers, and (c) showing relationship between the angles 2θ, α, β.

in Fig. 8.15(b). Over one period $T = 2\pi/\omega$ and we have

$$\langle \cos^2(\omega t - \beta) \rangle = \frac{\displaystyle\int_0^{2\pi/\omega} \cos^2(\omega t - \beta)\, dt}{\displaystyle\int_0^{2\pi/\omega} dt}$$

$$= \frac{1}{2}$$

Hence, the intensity of the scattered radiation at **R** due to the single electron at O is

$$I_\phi = \frac{1}{2}c\varepsilon_0 E_\phi^2 \tag{8.7-3}$$

It can be shown that intensity has units of $J\,m^{-2}\,s^{-1}$, that is, units of energy flux.

If Z electrons were concentrated at O, the total intensity at **R** would be Z times the value in eq. (8.7-3). However, suppose instead that we have a multielectronic atom centered at O. Consider a second electron, located at position **r** (Fig. 8.15(b)). We assume that $R \gg r$ so that the wave vectors of the scattered waves from the two electrons are essentially identical (equal to **k**′). From the geometry of the figure, it is seen that the wave incident upon P must travel the extra distance $\Delta\ell$

$$\Delta\ell = \frac{\mathbf{r}\cdot\mathbf{k}}{|\mathbf{k}|} - \frac{\mathbf{r}\cdot\mathbf{k}'}{|\mathbf{k}'|}$$

$$= \frac{\mathbf{r}\cdot(\mathbf{k}-\mathbf{k}')}{2\pi/\lambda} \tag{8.7-4}$$

where the wave vector **k** of the incident wave is parallel to the x axis. If $\mathbf{E}_\phi$ is the electric field at **R** of the wave scattered from O and $\mathbf{E}_1$ is the field at **R** of the wave scattered from P, we then have for the net electric field

$$\mathbf{E}_{tot} = \mathbf{E}_\phi + \mathbf{E}_1 = \mathbf{E}_\phi + \mathbf{E}_\phi e^{-i\mathbf{r}\cdot(\mathbf{k}-\mathbf{k}')}$$

where the exponential factor accounts for the shift in the phase of the field. Note that for convenience we have expressed the field in complex form, so that the intensity is now given by $I = c\varepsilon_0\langle\mathbf{E}\cdot\mathbf{E}^*\rangle$.

In general, for a collection of Z electrons belonging to one atom,

$$\mathbf{E}_{tot} = \sum_{j=1}^{Z}\mathbf{E}_j$$

$$= \mathbf{E}_\phi\sum_{j=1}^{Z}e^{-i\mathbf{r}_j\cdot(\mathbf{k}-\mathbf{k}')}$$

The configuration of the electrons is not stationary in time, however. Let $\rho(\mathbf{r})$ be the number density of electrons in a volume element d^3r situated at **r**. Then configurational averaging transforms the previous equation into

$$\overline{\mathbf{E}}_{tot} = \mathbf{E}_\phi\int\rho(\mathbf{r})e^{-i\mathbf{r}\cdot(\mathbf{k}-\mathbf{k}')}\,d^3r$$

$$= \mathbf{E}_\phi f \tag{8.7-5}$$

where f is termed the **atomic scattering factor** of the atom. It is a function of the scalar $|\mathbf{k} - \mathbf{k}'|$, and is tabulated for various atoms (MacGillavry and Rieck, 1968). The function $\rho(\mathbf{r})$, of course, has the normalization $\int \rho(\mathbf{r}) \, d^3r = Z$.

Now we consider that in addition to an atom at O there are several other *identical* atoms distributed about the origin. If $\mathbf{q}_j$ is the location of any such atom, then the scattered wave from this atom has an electric vector $\mathbf{E}_\phi f e^{-i\mathbf{q}_j \cdot (\mathbf{k}-\mathbf{k}')}$. For a collection of N atoms, the net electric vector $\mathbf{E}$ is

$$\mathbf{E} = \mathbf{E}_\phi f \sum_{j=1}^{N} e^{-i\mathbf{q}_j \cdot (\mathbf{k}-\mathbf{k}')}$$

The intensity at $\mathbf{R}$ is then

$$\begin{aligned}
I(\mathbf{R}) &= c\varepsilon_0 \langle \mathbf{E} \cdot \mathbf{E}^* \rangle \\
&= \frac{1}{2} c\varepsilon_0 E_\phi^2 f^2 \sum_{j=1}^{N} e^{-i\mathbf{q}_j(\mathbf{k}-\mathbf{k}')} \sum_{n=1}^{N} e^{+i\mathbf{q}_n \cdot (\mathbf{k}-\mathbf{k}')} \\
&= I_\phi f^2 \sum_{j=1}^{N} \sum_{n=1}^{N} e^{-i(\mathbf{k}-\mathbf{k}') \cdot (\mathbf{q}_j - \mathbf{q}_n)}
\end{aligned} \tag{8.7-6a}$$

We recall from eq. (8.6-4) that $p^{(2)}(\mathbf{q}_j, \mathbf{q}_n) \, d^3q_j \, d^3q_n$ is the probability of finding any atom in volume element d^3q_n and simultaneously any other atom in volume element d^3q_j. Making use of eq. (8.6-7), we can then transform eq. (8.7-6a) into the configurationally averaged quantity

$$\overline{I(\mathbf{R})} = I_\phi f^2 \left(N + \iint \rho_0^2 g^{(2)}(\mathbf{q}_j, \mathbf{q}_n) e^{-i(\mathbf{k}-\mathbf{k}') \cdot (\mathbf{q}_j - \mathbf{q}_n)} \, d^3q_j \, d^3q_n \right)$$

The term N arises from those terms in eq. (8.7-6a) where $j = n$; ρ_0 is the bulk number density of the sample. Let α be the angle between the vectors $(\mathbf{q}_j - \mathbf{q}_n)$ and $(\mathbf{k} - \mathbf{k}')$ (Fig. 8.15c). Making the usual substitutions $g^{(2)}(\mathbf{q}_j, \mathbf{q}_n) \to g(q_{jn})$, $d^3q_j \, d^3q_n \to d^3q_j \, d^3q_{jn}$ and integrating over the coordinate $\mathbf{q}_j$, we obtain

$$\overline{I(\mathbf{R})} = I_\phi f^2 N \left(1 + \int_V \rho_0 g(q_{jn}) e^{-isq_{jn}\cos\alpha} \, d^3q_{jn} \right) \tag{8.7-6b}$$

Note that we have taken the z axis of the polar coordinate system to be parallel to the difference of wave vectors $\mathbf{k} - \mathbf{k}'$. The quantity s is related to the scattering angle 2θ according to

$$\begin{aligned}
s &= |\mathbf{k} - \mathbf{k}'| \\
&= \frac{4\pi \sin\theta}{\lambda}
\end{aligned} \tag{8.7-7}$$

The volume element d^3q_{jn} can be written as $q_{jn}^2 \sin \alpha \, dq_{jn} \, d\alpha \, d\beta$. We make this substitution in eq. (8.7-6b) and integrate over the polar angles α, β to obtain

$$\overline{I(\mathbf{R})} = I_\phi f^2 N \left[1 + \int_0^{q_{max}} 4\pi q^2 \rho_0 g(q) \left(\frac{\sin sq}{sq} \right) dq \right] \qquad (8.7\text{-}8a)$$

or

$$\frac{\overline{I(\mathbf{R})}}{I_\phi N f^2} - 1 = \int_0^{q_{max}} 4\pi q^2 \rho_0 (g(q) - 1) \left(\frac{\sin sq}{sq} \right) dq$$

$$+ \int_0^{q_{max}} 4\pi q^2 \rho_0 \left(\frac{\sin sq}{sq} \right) dq \qquad (8.7\text{-}8b)$$

Equation (8.7-8a) gives a relation between the intensity $\overline{I(\mathbf{R})}$ and the pair correlation function $g(q)$; inversion of the equation would, in principle, provide $g(q)$ from knowledge of the experimental intensities. Greater accuracy is generally obtained (Paalman and Pings, 1963) if one first rearranges eq. (8.7-8a) to eq. (8.7-8b). The second integral in eq. (8.7-8b) is simply

$$\int_0^{q_{max}} 4\pi q^2 \rho_0 \left(\frac{\sin sq}{sq} \right) dq = 4\pi \rho_0 q_{max}^3 \left[\frac{\sin x - x \cos x}{x^3} \right] \qquad (x = sq_{max})$$

The limit of the expression in brackets as $x \to 0$ is $\frac{1}{3}$; for $x > 0$ this expression behaves like a damped cosine curve. The first zero of the function occurs at the point where $x = \tan x$, and is $x = 1.430\pi$. In a typical case, let $\lambda = 1.54$ Å (copper K_α radiation) and let $q_{max} = 1$ mm. Using eq. (8.7-7), we find that the function then has appreciable nonzero values only if $\theta < 0.011''$ of arc. Hence, except at the very smallest of scattering angles, the second integral in eq. (8.7-8b) may be neglected. Designating the left side of eq. (8.7-8b) by $i(s)$, we then write (with a slight change in notation)

$$si(s) = \int_0^\infty 4\pi r^2 \rho_0 [g(r) - 1] \left(\frac{\sin sr}{r} \right) dr \qquad (8.7\text{-}9)$$

where now the upper limit has been taken to be ∞, and is justified by the fact that the pair correlation function $g(r)$ approaches 1 very rapidly as r reaches a few atomic diameters.

We now invert eq. (8.7-9). According to the **Fourier Integral Theorem** (Tolstov, 1976), if the Fourier sine transform $F(\omega)$ exists,

$$F(\omega) = \sqrt{\frac{2}{\pi}} \int_0^\infty \sin(\omega t) f(t) \, dt$$

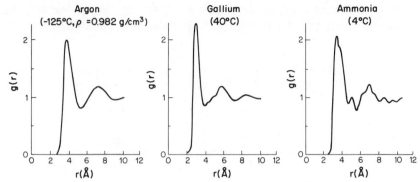

FIGURE 8.16. Pair correlation function for (a) liquid argon [from P. G. Mikolaj and C. J. Pings, *J. Chem. Phys.*, **46**, 1401 (1967)] (b) liquid gallium [from S. E. Rodriguez and C. J. Pings, *J. Chem. Phys.*, **42**, 2435 (1965)], and (c) liquid ammonia [from A. H. Narten, *J. Chem. Phys.*, **66**, 3117 (1977)] (all redrawn with permission).

then the function $f(t)$ is given by

$$f(t) = \sqrt{\frac{2}{\pi}} \int_0^\infty \sin(\omega t) F(\omega)\, d\omega$$

provided $\lim_{\omega \to \infty} F(\omega) = 0$. Applying this theorem to eq. (8.7-9), we obtain finally

$$g(r) = 1 + \frac{1}{2\pi^2 r \rho_0} \int_0^\infty si(s)\sin sr\, ds \qquad (8.7\text{-}10)$$

This relation is the basis of all modern X-ray studies of monatomic liquids. It was first used by Debye and Menke in 1930 to study the scattering of copper and molybdenum K_α radiation by mercury.

Note that eq. (8.7-10) supposes all the scattered radiation to be *coherent*. However, some scattered radiation is **incoherent** because of the Compton effect. The X-ray detector picks up both coherent and incoherent radiation, and the contribution due to the latter must be subtracted out before using eq. (8.7-10). Various approximate formulas are available for estimating the intensity of incoherently scattered radiation (James, 1965).

In Fig. 8.16 we show the experimental pair correlation functions as deduced from eq. (8.7-10) for three different liquids. The curves are essentially interpolations between the two types of curves shown in Fig. 8.14.

8.8   CONNECTION BETWEEN THE PAIR CORRELATION FUNCTION AND THERMODYNAMIC PROPERTIES

We now establish the connection between the pair correlation function $g(r)$ and thermodynamic properties of a fluid. From eqs. (6.6-9) and (8.5-2) we have for a mole of monatomic fluid

$$\tilde{E} = k_B T^2 \left(\frac{\partial \ln Q}{\partial T} \right)_{N,V}$$

$$= \frac{3}{2} RT + k_B T^2 \left(\frac{\partial \ln Z}{\partial T} \right)_{N,V}$$

where Z is the configuration integral (8.5-3). Now let us introduce the key approximation upon which everything in this section depends. The potential-energy function U is assumed to be a sum of scalar pair potentials, that is,

$$U(\mathbf{r}_1, \mathbf{r}_2, \ldots, \mathbf{r}_N) \simeq \sum_i \sum_{j>i} u(r_{ij}) \qquad (8.8\text{-}1)$$

We actually used this approximation in eq. (8.5-8a) although it was not stressed there. For gases, eq. (8.8-1) should be reliable, but at normal liquid densities there is evidence that three-body forces are important (Barker et al, 1969; Croxton, 1974).

However, if we employ eq. (8.8-1) and then differentiate under the integral sign,

$$\left(\frac{\partial \ln Z}{\partial T} \right)_{N,V} = \frac{1}{k_B T^2} \frac{\int \cdots \int \left[\sum_i \sum_{j>i} u(r_{ij}) \exp\left(-\frac{U}{k_B T} \right) \right] \prod_{i=1}^{N} d^3 r_i}{Z}$$

In the brackets there are $N(N-1)/2$ equivalent terms since the ith molecule can be chosen in N ways and the jth in $(N-1)$ ways; the factor of $\frac{1}{2}$ prevents us from counting the same interactions (e.g., U_{12} and U_{21}) twice. Therefore, we have

$$\left(\frac{\partial \ln Z}{\partial T} \right)_{N,V}$$

$$= \frac{1}{k_B T^2} \frac{N(N-1)}{2} \frac{\int\int u(r_{12}) \left[\int \cdots \int \exp\left(\frac{-U}{k_B T} \right) \prod_{i=3}^{N} d^3 r_i \right] d^3 r_1 d^3 r_2}{Z}$$

Now observe that the expression in brackets times $N(N-1)/Z$ is just the two-body generic configurational distribution function of eq. (8.6-4). Hence, we have

$$\left(\frac{\partial \ln Z}{\partial T}\right)_{N,V} = \frac{1}{2k_B T^2} \iint u(r_{12}) p^{(2)}(\mathbf{r}_1, \mathbf{r}_2)\, d^3 r_1\, d^3 r_2$$

$$= \frac{\rho_0^2}{2k_B T^2} \iint u(r_{12}) g^{(2)}(\mathbf{r}_1, \mathbf{r}_2)\, d^3 r_1\, d^3 r_2$$

$$= \frac{N_0 \rho_0}{2k_B T^2} \int 4\pi r_{12}^2 g(r_{12}) u(r_{12})\, dr_{12} \qquad (8.8\text{-}2)$$

from eq. (8.6-8). Finally, we have for the molar internal energy

$$\tilde{E} = \frac{3}{2} RT + 2\pi N_0 \rho_0 \int_0^\infty r^2 g(r) u(r)\, dr \qquad (8.8\text{-}3)$$

It is worth emphasizing that eq. (8.8-3) is valid only if the approximation (8.8-1) holds. Where this is true, the pair correlation function $g(r)$ is of dominant importance. If three-body forces were taken into account, then the triplet correlation function $g^{(3)}$ would appear in additional terms in eq. (8.8-3).

Another thermodynamic function conveniently expressible in terms of the pair correlation function is the pressure. From Exercise 6.17 we recall that

$$\langle P \rangle = -\left(\frac{\partial \langle A \rangle}{\partial V}\right)_{T,N}$$

$$= k_B T \left(\frac{\partial \ln Q}{\partial V}\right)_{T,N}$$

$$= k_B T \left(\frac{\partial \ln Z}{\partial V}\right)_{T,N} \qquad (8.8\text{-}4)$$

from eq. (8.5-2). The configuration integral Z is an implicit function of the volume. In order to carry out the differentiation indicated in eq. (8.8-4), we make a change of variable that introduces the volume explicitly. Let $\{x_i\} = \{r_i V^{-1/3}\}$ be the new variables of integration. Then one has for Z on the assumption of pairwise interactions

$$Z = V^N \int \cdots \int \exp\left(-\sum_i \sum_{j>i} \frac{u\left(x_{ij} V^{1/3}\right)}{k_B T}\right) \prod_{i=1}^N d^3 x_i \qquad (8.8\text{-}5)$$

Now make use of the chain rule to write

$$\frac{\partial u\left(x_{ij}V^{1/3}\right)}{\partial V} = \frac{\partial u\left(x_{ij}V^{1/3}\right)}{\partial \left(x_{ij}V^{1/3}\right)}\frac{d\left(x_{ij}V^{1/3}\right)}{dV}$$

$$= \frac{r_{ij}}{3V}\frac{du(r_{ij})}{dr_{ij}}$$

so that

$$\left(\frac{\partial U}{\partial V}\right)_{T,N} = \frac{1}{3V}\sum_i\sum_{j>i} r_{ij}\frac{du}{dr_{ij}}$$

Combination of this with eqs. (8.8-4) and (8.8-5) yields

$$\langle P\rangle = \frac{Nk_BT}{V} - \frac{1}{3V}\frac{\int\cdots\int e^{-U/k_BT}\left(\sum_i\sum_{j>i} r_{ij}\dfrac{du}{dr_{ij}}\right)\prod_{i=1}^{N} d^3x_i}{ZV^{-N}}$$

$$= \frac{Nk_BT}{V} - \frac{1}{3V}\frac{\int\cdots\int e^{-U/k_BT}\left(\sum_i\sum_{j>i} r_{ij}\dfrac{du}{dr_{ij}}\right)\prod_{i=1}^{N} d^3r_i}{Z}$$

$$= \frac{Nk_BT}{V}$$

$$- \frac{1}{6V}\frac{\iint r_{12}\dfrac{du(r_{12})}{dr_{12}}\left(N(N-1)\int\cdots\int e^{-U/k_BT}\prod_{i=3}^{N} d^3r_i\right)d^3r_1\, d^3r_{12}}{Z}$$

$$= \frac{Nk_BT}{V} - \frac{1}{6V}\iint r_{12}\frac{du(r_{12})}{dr_{12}}p^{(2)}(\mathbf{r}_1,\mathbf{r}_{12})\,d^3r_1\,d^3r_{12} \qquad (8.8\text{-}6)$$

from eq. (8.6-4). Finally, calling upon eq. (8.6-6) again, we obtain

$$\langle P\rangle = \frac{Nk_BT}{V} - \frac{\rho_0^2}{6}\int_0^{\infty} 4\pi r^3\left(\frac{du}{dr}\right)g(r)\,dr \qquad (8.8\text{-}7)$$

The first term on the right-hand side of eq. (8.8-7) is the contribution to the pressure due simply to momentum transport; it is identical to that of the ideal gas, in which intermolecular forces are absent. The second term is the

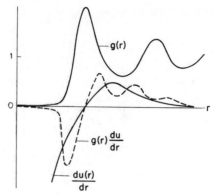

FIGURE 8.17. Plots of the pair correlation function $g(r)$, the pair intermolecular potential $u(r)$, and the product $g(r)(du/dr)$ for a liquid (from C. Croxton, *Introduction to Liquid State Physics*, Wiley, London, 1975, p. 61, with permission).

contribution due to intermolecular forces. This term is approximated by the van der Waals a constant of the van der Waals equation of state. Indeed, eq. (8.8-7) is an equation of state for a fluid subject only to the assumption of spherically symmetric pairwise interactions. It is a very sensitive test of any potential function or any pair correlation function $g(r)$ since the product $[du(r)/dr]g(r)$ varies rapidly in the vicinity of the first maximum of $g(r)$ (Fig. 8.17). An interesting comparison of various intermolecular potentials has been made using an experimental $g(r)$ for argon near its triple point (Chen and Present, 1971). Some of the results are given in Table 8.8. The large variation in value of the calculated vapor pressure of argon clearly shows the great sensitivity of eq. (8.8-7).

Table 8.8.   Test of Various Pair Potentials for Liquid Argon at 84.3 K

Potential[a]	σ (Å)[b]	ε/k_B (K)[c]	P (atm)[d]
LJ (12, 6)	3.410	116.0	460
LJ (12, 6)	3.448	121.1	1130
Kihara	3.314	147.2	−780
Kihara	3.381	138.0	230
Alder	3.280	138.2	−340

[a] For details of these potentials see the cited references in C. T. Chen and R. D. Present, *J. Chem. Phys.*, **54**, 58 (1971).
[b] Slow-collision diameter of argon atom.
[c] Depth of potential well.
[d] Calculated using the pair correlation function in B. A. Dasannacharya and K. R. Rao, *Phys. Rev.*, **137**, A417 (1965); the experimental vapor pressure is 0.71 atm!

Exercises

8.29. Write a general equation that connects the k-body correlation function $g^{(k)}(\mathbf{r}_1, \mathbf{r}_2, \ldots, \mathbf{r}_k)$ with the $(k + 1)$-body correlation function.

8.30. The probability that any three particles of a system will be found in the volume element $d^3r_1\, d^3r_2\, d^3r_3$ centered at $(\mathbf{r}_1, \mathbf{r}_2, \mathbf{r}_3)$ is given from the three-body generic configurational distribution function as $p^{(3)}(\mathbf{r}_1, \mathbf{r}_2, \mathbf{r}_3)\, d^3r_1\, d^3r_2\, d^3r_3$. In the so-called **superposition approximation**, this probability is set proportional to the product of the three separate pair probabilities.

$$
\begin{aligned}
p^{(3)}&(\mathbf{r}_1, \mathbf{r}_2, \mathbf{r}_3)\, d^3r_1\, d^3r_2\, d^3r_3 \\
&= \left[p^{(2)}(\mathbf{r}_1, \mathbf{r}_2)\, d^3r_1\, d^3r_2 \right]\left[p^{(2)}(\mathbf{r}_1, \mathbf{r}_3)\, d^3r_1\, d^3r_3 \right] \\
&\quad \times \left[p^{(2)}(\mathbf{r}_2, \mathbf{r}_3)\, d^3r_2\, d^3r_3 \right] C
\end{aligned}
$$

The proportionality constant C is chosen so that this expression reduces to the correct limiting form when r_{12}, r_{13}, r_{23} are all very great. Determine C. According to the superposition approximation, how is $g^{(3)}(\mathbf{r}_1, \mathbf{r}_2, \mathbf{r}_3)$ related to the pair correlation functions? The superposition approximation has been widely used in liquid studies since it was first proposed by Kirkwood in 1935.

8.31.[C] The following data (Narten, 1977) give the value of $g(r)$, where r is measured from nitrogen to nitrogen, for liquid ammonia at 4°C [$\rho_0 = 0.02237$ molecules Å^{-3}]. Estimate by graphical integration the number of nearest neighbors an ammonia molecule has in the liquid. Compare with the value of 12 in the solid, where there are 6 hydrogen-bonded neighbors at 3.4 Å and 6 nonbonded ones at 3.9 Å.

r (Å)	$g(r)$	r (Å)	$g(r)$	r (Å)	$g(r)$
2.88	0.04	3.68	1.85	4.48	0.92
2.95	0.16	3.74	1.86	4.54	0.96
3.01	0.38	3.80	1.84	4.60	0.99
3.07	0.69	3.87	1.78	4.66	0.99
3.13	1.06	3.93	1.68	4.72	0.96
3.19	1.43	3.99	1.55	4.79	0.91
3.25	1.75	4.05	1.41	4.85	0.86
3.31	1.97	4.11	1.27	4.91	0.83
3.37	2.06	4.17	1.14	4.97	0.81
3.44	2.04	4.23	1.03	5.03	0.80
3.50	1.97	4.30	0.94	5.09	0.79
3.56	1.90	4.36	0.90	5.15	0.78
3.62	1.86	4.42	0.89	5.22	0.76

8.32.[C] Elemental iron is a body-centered cubic (bcc) crystal at normal temperatures; the unit cell dimension is 2.8644 Å. Prepare a plot analogous to that in Fig. 8.14(b), using the same Gaussian representation for the Dirac delta functions.

8.33. Verify eq. (8.7-8a). Then check that the dimensions of the left-hand side are those of energy flux.

8.34. The **pressure equation** [eq. (8.8-7)] can be derived in another way from that given in the text. We develop this in the next few exercises. Consider the quantity

$$A = \sum_{i=1}^{N} \mathbf{p}_i \cdot \mathbf{r}_i$$

defined for a system of N identical particles. Suppose over any time interval of interest A remains finite. The time-averaged value of the time derivative of A is from eq. (6.3-1)

$$\left\langle \frac{dA}{dt} \right\rangle = \frac{1}{\Delta t} \int_0^{\Delta t} \left(\frac{dA}{dt} \right) dt$$

Hence, for a sufficiently long time interval Δt we obtain

$$\langle T \rangle = -\frac{1}{2} \left\langle \sum_{i=1}^{N} \mathbf{F}_i \cdot \mathbf{r}_i \right\rangle$$

where T is the kinetic energy of the system and $\mathbf{F}_i$ is the force on the ith particle at $\mathbf{r}_i$. Prove this result, which is known as the **virial theorem** (not to be confused with the virial equation). The virial theorem was first given by the German physicist Rudolph Clausius (1822–1888).

8.35. The left side of the virial theorem is just $\langle \sum_{i=1}^{N} \mathbf{p}_i \cdot \mathbf{p}_i / 2m \rangle$. We now invoke SM Postulate 2 and then use eq. (8.6-1) to carry out the ensemble averaging. Show that this leads to

$$\left\langle \sum_{i=1}^{N} \frac{\mathbf{p}_i \cdot \mathbf{p}_i}{2m} \right\rangle = \frac{3Nk_BT}{2}$$

8.36. In the right side of the virial theorem, assume that the forces are derivable from a potential energy function

$$\mathbf{F}_i = -\frac{\partial U_{\text{sys}}}{\partial \mathbf{r}_i}$$

Let the function U_{sys} consist of interactions of the particles with the walls of the vessel plus all of the interparticle interactions.

$$U_{sys} = U_{wall} + U_{int}$$

By Newton's Third Law the contribution to $\sum_{i=1}^{N}(-\partial U_{wall}/\partial \mathbf{r}_i)$ due to the element of the area $\mathbf{n}\,dS$ of the container wall is $-P\mathbf{n}\,dS$, where $\mathbf{n}$ is a unit vector in the direction of the outward normal to dS. Using the **divergence theorem** of Gauss (Spiegel, 1959), show that the contribution to the right side of the virial theorem made by the potential U_{wall} is $\frac{3}{2}PV$.

8.37. The virial theorem now reads from Exercise 8.34–8.36,

$$3Nk_BT = 3PV + \left\langle \sum_{i=1}^{N} \mathbf{r}_i \cdot \frac{\partial U_{int}}{\partial \mathbf{r}_i} \right\rangle$$

Assume U_{int} is a sum of pair potentials and complete the derivation of eq. (8.8-7).

8.38. The next few exercises lead to the derivation of an equation supplementary to eq. (8.8-7) that is often used to test proposed pair correlation functions. Return to the grand canonical formalism developed in Section 8.2. First show that the probability of finding a system to contain n particles is

$$p_n = \frac{e^{n\mu/k_BT}Q_n}{\Xi}$$

where Q_n is defined in eq. (8.4-1). Using this expression for p_n, next we show that the derivative of the average number of particles with respect to the chemical potential is given by

$$\left(\frac{\partial \langle n \rangle}{\partial \mu}\right)_{V,T} = \frac{1}{k_BT}(\langle n^2 \rangle - \langle n \rangle^2)$$

8.39. Now we establish a connection with thermodynamics.
(a) First prove that $(\partial \langle P \rangle/\partial \mu)_{V,T} = \langle n \rangle/V$.
(b) Next, we argue that $(\partial \langle \rho \rangle/\partial P)_{V,T} = (\partial \langle \rho \rangle/\partial P)_{n,T}$ because $\rho = n/V$ is an intensive quantity. Hence, show that

$$\left(\frac{\partial \langle \rho \rangle}{\partial P}\right)_{V,T} = \frac{\langle n \rangle}{V}\beta_T$$

where $\beta_T = -V^{-1}(\partial V/\partial P)_{n,T}$ is the coefficient of isothermal compressibility.

(c) Third, indicate how $(\partial\langle n\rangle/\partial P)_{V,T} = \langle n\rangle\beta_T$ follows from part (b).

(d) Combine the results of Exercises 8.38 and 8.39(a), (b), and (c) to yield the equation

$$\frac{\langle n^2\rangle - \langle n\rangle^2}{\langle n\rangle^2} = \frac{k_BT}{V}\beta_T$$

This interesting equation shows that the particle (or equivalently, the density) fluctuation is related to the isothermal compressibility.

8.40. Let the **kth-order grand canonical generic configurational distribution function** be designated $p_G^{(k)}(\mathbf{r}_1, \mathbf{r}_2, \ldots, \mathbf{r}_k)$. It should be given by the weighted average

$$p_G^{(k)}(\mathbf{r}_1, \mathbf{r}_2, \ldots, \mathbf{r}_k) = \sum_{n=k}^{\infty} p_n p_n^{(k)}(\mathbf{r}_1, \mathbf{r}_2, \ldots, \mathbf{r}_k)$$

where p_n is defined in Exercise 8.38, and $p_n^{(k)}(\mathbf{r}_1, \mathbf{r}_2, \ldots, \mathbf{r}_k)$ is the kth-order canonical generic configurational distribution function for a system of n particles [see eq. (8.6-4)].

(a) Show that

$$\int \cdots \int p_G^{(k)}(\mathbf{r}_1, \mathbf{r}_2, \ldots, \mathbf{r}_k)\, d^3r_1\, d^3r_2 \cdots d^3r_k = \left\langle \frac{n!}{(n-k)!} \right\rangle$$

(b) Now be specific and let $k = 2$. From part (a) deduce

$$\langle n\rangle\rho \int_0^\infty 4\pi r^2 g(r)\, dr = \langle n^2\rangle - \langle n\rangle$$

(c) Finally, combine the results of Exercise 8.39(d) and 8.40(b) to yield

$$1 + \rho \int_0^\infty 4\pi r^2 h(r)\, dr = \langle n\rangle\frac{k_BT}{V}\beta_T$$

where $h(r) = g(r) - 1$ is known as the **total pair correlation function**. The result of part (c) is commonly referred to as the **compressibility equation of state**, in contrast to eq. (8.8-7), which is the **pressure equation of state**. Note that the compressibility equation has not required us to assume the potential energy function is a sum or pairwise potentials.

8.9 INTEGRAL EQUATION FOR THE PAIR CORRELATION FUNCTION

In Section 8.8 it was shown how to calculate the standard thermodynamic functions from the pair correlation function $g(r)$. To complete the theoretical description of liquids we need to find an equation for $g(r)$. Differing viewpoints and approximations have led to a variety of **integral equations** for $g(r)$. We shall not attempt to summarize them in the limited space here, as reviews appear frequently (Rowlinson, 1968; Watts, 1973; Barker and Henderson, 1976). However, to illustrate the methodology, we develop one such equation.

Into eq. (8.6-4) let us substitute the pair potential approximation, eq. (8.8-1)

$$p^{(2)}(\mathbf{r}_1, \mathbf{r}_2) = n(n-1) \frac{\int \cdots \int \exp\left[-\sum_i \sum_{j>i} \frac{u(r_{ij})}{k_B T}\right] \prod_{i=3}^{n} d^3 r_i}{Z} \tag{8.9-1}$$

Next, we take the gradient of both sides of eq. (8.9-1) with respect to the position of particle 1,

$$\frac{\partial p^{(2)}(\mathbf{r}_1, \mathbf{r}_2)}{\partial \mathbf{r}_1}$$

$$= n(n-1) \frac{-\frac{1}{k_B T} \sum_{j=2}^{n} \int \cdots \int \frac{\partial u(r_{1j})}{\partial \mathbf{r}_1} \exp\left[-\sum_i \sum_{j>i} \frac{u(r_{ij})}{k_B T}\right] \prod_{i=3}^{n} d^3 r_i}{Z} \tag{8.9-2a}$$

where $\partial/\partial \mathbf{r}_1$ signifies the gradient ∇_1. Singling out the $j = 2$ term, we can rewrite eq. (8.9-2a) as

$$\frac{\partial p^{(2)}(\mathbf{r}_1, \mathbf{r}_2)}{\partial \mathbf{r}_1} = -\frac{1}{k_B T} \frac{\partial u(r_{12})}{\partial \mathbf{r}_1} p^{(2)}(\mathbf{r}_1, \mathbf{r}_2)$$

$$-\frac{1}{k_B T} \frac{n(n-1)}{Z} \sum_{j=3}^{n} \int \cdots \int \frac{\partial u(r_{1j})}{\partial \mathbf{r}_1}$$

$$\times \exp\left[-\sum_i \sum_{j>i} \frac{u(r_{ij})}{k_B T}\right] \prod_{i=3}^{n} d^3 r_i \tag{8.9-2b}$$

Now look at the outer summation of eq. (8.9-2b). For the case of $j = 3$ we can write the corresponding term as

$$\int \frac{\partial u(r_{13})}{\partial \mathbf{r}_1} \left(\int \cdots \int e^{-U/k_B T} \prod_{i=4}^{n} d^3 r_i \right) d^3 r_3$$

$$= \int \frac{\partial u(r_{13})}{\partial \mathbf{r}_1} \left[p^{(3)}(\mathbf{r}_1, \mathbf{r}_2, \mathbf{r}_3) \frac{Z}{n(n-1)(n-2)} \right] d^3 r_3$$

after referring to eq. (8.6-4). The choice of $j = 3$ is arbitrary, however, and so a quantity identical to that on the right side occurs a total of $n - 2$ times. It follows that eq. (8.9-2b) can be rewritten

$$\frac{\partial p^{(2)}(\mathbf{r}_1, \mathbf{r}_2)}{\partial \mathbf{r}_1} = -\frac{1}{k_B T} \frac{\partial u(r_{12})}{\partial \mathbf{r}_1} p^{(2)}(\mathbf{r}_1, \mathbf{r}_2)$$

$$-\frac{1}{k_B T} \int \frac{\partial u(r_{13})}{\partial \mathbf{r}_1} p^{(3)}(\mathbf{r}_1, \mathbf{r}_2, \mathbf{r}_3) \, d^3 r_3$$

or upon multiplication of both sides by $k_B T / p^{(2)}(\mathbf{r}_1, \mathbf{r}_2)$,

$$k_B T \frac{\partial \ln p^{(2)}(\mathbf{r}_1, \mathbf{r}_2)}{\partial \mathbf{r}_1} = -\frac{\partial u(r_{12})}{\partial \mathbf{r}_1} - \int \frac{\partial u(r_{13})}{\partial \mathbf{r}_1} \frac{p^{(3)}(\mathbf{r}_1, \mathbf{r}_2, \mathbf{r}_3)}{p^{(2)}(\mathbf{r}_1, \mathbf{r}_2)} \, d^3 r_3 \quad (8.9\text{-}3)$$

Equation (8.9-3) is known as the **Born–Green equation**, or sometimes as the **Born–Green–Bogoliubov–Yvon (BGBY) equation**. It is an **integrodifferential equation**, and is *exact* within the approximation of pairwise interactions. The generic pair configurational distribution function is found to depend upon the generic triplet configurational distribution function. Unfortunately, if we do not know $p^{(2)}$ *a priori*, then neither do we know $p^{(3)}$. The triplet distribution function $p^{(3)}$ can be related to $p^{(4)}$, however, and so on through successively higher orders. To close this infinite hierarchy, Kirkwood suggested that the triplet distribution function depends principally upon the distribution of pairs of molecules and that to a first approximation it contains no contributions from three-molecule correlations. This **superposition approximation** was given in Exercise 8.30.

$$p^{(3)}(\mathbf{r}_1, \mathbf{r}_2, \mathbf{r}_3) = \frac{p^{(2)}(\mathbf{r}_1, \mathbf{r}_2) p^{(2)}(\mathbf{r}_1, \mathbf{r}_3) p^{(2)}(\mathbf{r}_2, \mathbf{r}_3)}{\rho_0^3}$$

Its insertion into eq. (8.9-3) yields an equation that contains only pair distribution functions.

$$k_B T \frac{\partial \ln p^{(2)}(\mathbf{r}_1, \mathbf{r}_2)}{\partial \mathbf{r}_1} = -\frac{\partial u(r_{12})}{\partial \mathbf{r}_1}$$

$$-\frac{1}{\rho_0^3} \int \frac{\partial u(r_{13})}{\partial \mathbf{r}_1} p^{(2)}(\mathbf{r}_1, \mathbf{r}_3) p^{(2)}(\mathbf{r}_2, \mathbf{r}_3) \, d^3 r_3 \quad (8.9\text{-}4)$$

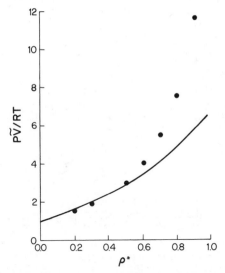

FIGURE 8.18. Plot of compressibility factor versus reduced density for hard spheres as determined by solution of the Born–Green equation (—) and by computer simulation ($\cdots$) [from J. A. Barker and D. Henderson, *Rev. Mod. Phys.*, **48**, 587 (1976), with permission].

In general, eq. (8.9-4) must be solved numerically on a computer. To test the numerical method, one must first choose a pair potential $u(r_{12})$. Since real liquids are not strictly described by pair potentials, the comparison is more appropriate for some *model* liquid, which can be simulated on computers. In Fig. 8.18 we show a plot of the compressibility factor $Z = P\tilde{V}/RT$ for a fluid of hard spheres of diameter d versus the reduced density $\rho^* = \rho d^3$. The curve was obtained by solving eq. (8.9-4), and then using the pair correlation function, $g(r_{12}) = \rho_0^{-2} p^{(2)}(r_{12})$, therefrom to calculate Z from the pressure equation [eq. (8.8-7)]. The dots in Fig. 8.18 are the "exact" results determined

Table 8.9.   Critical Constants for the Lennard-Jones 6-12 Potential

	T_{cr}^{*b}	ρ_{cr}^{*b}	P_{cr}^{*b}	Z_{cr}
Born–Green[a]	1.45	0.40	0.26	0.44
Percus–Yevick[a]	1.25	0.29	0.11	0.30
Computer simulation[c]	1.32–1.36	0.32–0.36	0.13–0.17	0.30–0.36
Experimental[d]	1.26	0.316	0.117	0.293

[a] Calculations done via the pressure equation [eq. (8.8-7)].
[b] $T_{cr}^* = k_B T_{cr}/\varepsilon$, $\rho_{cr}^* = \rho_{cr}\sigma^3$, $P_{cr}^* = P_{cr}\sigma^3/\varepsilon$.
[c] All entries in this table taken from Table VII of J. A. Barker and D. Henderson, *Rev. Mod. Phys.*, **48**, 587 (1976). The LJ parameters are those roughly appropriate to argon: $\varepsilon/k_B = 119.8$ K and $\sigma = 3.405$ Å.
[d] The actual experimental values for argon.

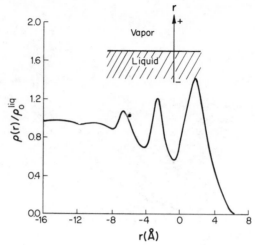

FIGURE 8.19. Density profile in the vicinity of the gas–liquid interface [from G. M. Nazarian, *J. Chem. Phys.*, **56**, 1408 (1972), with permission].

by computer simulation. One sees that agreement is satisfactory at moderate densities, but that divergence becomes pronounced above roughly $\rho^* = 0.6$. This has generally been attributed to breakdown of the superposition approximation (Barker and Henderson, 1976).

Another test of eq. (8.9-4) is to calculate the critical constants for a fluid with some specified potential. In Table 8.9 are given the results of computer simulation, of numerical solution of eq. (8.9-4), and of solution of a more recent integral equation (the **Percus—Yevick equation**) for a fluid with a Lennard-Jones 6-12 potential. The BGBY equation does not work very well here, whereas the newer integral equation does quite well.

An interesting application of the BGBY equation has been made to the calculation of the density profile of a fluid in the vicinity of its liquid-vapor interface (Nazarian, 1972). The fluid studied was argon at 90 K for which a Lennard-Jones potential was used. The result is shown in Fig. 8.19. Attention is called to the fluctuations in liquid density that are predicted to exist a few angstroms beneath the surface.

As just implied, integral equations more recent than the BGBY equation have proved more successful. However, the BGBY equation has been very important in stimulating research in the problem of phase transitions. From Exercise 8.40(c) one has

$$4\pi \int_0^\infty r^2 h(r) \, dr = k_B T \beta_T - \frac{1}{\rho}$$

and thus any solution $p^{(2)}(\mathbf{r}_1, \mathbf{r}_2)$ of eq. (8.9-4) should yield a total pair correlation function $h(r)$ whose integral is finite. It was found early that at reduced densities greater than about 0.95 the solutions to the BGBY equation for hard spheres lead to $h(r)$ functions whose integrals diverge. There was considerable excitement that this might be indicative of a phase transition from liquid to solid. Hence, confirmation was sought in computer "experiments" (Alder and Wainwright, 1957). Indeed, the computer experiments indicated that for $\rho^* > 0.88$ a fork in the plot of Z versus ρ^* appears, leading to equations of state for two distinct phases. Subsequent study by many workers in pursuit of a theory of melting ensued (Longuet-Higgins and Widom, 1964). It has been a matter of debate whether the behavior of the BGBY equation at high density is actually symptomatic of a phase change or merely represents a breakdown of the theory (Croxton, 1974). Most seem to agree that the computer experiments indicate a genuine phase transition (although, is it truly a liquid-to-solid phase transition?) for a collection of hard spheres. In any event, the early results with the BGBY equation have spawned considerable activity, which continues even today (Chandler et al., 1983).

8.10   COMPUTER SIMULATIONS OF LIQUIDS

The use of the computer to simulate the behavior of real liquids has probably been the single most important development in advancing our understanding of liquids. Two general approaches are commonly used: the **molecular dynamics** method and the **Monte Carlo** method (McDonald and Singer, 1973).

In the molecular dynamics method, the system is studied by solving Newton's equations of motion, which are all interconnected by the potential-energy function for the system (see Section 1.1). To keep the mathematics manageable, the potential-energy function is usually assumed to be a sum of pair potentials. The size of a system that can be studied is limited by the capacity of present-day computers. Very early molecular-dynamics calculations employed only a few tens of particles; currently, the limit is several thousands. The set of Newtonian equations is solved numerically to give the position and velocity of each particle as a function of time; this is done by means of a straightforward extension of the one-particle treatment discussed in Section 1.5.

Let $\mathbf{r}_i(t)$ and $\mathbf{v}_i(t)$ be the position and velocity of particle i at time t. The force on this particle at time t is

$$\mathbf{F}_i(t) = -\frac{\partial U}{\partial \mathbf{r}_i}$$

$$= -\sum_{j \neq i}^{n} \frac{\partial u(r_{ij})}{\partial \mathbf{r}_i}$$

Thus, from Newton's Second Law of Motion the acceleration is $\mathbf{a}_i(t) =$

$F_i(t)/m_i$. We choose a small time interval, Δt, typically about 10^{-15} s. If the acceleration is approximately constant during this interval, then the new position is, from eq. (1.5-6),

$$\mathbf{r}_i(t + \Delta t) = 2\mathbf{r}_i(t) - \mathbf{r}_i(t - \Delta t) - m_i^{-1}\Delta t^2 \mathbf{F}_i(t) \qquad (8.10\text{-}1a)$$

The velocity is

$$\mathbf{v}_i(t) = [\mathbf{r}_i(t + \Delta t) - \mathbf{r}_i(t - \Delta t)]/2\Delta t \qquad (8.10\text{-}1b)$$

At time $t = 0$ the particles are assigned arbitrary velocities and positions. As motion proceeds, one monitors the quantity

$$T = \frac{1}{3k_B} \sum_{i=1} m_i v_i^2 \qquad (8.10\text{-}2)$$

This quantity fluctuates strongly at the beginning of the simulation, but after several collisions (typically, a hundred or so) thermal equilibrium is achieved and T then represents the temperature of the system. Subsequently, macroscopic properties of the system are evaluated as *time* averages. For example, the mean potential energy is

$$\bar{U} = \sum_i \sum_{j>i} u(r_{ij})$$

where the overbar connotes the time average for a period 10^{-11} s or longer. The pressure is evaluated by means of the virial theorem, which was given in Exercise 8.37.

$$\bar{P} = \frac{Nk_BT}{V} - \frac{1}{3V} \overline{\sum_{i=1} \mathbf{r}_i \cdot \sum_{j \neq i} \frac{\partial u(r_{ij})}{\partial \mathbf{r}_i}} \qquad (8.10\text{-}3)$$

Note that in the molecular dynamics method the volume is fixed by choice (so as to fix the density), and the total energy is constant because the system is isolated.

In the Monte Carlo method one does not allow the system to evolve in real time according to the true classical laws of motion. Rather, one generates a sequence of configurations $\{(\mathbf{r}_1, \mathbf{r}_2, \ldots, \mathbf{r}_n)\}$ of the system by random displacements of the particles. To create a new configuration from a previous one, a particle is selected at random and is moved in a random direction by a tiny amount $\delta \mathbf{r}$. If the change in potential energy, δU, is negative, the new configuration is accepted. But if δU is positive, the new configuration is accepted only with the probability

$$p = \exp\left(-\frac{\delta U}{k_B T}\right) \tag{8.10-4}$$

This is effected by choosing a random number between 0 and 1 (the limits of the range for a probability); if the exponential is larger than the random number, the new configuration is accepted, otherwise it is rejected and the old configuration is counted a second time.

After a great many configurational moves, one obtains a nearly continuous probability distribution function

$$p(\mathbf{r}_i, \mathbf{r}_2, \ldots, \mathbf{r}_n) \simeq \exp\left(-\frac{U(\mathbf{r}_1, \mathbf{r}_2, \ldots, \mathbf{r}_n)}{k_B T}\right) \tag{8.10-5}$$

Table 8.10.   Some Substances Studied by Computer Simulation

Substance	Method[a]	References
CH_4	MD	S. Murad et al., *Mol. Phys.*, **37**, 725 (1979)
NH_3	MC	W. L. Jorgensen and M. Ibrahim, *J. Am. Chem. Soc.*, **102**, 3309 (1980)
CCl_4	MD	O. Steinhauser and M. Neumann, *Mol. Phys.*, **40**, 115 (1980)
VCl_4	MD	S. Murad and K. E. Gubbins, *Mol. Phys.*, **39**, 271 (1980)
C_6H_6	MC	F. Serrano Adan, A. Bañon, and J. Santamaria, *Chem. Phys.*, **86**, 433 (1984)
H_2O (ℓ)	MC	R. O. Watts, *Mol. Phys.*, **28**, 2069 (1974)
	MD	I. R. McDonald and M. L. Klein, *J. Chem. Phys.*, **68**, 4875 (1978)
	MC	W. L. Jorgensen, *Chem. Phys. Lett.*, **70**, 326 (1980)
H_2O (ice melting)	MD	T. A. Weber and F. H. Stillinger, *J. Phys. Chem.*, **87**, 4277 (1983)
H_2O (on a silicate surface)	MD	D. J. Mulla et al., *J. Colloid Interface Sci.*, **100**, 576 (1984)

[a] MC = Monte Carlo method; MD = molecular dynamics method.

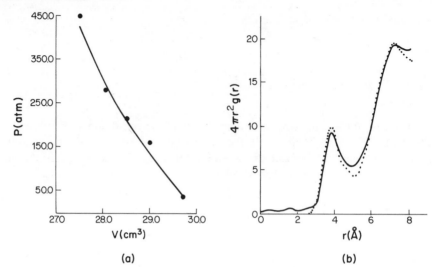

(a)   (b)

FIGURE 8.20. (*a*) Comparison of the Monte Carlo calculation (· · ·) of the pressure of argon at 97 K with the experimental value (—) [from I. R. McDonald and K. Singer, *Disc. Faraday Soc.*, **43**, 40 (1967)], (*b*) comparison of the molecular dynamics calculation (· · ·) of the radial distribution function for liquid sodium at 373 K with the experimental function [from A. Paskin and A. Rahman, *Phys. Rev. Lett.*, **16**, 300 (1966)] (both with permission).

Then the average of any classical quantity f expressible as a function of the coordinates,

$$\langle f \rangle = \frac{\int \cdots \int f(\mathbf{r}_1, \mathbf{r}_2, \ldots, \mathbf{r}_n)\, p(\mathbf{r}_1, \mathbf{r}_2, \ldots, \mathbf{r}_n)\, d^3r_1\, d^3r_2 \cdots d^3r_n}{\int \cdots \int p(\mathbf{r}_1, \mathbf{r}_2, \ldots, \mathbf{r}_n)\, d^3r_1\, d^3r_2 \cdots d^3r_n}$$

is approximated by the *ensemble* average. Typically, one generates about 700,000 configurations, of which the first 200,000 are discarded in order to remove any influence of the choice of initial configuration, and the remaining 500,000 are employed to calculate $\langle f \rangle$.

In cases where calculations have been performed using both Monte Carlo and molecular-dynamics methods, essentially identical results are obtained. This provides a computational corroboration of SM Postulate 2 (see Section 6.3). Table 8.10 lists a small sampling of some of the particular substances that have been studied by either the Monte Carlo or molecular-dynamics method. In Fig. 8.20*a*, *b* we show some of the results that can be obtained in a typical computer "experiment" with realistic potentials. The great utility of the computer is clearly evident.

8.11   COMPUTER HIGHLIGHT: NUMERICAL SOLUTION OF AN INTEGRAL EQUATION

We have seen in Section 8.7 the form of a typical pair correlation function for a monatomic liquid [Fig. 8.16(a)]. It is interesting that most of the features of $g(r)$ can be reproduced by an integral equation based on an heuristic argument on the packing of rigid spheres (e.g., ball bearings) [Kirkwood (1939)]. We first "derive" this equation and then consider its numerical solution.

Imagine a large collection of impenetrable spheres. To this collection we wish to add one more sphere; in order to do so, there must be a hole in the collection large enough to accommodate the added sphere. Let us take one member of the collection as a reference sphere, and let $P_n(R)$ be the probability that a point can be found on the spherical surface of radius R about this reference sphere such that a hole exists at the point that is large enough to admit the added sphere. Once the new sphere is added, it is like all others in the collection, and so $P_n(R)$ must at the same time be proportional to the probability of finding *any* sphere a distance R from the central one (i.e., the reference sphere).

Now consider a spherical region Γ of diameter d, whose center is R units away from the central sphere (Fig. 8.21). On a time average, the number of spheres whose centers lie inside Γ is

$$n_s(R) = \int_\Gamma \rho(r)\, d^3r$$

where $\rho(r)$ is the number density of spheres in volume element d^3r located r units from the center of the reference sphere. Now if d is at least as large as the sphere's diameter, $n_s(R)$ cannot be greater than $\rho_c\Gamma$, where ρ_c is the density under close-packed conditions. If $n_s(R)$ is "normalized" by dividing

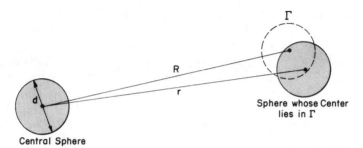

FIGURE 8.21. Showing a spherical region Γ that is R units away from a central sphere, where a hole in the liquid may exist.

by $\rho_c \Gamma$, we can interpret the normalized expression

$$P_s(R) = \frac{1}{\rho_c \Gamma} \int_\Gamma \rho(r) \, d^3r \qquad (8.11\text{-}1)$$

as the *probability* that a sphere's center lies within Γ, and so $1 - P_s(R)$ is then the probability that a hole exists in Γ.

$$P_n(R) = 1 - P_s(R)$$

However, we expect P_n to be proportional to $\rho(R)$ because $P_n(R)$ is also proportional to the probability of finding any sphere a distance R from the reference sphere. Thus, we can write

$$\rho(R) = C[1 - P_s(R)] \qquad (8.11\text{-}2a)$$

To evaluate the constant C we use the fact that at large R the function $\rho(R)$ must approach the bulk number density ρ_0

$$\rho_0 = C\left(1 - \frac{\rho_0}{\rho_c}\right) \qquad (8.11\text{-}2b)$$

Combination of eqs. (8.11-1) and (8.11-2) then yields

$$\frac{\rho(R)}{\rho_0} = \frac{1}{1 - \rho_0/\rho_c}\left(1 - \frac{1}{\rho_c \Gamma} \int \rho(r) \, d^3\mathbf{r}\right)$$

or

$$g(R) = \frac{1}{1 - \rho_0/\rho_c}\left(1 - \frac{\rho_0}{\rho_c \Gamma} \int g(r) \, d^3\mathbf{r}\right) \qquad (8.11\text{-}3)$$

where the second line follows from the results of Section 8.6. Note that this intuited integral equation is very different in nature from the approximate BGBY equation (8.9-4). Most important for our purposes is that eq. (8.11-3) is *linear* in $g(R)$, which makes it far easier to solve.

While the volume Γ of integration in eq. (8.11-3) is spherical, the origin of the integration variable r is not the center of Γ. Hence spherical polar coordinates are not useful in this instance. Instead we introduce **bipolar coordinates**, which are depicted in Fig. 8.22. A point P in Γ is specified either in terms of the Cartesian coordinates (x, y, z) with respect to the central (reference) sphere as origin O, or in terms of bipolar coordinates (r, r', ϕ), where r is the distance from O to P, r' is the distance of P from the center of Γ and ϕ is the angle of rotation in the plane containing P and parallel to the

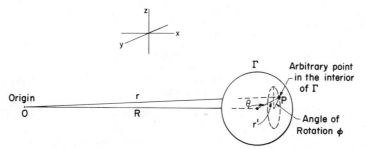

FIGURE 8.22. Illustration of the bipolar coordinate system.

yz plane. Elementary trigonometry yields the transformation equations:

$$\begin{cases} x = \dfrac{1}{2}R - \dfrac{r'^2}{2R} + \dfrac{r^2}{2R} \\[2mm] y = (r^2 - x^2)^{1/2}\sin\phi \\[2mm] z = (r^2 - x^2)^{1/2}\cos\phi \end{cases} \qquad (8.11\text{-}4)$$

The volume element is determined by evaluating the Jacobian of the transformation (Dence, 1975):

$$d^3r = \left(\frac{rr'}{R}\right) dr\, dr'\, d\phi$$

Then eq. (8.11-3) can be cast explicitly as

$$g(R) = \frac{1}{1 - (\rho_0/\rho_c)}\left(1 - \frac{\rho_0}{\rho_c\Gamma}\int_{R-d/2}^{R+d/2}\int_{|R-r|}^{d/2}\int_0^{2\pi}\frac{g(r)rr'\,d\phi\,dr'\,dr}{R}\right)$$

$$= \frac{1}{1 - (\rho_0/\rho_c)}\left[1 - \frac{\rho_0\pi}{\rho_c\Gamma R}\int_{R-d/2}^{R+d/2}\left(\frac{d^2}{4} - (R - r)^2\right)r\,g(r)\,dr\right]$$

$$(8.11\text{-}5)$$

For the purpose of solving eq. (8.11-5) numerically, it is convenient to measure distance in units of the diameter d, which we now also take to be the sphere's diameter. Thus, making the change of variables

$$x = R/d \qquad y = r/d$$

we can rewrite eq. (8.11-5) as

$$g(x) = \frac{1}{1 - \eta} - \frac{6\eta}{(1 - \eta)}\frac{1}{x}\int_{x-1/2}^{x+1/2}\left[\frac{1}{4} - (x - y)^2\right]y\,g(y)\,dy \quad (8.11\text{-}6)$$

where

$$\eta = \frac{\rho_0}{\rho_c}$$

Note that the only disposable parameter in the reduced integral equation [eq. (8.11-6)] is η, the *ratio* of the bulk density to close-packed density. The maximum value of η is 1, attained in the ideal solid where the spheres are closely packed.

Although eq. (8.11-6) is a good deal simpler in form than the BGBY equation, it nevertheless defies analytical methods of solution. Moreover, the numerical solution of eq. (8.11-6) poses a difficult computational task, especially if one is restricted to a microcomputer. One possibility is iteration to self consistency, as was done in Section 4.4 in order to solve the Hartree–Fock equations. We guess a solution $g(y)$ and use it to evaluate the integral on the right side of eq. (8.11-6), thus obtaining a first iterate $g(x)$. This, or perhaps some combination of it and the initial guess, is then substituted into the right side of eq. (8.11-6) to obtain a second iterate, and so on. When successive iterates differ (in the least-squares sense) by less than a prescribed value, the solution has converged. In practice, the iterative procedure is very sensitive to the various parameters involved, for example, the initial guess, and requires "fine tuning" in order to work. Indeed, it is an active area of research today in the theory of liquids.

The fact that eq. (8.11-6) is linear in g permits a more direct method than iteration. We discretize g, by representing it as a set of discrete values $\{g_k\}$ on a grid (see Fig. 8.23). From Section 8.6, we recall the limiting behavior of $g(r)$ at small r and large r. This translates into the requirements $g(x) = 0$ for $x \le 1$ and $g(x) \to 1$ as $x \to \infty$. These conditions are incorporated in our computation as follows. We set $x_0 = 1$ and $g_k = 0$ for $k \le 0$. We take a sufficiently large number N of equally spaced grid points such that $x_N (= 1 + Nh)$ is large and then we set $g_k = 1$ for $k \ge N + 1$.

Now we write the discrete analog of eq. (8.11-6) at each point of the grid:

$$g_k = \frac{1}{1 - \eta} - \frac{6\eta}{(1 - \eta)} \cdot \frac{1}{(1 + kh)} \int_{x_k - 1/2}^{x_k + 1/2} \left[\frac{1}{4} - (x_k - y)^2 \right] y g(y) \, dy$$

$$k = 1, 2, \ldots, N \quad (8.11\text{-}7)$$

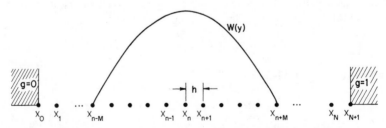

FIGURE 8.23. Grid used to discretize the integral equation (8.11-6).

where h is the grid spacing. The integral is approximated by the extended trapezoidal rule as

$$\int_{x_k-1/2}^{x_k+1/2} W(y)g(y)\,dy = h\left[\frac{1}{2}W_{k-M}g_{k-M}\right.$$

$$\left. + \sum_{j=1}^{2M-1} W_{k-M+j}g_{k-M+j} + \frac{1}{2}W_{k+M}g_{k+M}\right]$$

where

$$W(y) \equiv \left[\frac{1}{4} - (x_k - y)^2\right]y$$

According to Fig. 8.23 we have

$$x_k = 1 + kh$$
$$y_j = 1 + (k - M + j)h$$
$$W_{k-M+j} = \left[\frac{1}{4} - (M-j)^2h^2\right][1 + (k - M + j)h] \qquad 0 \le j \le 2M$$

Also, since $W(y)$ vanishes at $y = x_k \pm \frac{1}{2}$, we choose M and h such that $Mh = \frac{1}{2}$. The discretized integral equation (8.11-7) can then be rewritten

$$g_k = \frac{1}{1 - \eta} - \frac{6\eta}{1 - \eta}\sum_{j=1}^{2M-1} \Omega_{k-M+j}g_{k-M+j} \qquad k = 1, 2, \ldots, N \quad (8.11\text{-}8)$$

where

$$\Omega_{k-M+j} \equiv h\left[\frac{1}{4} - (M-j)^2h^2\right]\left(1 + \frac{(j-M)h}{1+kh}\right)$$

Multiplying eq. (8.11-8) by $(1 - \eta)/6\eta$ and rearranging, we obtain

$$\left(\frac{(1-\eta)}{6\eta} + \Omega_k\right)g_k + \sum_{j=1}^{2M-1}{}' \Omega_{k-M+j}g_{k-M+j} = \frac{1}{6\eta}$$

$$k = 1, 2, \ldots, N \quad (8.11\text{-}9)$$

where the prime on Σ' signifies $j \ne M$.

Equations (8.11-9) constitute N simultaneous linear equations in the N unknowns, $\{g_k\}$. Observe, however, that for $k \le M - 1$, the summation in eq. (8.11-9) involves terms g_k for $k \le 0$. These make no contribution on account of the boundary condition $g(x) = 0$ for $x \leq 1$. Likewise, for $k \leq N - M + 2$ the summation contains terms g_k for $k \geq N + 1$. Invoking the boundary condition $g(x) = 1$ for $x \geq x_{N+1}$, we set these g_k equal to one and transfer the resulting constant to the right side of eq. (8.11-9).

To obtain an accurate solution of eqs. (8.11-9), h must be sufficiently small so that the integration is accurate and N sufficiently large so that the boundary condition for $x > x_N = 1 + Nh$ is appropriate. This dictates, in general, a large number of equations, say several hundred. While well-developed computer algorithms for solving such large sets of simultaneous equations are available, they are difficult to implement on a microcomputer and extremely time consuming. It appears that this problem is beyond the reach of most contemporary microcomputers.

Even on a mainframe computer, the solution of large systems of simultaneous equations poses a serious problem on account of the large memory requirements. The structure of eqs. (8.11-9) is, however, propitious in that the matrix of coefficients of the unknowns $\{g_n\}$ is nonzero only in a band of width $(2M + 1)$ about the main diagonal. This structure is due to the fact that the equation for point k involves only the M points immediately preceding it and the M points immediately following it. A matrix having this structure is said to be **banded** and subroutines for simultaneous equations involving banded matrices are available in the system's library of many computing facilities. On account of the variable nature of such routines, it is not appropriate to go into details here.

As an example, we have utilized a FORTRAN main program that calls a banded-matrix solver based on the technique of Gaussian elimination (Deif, 1982). A plot of $g(x)$ for the case $\eta = 0.918$ is shown in Fig. 8.24. As mentioned at the beginning of this section, $g(x)$ exhibits the main features of $g(r)$ noted in Fig. 8.16(a), namely, oscillations that decay with increasing r (or x). In the present case there are more oscillations and these appear more distinct. This depends, however, on the parameter η; as η decreases, the oscillations wash out, as might be expected.

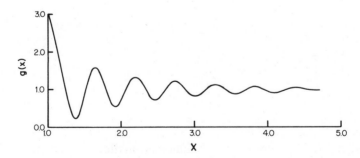

FIGURE 8.24. Numerical solution $g(x)$ of eq. (8.11-6).

Exercises

8.41. The solving of integral equations is a very specialized part of mathematics, and as with differential equations, a great many techniques are available. For example, the first-order *differential* equation $d\Phi/dx = F(x, \Phi)$, $a \le x \le b$, can be reexpressed as the *integral* equation

$$\Phi(x) = \Phi(a) + \int_a^x F[s, \Phi(s)]\, ds$$

which is an example of the **Volterra equation of the second kind**. Solve the Volterra equation

$$\Phi(x) = \frac{1}{2} + 2\int_0^x s\,\Phi(s)\, ds$$

8.42. Suppose the **kernel** of an integral equation can be written in the form $K(x, s) = \lambda u(x)v(s)$. Then in the **Fredholm equation of the second kind**,

$$\Phi(x) = f(x) + \int_a^b K(x, s)\Phi(s)\, ds \qquad a \le x \le b$$

if we insert the form for the kernel, multiply both sides by $v(x)$, and integrate over the interval $[a, b]$, we obtain

$$\int_a^b v(x)\Phi(x)\, dx = \frac{\int_a^b v(x)f(x)\, dx}{1 - \lambda \int_a^b v(x)u(x)\, dx}$$

Show this. Then show that the solution to the integral equation can be written in the form

$$\Phi(x) = f(x) + \lambda \int_a^b R(x, s, \lambda)f(s)\, ds$$

where

$$R(x, s, \lambda) = \frac{u(x)v(s)}{1 - \lambda \int_a^b v(t)u(t)\, dt}$$

Solve the integral equation

$$\Phi(x) = x + \lambda \int_0^\pi \sin nx \sin ns\,\Phi(s)\, ds$$

for the case where n is integral.

8.43. Write the equation, analogous to eq. (8.9-3), that would be the next equation in the BGBY hierarchy. How might this equation be simplified?

8.44. In an important paper (Carnahan and Starling, 1969), an analytical equation of state was postulated for a hard-sphere fluid, which yields virial coefficients and other properties in superb agreement with computer simulation. The equation of state is

$$Z = \frac{PV}{Nk_BT} = \frac{1 + \lambda + \lambda^2 - \lambda^3}{(1 - \lambda)^3}$$

$$\lambda = \frac{\pi N d^3}{6V}$$

Prepare a plot of Z versus reduced density ρ^* up to the limiting value of ρ^* where the phase transition occurs. Why is Z never less than one? Calculate the second, third, and fourth virial coefficients of eq. (8.4-11) for a hard-sphere fluid.

8.45. If $U(r)$ is the hard-sphere pair potential, rationalize the relation

$$\frac{dU}{dr} = -k_BT\delta(r - d)$$

where the right side is a Dirac delta function. Now use eq. (8.8-7) and derive the result

$$\frac{PV}{Nk_BT} = 1 + \frac{2\pi d^3}{3}\rho_0 g(d)$$

as an alternative form for the equation of state.

8.46. Following Guggenheim (1965) let us write for a fluid with attractive forces a "van der Waals type" of equation of state:

$$\frac{PV}{Nk_BT} = Z_{hs} - \frac{Na}{Vk_BT}$$

where Z_{hs} is the hard-sphere (repulsive) term and N^2a/V is the cohesive energy for N molecules of the substance, in the liquid state if V is the volume of the liquid, or in the solid state if V is the volume of the solid. At the triple point, the left side would be expected to be negligible. For argon at its triple point, the molar energy of vaporization is $8.56\,RT_{tr}$ and the molar energy of fusion is $1.883\,RT_{tr}$. Calculate $\tilde{V}_\ell/\tilde{V}_s$, the ratio of the molar volumes of the liquid and the solid at T_{tr}, and compare with the experimental value of 1.14.

8.47. Show that the absolute value of the Jacobian of the transformation (8.11-4) is rr'/R.

8.48. Verify eqs. (8.11-5) and (8.11-6).

8.49. Show that eq. (8.11-5) satisfies the constraint $\lim_{R \to \infty} g(R) = 1$.

REFERENCES

Abramowitz, M. and Stegun, I. A., *Handbook of Mathematical Functions*, Dover, New York, 1965, pp. 227–251. Tables of exponential, sine, and cosine integrals.

Alder, B. J. and Pople, J. A., "Third Virial Coefficient for Intermolecular Potentials with Hard Sphere Cores," *J. Chem. Phys.*, **26**, 325 (1957). The Lennard-Jones potential is illustrated in this paper.

Alder, B. J. and Wainwright, T. E., "Phase Transition for a Hard Sphere System," *J. Chem. Phys.*, **27**, 1208 (1957). An important early paper. Note that in the accompanying paper by Wood and Jacobson (p. 1207), the compressibility factor is plotted against reduced volume, V/V_0 where V_0 is the volume of fluid at closest packing. From the authors' Fig. 1, a fork appears at $V/V_0 \simeq 1.7$; since for hard spheres $nd^3/V = \sqrt{2}\,(V_0/V)$, this corresponds to $\rho^* = \rho d^3 \simeq 0.83$.

Barker, J. A., Henderson, D., and Smith, W. R., "Pair and Triplet Interactions in Argon," *Mol. Phys.*, **17**, 579 (1969). Calculations of the contribution of three-body interactions to thermodynamic properties of dense gaseous and liquid argon.

Barker, J. A. and Henderson, D., "What is 'Liquid'? Understanding the States of Matter," *Rev. Mod. Phys.*, **48**, 587 (1976). This article is an entire course on liquids and is worth several readings; see pp. 622–638 for material on the integral equation approach.

Bohr, N., "The Quantum Postulate and the Recent Development of Atomic Theory," *Nature* (London), **121**, 580 (1928). Bohr's presentation of what is now called his Correspondence Principle.

Bosi, P., Cilloco, F., and Ricci, M. A., "Neutron Diffraction Study of Liquid Iodine," *Mol. Phys.*, **40**, 1285 (1980). A typical report of a contemporary study of liquids by neutron diffraction.

Carnahan, N. F. and Starling, K. E., "Equation of State for Nonattracting Rigid Spheres," *J. Chem. Phys.*, **51**, 635 (1969). Development and testing of an analytical equation of state for a hard-sphere fluid.

Chandler, D., Weeks, J. D., and Andersen, H. C., "Van der Waals Picture of Liquids, Solids, and Phase Transformations," *Science*, **220**, 787 (1983). Pushes the original van der Waals point of view that attractive interactions fix the volume of a system, but repulsive forces determine the arrangements and motions of molecules within that volume. See also Rigby, M., *Quart. Rev.*, **24**, 416 (1970).

Chen, C. T. and Present, R. D., "Vapor Pressure Test for Intermolecular Potentials," *J. Chem. Phys.*, **54**, 58 (1971). Authors use $g(r)$ functions obtained by X-ray and neutron diffraction experiments to test various potential energy functions for argon.

Compton, A. H. and Allison, S. K., *X-Rays in Theory and Experiment*, 2nd ed., Van Nostrand, New York, 1935, p. 237. The Klein–Nishina formula for the intensity of scattering by a free electron according to Dirac theory.

Croxton, C. A., *Liquid State Physics—A Statistical Mechanical Introduction*, Cambridge University Press, London, 1974, pp. 37 and 82. Brief discussion of the breakdown of the BGBY equation, and of nonadditive effects in interparticle interactions.

Davidson, N., *Statistical Mechanics*, McGraw-Hill, New York, 1962, pp. 248–257. See these pages for a supplementary discussion of the grand canonical ensemble.

Deif, A. S., *Advanced Matrix Theory for Scientists and Engineers*, Abacus Press, Tunbridge Wells, 1982, pp. 57–77. Discussion of Gauss–Jordan elimination and other techniques for the solution of a system of linear equations.

Dence, J. B., *Mathematical Techniques in Chemistry*, Wiley, New York, 1975, p. 106. Brief discussion of Jacobians.

Dieterici, C., *Ann. Physik. Chem.*, **69**, 685 (1899), as quoted in MacDougall, F. H., "The Equation of State for Gases and Liquids," *J. Am. Chem. Soc.*, **38**, 528 (1916).

Dymond, J. H. and Smith, E. B., *The Second Virial Coefficients of Pure Gases and Mixtures*, Oxford University, Press, London, 1980. Covers data up to 1979.

Fitts, D. D., "Statistical Mechanics: A Study of Intermolecular Forces," *Ann. Rev. Phys. Chem.*, **17**, 59 (1966). Discussion of various model potential energy functions.

Fulks, W., *Advanced Calculus*, Wiley, New York, 1969, p. 402. A statement, without proof, of the conditions for the existence of an inverse to a function defined by an infinite series.

Gingrich, N. S., "The Diffraction of X-Rays by Liquid Elements," *Rev. Mod. Phys.*, **15**, 90 (1943).

Green, H. S., *The Molecular Theory of Fluids*, Dover, New York, 1969, pp. 57–61. Our discussion in Section 8.7 is based in part on these pages.

Guggenheim, E. A., "The New Equation of State of Longuet-Higgins and Widom," *Mol. Phys.*, **9**, 43 (1965).

Guggenheim, E. A., *Applications of Statistical Mechanics*, Oxford University Press, London, 1966. See Chapter 11 on "Localized Monolayer and Multilayer Adsorption of Gases."

Hill, T. L., *An Introduction to Statistical Thermodynamics*, Addison-Wesley, Reading, 1960, pp. 261–266, 462–464. The first citation forms the basis of our Section 8.4, while the second presents Kirkwood's argument for the appearance of h in classical canonical ensemble partition functions.

Hirschfelder, J. O., Curtiss, C. F., and Bird, R. B., *Molecular Theory of Gases and Liquids*, Wiley, New York, 1954, p. 163. The explicit infinite series representation of the second-virial coefficient for an LJ potential is given. The series converges rapidly for reduced temperatures greater than 4.

James, R. W., *The Optical Principles of the Diffraction of X-Rays*, Cornell University Press, Ithaca, 1965, pp. 461–463. Brief discussion here and elsewhere of formulas for incoherent scattering. For the general formula see eq. (3.48) in this monograph.

Jones, J. E., "On the Determination of Molecular Fields. II.," *Proc. Roy. Soc. London, Ser. A*, **106**, 463 (1924). Lennard-Jones introduces his potential (the form had, in fact, been suggested as early as 1903 by Mie) and applies it to the calculation of gas viscosities.

Karnicky, J. F. and Pings, C. J., "Recent Advances in the Study of Liquids by X-Ray Diffraction," *Adv. Chem. Phys.*, **34**, 157 (1976). An updated review article similar to that of Gingrich.

Kirkwood, J. G., "Molecular Distribution in Liquids," *J. Chem. Phys.*, **7**, 919 (1939). This article is reprinted in B. J. Alder (Ed.), *John G. Kirkwood: Theory of Liquids*, Gordon and Breach, New York, 1968.

Kohler, F., *The Liquid State*, Verlag Chemie, Weinheim, 1972, pp. 47–49. Further discussion, supplementing that in Green, of the scattering of X-rays by liquids. This book is an excellent introduction to the study of liquids.

Kreglewski, A., *Equilibrium Properties of Fluids and Fluid Mixtures*, Texas A & M University Press, College Station, 1984, Chap. 6. Concerns empirical equations of state.

Longuet-Higgins, H. C. and Widom, B., "A Rigid Sphere Model for the Melting of Argon," *Mol. Phys.*, **8**, 549 (1964). Interesting approximate model in the spirit of van der Waals for melting argon, which works reasonably well. For follow-up comments, see Guggenheim, E. A., *Mol. Phys.*, **9**, 43 (1965).

MacGillavry, C. H. and Rieck, G. D. (Eds.), *International Tables for X-Ray Crystallography*, Vol. III, 2nd ed. Kynoch Press, Birmingham, 1968, p. 201.

Maitland, G. C., Rigby, M., Smith, E. B., and Wakeham, W. A., *Intermolecular Forces: Their Origin and Determination*, Clarendon Press, Oxford, 1981.

McDonald, I. R. and Singer, K., "Computer Experiments on Liquids," *Chem. Brit.*, **9**(2), 54 (1973). A strictly qualitative review; see also McDonald, I. R. and Singer, K., *Quart. Rev.*, **24**, 238 (1970). A more thorough discussion is in Chap. 5 of Croxton.

Mourits, F. M. and Rummens, F. H. A., "A Critical Evaluation of Lennard-Jones and Stockmayer Potential Parameters and of Some Correlation Methods," *Can. J. Chem.*, **55**, 3007 (1977).

Narten, A. H., "Liquid Ammonia: Molecular Correlations from X-Ray Diffraction," *J. Chem. Phys.*, **66**, 3117 (1977).

Nazarian, G. M., "Statistical Mechanical Calculation of the Density Variation through a Liquid-Vapor Interface," *J. Chem. Phys.*, **56**, 1408 (1972).

Paalman, H. H. and Pings, C. J., "Fourier Analysis of X-Ray Diffraction Data from Liquids," *Rev. Mod. Phys.*, **35**, 389 (1963). A discussion of the application of the Fourier Integral Theorem to X-ray data from liquids.

Rowlinson, J. S., "A Comparison of the Solutions of Integral Equations for the Distribution Functions with the Properties of Model and Real Systems," in H. N. V. Temperley, J. S. Rowlinson, and G. S. Rushbrooke (Eds.), *Physics of Simple Liquids*, North-Holland, Amsterdam, 1968, Chap. 3. Short, clearly written article.

Rybolt, T. R., "A Virial Treatment of the Joule and Joule-Thomson Coefficients," *J. Chem. Educ.*, **58**, 620 (1981).

Slater, J. C. and Kirkwood, J. G., "The van der Waals Forces in Gases," *Phys. Rev.*, **37**, 682 (1931). The authors' pair potential energy function for helium is shown in Fig. 1 of this paper; an interesting approximate relationship between the polarizability of an atom and the number of electrons and their principal quantum number in the valence shell of the atom is also derived.

Spiegel, M. R., *Vector Analysis*, Schaum, New York, 1959. See Chap. 6 on the vector integral theorems.

Stockmayer, W. H., "Second Virial Coefficients of Polar Gases," *J. Chem. Educ.*, **9**, 398 (1941).

Tolstov, G. P., *Fourier Series*, Dover, New York, 1976, pp. 181–193. Rigorous but clear discussion of the Fourier Integral Theorem.

Watts, R. O., "Integral Equation Approximations in the Theory of Fluids," in K. Singer (Ed.), *Specialist Periodical Reports—Statistical Mechanics*, Vol. 1, Chemical Society, London, 1973, p. 1.

Yourgrau, W., van der Merwe, A., and Raw, G., *Treatise on Irreversible and Statistical Thermophysics*, Dover, New York, 1982, pp. 107–112. Our discussion of the distribution law in the grand canonical ensemble is based on these pages.

Fundamental Physical Constants and Units

Table A1.   Constants of Nature[a]

Quantity	Symbol	Value	Units
1. Speed of light *in vacuo*	c	2.99792×10^8	$\mathrm{m\ s^{-1}}$
2. Permittivity of vacuum	ε_0	8.85419×10^{-12}	$\mathrm{C^2\ N^{-1}\ m^{-2}}$ $(\mathrm{F\ m^{-1}})$
	$4\pi\varepsilon_0$	$111.26501 \times 10^{-12}$	$\mathrm{C^2\ N^{-1}\ m^{-2}}$ $(\mathrm{F\ m^{-1}})$
3. Proton charge	e	1.60219×10^{-19}	C
4. Planck's constant	h	6.62618×10^{-34}	$\mathrm{J\ s}$
	$\hbar$	1.05459×10^{-34}	$\mathrm{J\ s}$
5. Avogadro's number	N_0	6.02205×10^{23}	$\mathrm{mol^{-1}}$
6. Electron rest mass	m_e	9.10953×10^{-31}	kg
7. Proton rest mass	m_p	1.67265×10^{-27}	kg
8. Neutron rest mass	m_n	1.67495×10^{-27}	kg
9. Faraday's constant	F	9.64846×10^4	$\mathrm{C\ mol^{-1}}$
10. Rydberg constant for infinite mass	$\mathscr{R}_\infty$	1.09737×10^7	$\mathrm{m^{-1}}$
11. Bohr radius	a_0	5.29177×10^{-11}	m
12. Bohr magneton	β	9.27408×10^{-24}	$\mathrm{J\ T^{-1}}$
13. Nuclear magneton	β_N	5.05082×10^{-27}	$\mathrm{J\ T^{-1}}$
14. Molar gas constant	R	8.31441	$\mathrm{J\ K^{-1}\ mol^{-1}}$
15. Boltzmann's constant	k_B	1.38066×10^{-23}	$\mathrm{J\ K^{-1}}$
16. Gravitational constant	G	6.67204×10^{-11}	$\mathrm{N\ m^2\ kg^{-2}}$

[a]Abstracted from a more extensive compilation in R. C. Weast and M. J. Astle (Eds.), *CRC Handbook of Chemistry and Physics*, 61st ed., CRC Press, Boca Raton, 1980, pp. *F*-246 to *F*-247.

Table A2. Energy Conversions

	J	eV	cm^{-1}	kcal mol^{-1}
J	1			
eV	1.60219×10^{-19}	1		
cm^{-1}	1.98648×10^{-23}	1.23985×10^{-4}	1	
kcal mol^{-1}	6.95246×10^{-21}	4.33934×10^{-2}	3.49989×10^{2}	1
au	4.35982×10^{-18}	27.21166	2.19474×10^{5}	627.5116

Table A3. Miscellaneous Conversions

$1 \text{ J} = 10^{7} \text{ erg}$
$\quad = 9.86923 \times 10^{-3} \text{ L atm}$
$1 \text{ N} = 10^{5} \text{ dyne}$
$1 \text{ Pa} = 0.1 \text{ dyne cm}^{-2}$
$\quad = 9.86923 \times 10^{-6} \text{ atm}$
$1 \text{ Å} = 10^{-10} \text{ m} = 10^{-8} \text{ cm}$

$1 \text{ T(Wb m}^{-2}) = 10^{4} \text{ G}$
$1 \text{ V} = 1 \text{ J C}^{-1}$
$1 \text{ C} = 2.99792 \times 10^{9} \text{ statcoulomb (esu)}$
$\quad = 0.1 \text{ abcoulomb (emu)}$
$1 \text{ D} = 3.33565 \times 10^{-30} \text{ C m}$
$\quad = 10^{-18} \text{ statcoulomb cm}$

Sample Final Examinations
(closed book)

EXAMINATION 1. (120 MINUTES)

1. (10 pts) Terminology
Give precise one-sentence definitions or statements of the following: (a) potential energy of a particle, (b) Hermitian operator, (c) Koopmans' Theorem, (d) quantum ergodic hypothesis, and (e) configuration integral.

2. (8 pts) Conservative Forces
A necessary and sufficient condition that a force $\mathbf{F}$ be conservative is that curl $\mathbf{F} = \mathbf{0}$. Show that the Coulomb force meets this test: $\mathbf{F} = (q_1 q_2 / 4\pi\varepsilon_0)\mathbf{r}/r^3$.

3. (15 pts) Angular Momentum Operator
The following is an energy eigenstate of the hydrogen atom.

$$\Psi = \frac{1}{8\sqrt{\pi}} a_0^{-5/2} r e^{-r/2a_0} \sin\theta\, e^{-i\phi}$$

After first writing down the classical definition for angular momentum ℓ, deduce the quantum mechanical operator for ℓ_z, and show that for this eigenstate the electron possesses a definite value of the z component of its angular momentum. You may need to make use of the following relations:

$$x = r\sin\theta\cos\phi \qquad \frac{d}{dx}\tan^{-1}u = \frac{1}{1+u^2}\frac{du}{dx}$$

$$y = r\sin\theta\sin\phi \qquad \frac{d}{dx}\cos^{-1}u = -\frac{1}{\sqrt{1-u^2}}\frac{du}{dx}$$

$$z = r\cos\theta$$

310

4. (13 pts) Hermitian Operators

What is the principal reason why operators for observables are postulated in quantum mechanics to be Hermitian? Prove the basic theorem involved here.

5. (12 pts) Discussion

Write down a Slater determinantal wave function for the beryllium atom. Then present a well-organized discussion of as many aspects of the wave function as you can.

6. (8 pts) Mulliken Population Analysis

A Roothaan–Hartree–Fock calculation on HF with a minimal basis set has yielded the following set of occupied molecular orbitals.

AO \ MO	1	2	3	4	5
$1s_H$	− 0.0046	− 0.1606	− 0.5761	0	0
$1s_F$	0.9963	0.2435	− 0.0839	0	0
$2s_F$	0.0163	− 0.9322	0.4715	0	0
$2pz_F$	0.0024	0.0907	− 0.6870	0	0
$2px_F$	0	0	0	1	0
$2py_F$	0	0	0	0	1

What electron population is associated with the $1s_F$ orbital? Is this reasonable?

7. (10 pts) Partition Function

Recall that for an ideal gas of rigid rotors, the molecular rotational partition function for a diatomic molecule with symmetry factor σ is

$$q = \frac{T}{\sigma\Theta_r}\left[1 + \frac{1}{3}\left(\frac{\Theta_r}{T}\right) + \frac{1}{15}\left(\frac{\Theta_r}{T}\right)^2 + \cdots + \right]$$

In the high-temperature limit derive the following expression for the molar Gibbs free energy: $\langle \tilde{G} \rangle = RT[1 - \ln(T/\sigma\Theta_r) - R\Theta_r/3$. Recall also that $\langle A \rangle = -k_BT \ln Q$.

8. (10 pts) Second-Virial Coefficient

The Redlich–Kwong equation of state is

$$P = \frac{nRT}{V - nb} - \frac{n^2a}{T^{1/2}V(V + nb)}$$

Deduce from this an expression for the second virial coefficient $B(T)$.

9. (14 pts) Computer Program

We have need of the value of the Debye function $D(1.8)$, where

$$D(u) = \frac{3}{u^3}\int_0^u \frac{x^3\,dx}{e^x - 1}$$

Write out a computer program that would yield an accurate value of $D(1.8)$.

EXAMINATION 2.   (135 MINUTES)

1. **(10 pts) Terminology**
   Give precise one-sentence definitions or statements of the following: (a) Lagrangian for an isolated system of particles, (b) Slater screening factor, (c) Fundamental Symmetry Principle, (d) spectroscopic dissociation energy, and (e) grand canonical ensemble.

2. **(10 pts) Matrix Representations**
   A certain quantum mechanical observable has the 3×3 matrix representation

$$\frac{1}{\sqrt{3}} \begin{bmatrix} 0 & 1 & 0 \\ 1 & 0 & 1 \\ 0 & 1 & 0 \end{bmatrix}$$

   Find the normalized eigenvectors and the corresponding eigenvalues.

3. **(8 pts) Orthogonality**
   For the operator $\hat{O}$, the eigenvalue O_n is twofold degenerate; the eigenfunctions are Ψ_1 and Ψ_2, and as initially obtained are not orthogonal. Construct two eigenfunctions of $\hat{O}$, with eigenvalue O_n, which are orthogonal.

4. **(8 pts) Discussion**
   The Poisson bracket of some pairs of quantities in classical mechanics may vanish. Discuss the physical significance of the quantum mechanical analog of this, and then given an illustration or two in the quantal case.

5. **(12 pts) Variation Theorem**
   A four-term trial wave function for H_2^+ has the form

$$\Psi_{\text{tr}} = c_1(1s_A + 1s_B) + c_2(2pz_A + 2pz_B)$$
$$= c_1\Phi_1 + c_2\Phi_2$$

   Show that the expected value of the energy is given by the expression

$$\mathscr{E} = \frac{N \pm \left[N^2 - D^2\left(H_{11}H_{22} - H_{21}^2 \right) \right]^{1/2}}{D}$$

   where $N = S_{11}H_{22} + S_{22}H_{11} - 2S_{21}H_{21}$, $D = 2(S_{11}S_{22} - S_{21}^2)$, and the H and S symbols stand for $H_{21} = \int \Phi_2 \hat{H} \Phi_1 \, d\tau$, and so on. Which sign should be taken in order to yield the ground-state energy?

6. (10 pts) *h*-Theorem

State Boltzmann's *h*-Theorem, and then trace the reasoning that leads from it to an expression for the entropy of an isolated system.

7. (14 pts) Pauli Spin Matrices

In the theory of electron spin angular momentum, the following commutation relations are postulated:

$$\hat{s}_x\hat{s}_y - \hat{s}_y\hat{s}_x = i\hbar\hat{s}_z$$

$$\hat{s}_y\hat{s}_z - \hat{s}_z\hat{s}_y = i\hbar\hat{s}_x$$

$$\hat{s}_z\hat{s}_x - \hat{s}_x\hat{s}_z = i\hbar\hat{s}_y$$

Further, the operator $\hat{s}_z$ is stipulated to have just two eigenfunctions, with eigenvalues of $+\frac{1}{2}\hbar$ and $-\frac{1}{2}\hbar$. Show how to set up 2×2 matrix representations of $\hat{s}_x$, $\hat{s}_y$, and $\hat{s}_z$ in the basis of the eigenfunctions of $\hat{s}_z$.

8. (10 pts) Rotational Partition Function

The nuclei of a homonuclear diatomic molecule A_2 have spin quantum number $\frac{3}{2}$. The characteristic rotational temperature is 10 K. Calculate the percentage of A_2 molecules that are in the *para* form at the *low* temperature of 8 K in an equilibrium mixture.

9. (8 pts) Law of Corresponding States

Debye's formula for the heat capacity of a crystal is said to obey a Law of Corresponding States. Discuss Debye's model in the context of this statement.

10. (10 pts) Computer Program

In quantum chemical calculations on diatomic molecules, we have need of the functions $A_n(x)$,

$$A_n(x) = \frac{n!e^{-x}}{x^{n+1}} \sum_{k=0}^{n} \frac{x^k}{k!}$$

Write out a computer program to calculate the higher $A_n(x)$ functions from knowledge of $A_0(x)$.

EXAMINATION 3.   (120 MINUTES)

1. (10 pts) Terminology

Give precise one-sentence definitions or statements of the following: (a) classical Hamiltonian, (b) commutator, (c) canonical ensemble, (d) Boyle temperature, and (e) second-order specific configurational distribution function.

2. (12 pts) Semiclassical Hydrogen Atom

The Lagrangian for relative motion in a two-particle system of reduced mass μ is

$$L_{rel} = \frac{1}{2}\mu(\dot{r}^2 + r^2\dot{\phi}^2) - V(r)$$

Let the particle be restricted to a circle. Find the radius of the smallest such circle for a hydrogen atom for the case of the Coulomb potential [$V(r) = q_1q_2/4\pi\varepsilon_0|\mathbf{r}|$] if the angular momentum ℓ of the electron is quantized in the manner of Bohr. Evaluate this radius, given that $\varepsilon_0 = 8.85419 \times 10^{-12}$ $C^2\ N^{-1}\ m^{-2}$, $e = 1.60219 \times 10^{-19}$ C, $h = 6.62618 \times 10^{-34}$ J Hz^{-1}, $m_e = 9.10953 \times 10^{-31}$ kg, and $m_p = 1.67265 \times 10^{-27}$ kg.

3. (14 pts) Virial Theorem

Let $\hat{A}$ be a Hermitian operator and suppose the Hamiltonian $\hat{H}$ and possibly also $\hat{A}$ depend on a common parameter λ. Prove the relation

$$\left\langle \frac{\hat{A}\,\partial\hat{H}}{\partial\lambda} \right\rangle = \langle\hat{A}\rangle\frac{\partial E_i}{\partial\lambda} - \int \Psi_i^*[\hat{A}, \hat{H}]\frac{\partial\Psi_i}{\partial\lambda}\,dq$$

where E_i is the eigenvalue of $\hat{H}$ corresponding to the eigenfunction Ψ_i. Apply the formula to the hydrogen atom ($E_i = -2\pi^2\mu e^4/h^2i^2$) and thereby demonstrate the validity of the virial theorem: $\langle\hat{T}\rangle = -\langle\hat{V}\rangle/2$.

4. (10 pts) Discussion

The Hartree–Fock electronic energy for a closed-shell molecule of $2N$ electrons in its ground state is

$$E = \sum_{n=1}^{N}\left(2H_{nn} + \sum_{m=1}^{N}(2J_{nm} - K_{nm})\right)$$

Explain in detail the meaning of all terms of this equation. What approximations, if any, have gone into the derivation of the equation?

5. (9 pts) Dipole Moment

The molecular dipole moment operator (in the SI system) is

$$\hat{\mu} = \sum_{\alpha=1}^{M} Z_\alpha\mathbf{R}_\alpha - \sum_{i=1}^{N}\mathbf{r}_i$$

Rationalize the form of this operator and explain all the symbols.

6. (10 pts) Tensors

A molecule is oriented in a certain Cartesian coordinate system. The polarizability tensor is then calculated to be

$$\begin{bmatrix} 49.5 & 10.2 & 9.00 \\ 10.2 & 31.6 & 11.1 \\ 9.00 & 11.1 & 23.1 \end{bmatrix}$$

where each element is in units of a_0^3. An electric field

$$\mathbf{E} = 4.20\mathbf{e}_x + 13.55\mathbf{e}_y + 12.10\mathbf{e}_z$$

(units of kilovolts per centimeter) is now applied. Calculate the magnitude of the induced dipole moment in Debyes along the x direction, where $1 \text{ D} = 3.336 \times 10^{-30}$ C m. You will also need $a_0 = 5.292 \times 10^{-11}$ m, $4\pi\varepsilon_0 = 1.113 \times 10^{-10}$ C^2 N^{-1} m^{-2}.

7. (12 pts) NDDO Theory

The first three approximations in Pople's NDDO theory of electronic structure may be referred to as (a) the core approximation, (b) orthonormality of the basis set, and (c) neglect of differential diatomic overlap. Explain fully the physical content of these three approximations.

8. (9 pts) Boyle Temperature

Prove that the definition of the Boyle temperature is equivalent to the statement that T_B is that temperature at which the second virial coefficient $B(T)$ vanishes.

9. (14 pts) Vibrational Contribution to Entropy

(a) Let q_{vib} stand for the molecular vibrational partition function of a molecule. Starting from $\langle A_{\text{vib}} \rangle = -k_B T \ln Q_{\text{vib}}$, derive the vibrational contribution to the molar entropy in terms of q_{vib} for an independent collection of molecules.

(b) Now consider the molecules to be anharmonic oscillators having the potential energy $V(x) = \frac{1}{2}\mu\omega^2 x^2 + Dx^4$. Time-independent perturbation theory yields for the approximate eigenvalues

$$\varepsilon_n = \left(n + \tfrac{1}{2}\right)\hbar\omega + \tfrac{3}{4}(2n^2 + 2n + 1)\left(\frac{\hbar^2}{\mu^2\omega^2}\right)D$$

Now reexpress in more detail your answer in part (a).

(c) In a particular case, let μ, ω, D be given. Write a computer program that could then be used to calculate $\langle \tilde{S}_{\text{vib}} \rangle$.

Hints and Answers for Selected Exercises in the Text

CHAPTER 1

1.1. $\mathbf{F}_{12} = (q_1 q_2 / 4\pi\varepsilon_0)(\mathbf{r}_1 - \mathbf{r}_2)/|\mathbf{r}_1 - \mathbf{r}_2|^3$

1.2. $x = A \sin\sqrt{k/m}\, t + B \cos\sqrt{k/m}\, t$; $\quad T + V(x) = \frac{1}{2}k(A^2 + B^2) + c = $ constant

1.3. $v(t)_{\text{term}} = qE/\beta$

1.8. $m_H R^2 \ddot{\theta} = -k(\theta - \theta_0)$; if $\theta(0) = \theta_0$, $\dot{\theta}(0) = v_0 \neq 0$, then $\theta(t) = \theta_0 + v_0 R\sqrt{m/k} \sin(\sqrt{k/m}\, t/R)$

1.11. Make use of the relation $\dot{\mathbf{r}} = \boldsymbol{\omega} \times \mathbf{r}$ between linear velocity and angular velocity; also make use of the vector identity $\mathbf{A} \times (\mathbf{B} \times \mathbf{C}) = (\mathbf{A} \cdot \mathbf{C})\mathbf{B} - (\mathbf{A} \cdot \mathbf{B})\mathbf{C}$.

1.12. In the case of the Coulombic interaction $r = -4\pi\varepsilon_0 \ell^2 / \mu q_1 q_2$.

1.15. Use mathematical induction.

1.16. Let $a(t) = \omega m x(t) + i p(t)$ and show that this yields $\omega m \dot{x}(t) + i \dot{p}(t) = \omega p(t) - i m\omega^2 x(t)$.

1.17. To show that H is constant, express it as a function of r, ϕ, p_r, p_ϕ, and then look at $\dot{H}$.

1.19. For total energy greater than zero, the system is not a bound system and $x(\tau)$ will increase with τ instead of oscillating. A typical result follows, where we have taken $x(0) = 1$, $\dot{x}(0) = \sqrt{2}$, $\Delta\tau = 0.01$, and $\tau_{\max} = 10$:

τ	0	1	2	5	8	9	10
$x(\tau)$	1	2.233	3.259	6.024	8.712	9.606	10.500
$H(\tau)$	0.39958	0.39957	0.39957	0.39957	0.39957	0.39957	0.39957

CHAPTER 2

2.3. (a) and (d) are linear.

2.6. Only the second matrix is Hermitian.

2.7. $E_1 = \pi^2\hbar^2/2mL^2$

2.8.
$$\mathbf{p}_x = i\hbar \begin{bmatrix} 0 & -\frac{1}{6}\sqrt{6} & 0 \\ \frac{1}{6}\sqrt{6} & 0 & -\frac{1}{10}\sqrt{10} \\ 0 & \frac{1}{10}\sqrt{10} & 0 \end{bmatrix}$$

2.9. (a) $\mathbf{H} = \begin{bmatrix} E_1 & 0 \\ 0 & E_2 \end{bmatrix}$; (c) If A is an observable, then a is real.

2.11. Assume $\mathbf{x} = \sum_i A_i \mathbf{a}_i = \sum_i B_i \mathbf{a}_i$, and for an arbitrary j look at $\mathbf{a}_j \cdot \mathbf{x}$.

2.12. $\phi(x) = 1 = \dfrac{2}{\pi} \sum_{n=1}^{\infty} \dfrac{1}{n}(1 - \cos n\pi)\sin\dfrac{n\pi x}{L}, \qquad 0 \le x \le L.$

2.13. For a 15-term expansion, one obtains the following:

x/L	0.05	0.15	0.25	0.35	0.55	0.65	0.75	0.95
$\phi(x)$	1.1353	0.9644	0.9443	0.9876	1.0327	0.9876	0.9443	1.1353

2.14. (a) 1;   (b) 0;   (c) $\frac{1}{2}e^{-1/4}$;   (d) $\delta(x - x_2)$.

2.15. Show that $\lim_{n\to\infty} \int_{-1/n}^{1/n} \delta_n(t)f(t)\, dt = f(0)$, if $f(t)$ is continuous.

2.16.
$$\lim_{n\to\infty} \delta_n(x) = \begin{cases} \infty & x = 0 \\ 0 & x \ne 0 \end{cases}; \qquad \int_{-\infty}^{\infty} \delta_n(x)\, dx = 1$$
$$\lim_{n\to\infty} \int_{-\infty}^{\infty} \delta_n(x)f(x)\, dx = f(0)$$

2.17. In first sequence, plots approach that of a Heaviside step function:
$$H(x) = \begin{cases} 1 & x > 0 \\ 0 & x < 0 \end{cases};$$
derivatives form a delta sequence. In second sequence, plots approach that of $H(x) - 1$, and derivatives also form a delta sequence.

2.18. $\hat{x}$ and $\hat{p}$ are separately Hermitian; write the equations that express this and then combine them to look at $\hat{a}$ or $\hat{a}^\dagger$.

2.20. Careful of the order in performing the multiplications of the operators.

2.21. Make use of an integration by parts and recall that $\hat{p}$ is Hermitian.

2.23. Draw on the result of Exercise 2.21.

2.24.

$$x_{mn} = \left(\frac{\hbar}{2m\omega}\right)^{1/2} \begin{cases} (n+1)^{1/2} & m = n + 1 \\ n^{1/2} & m = n - 1 \\ 0 & \text{otherwise} \end{cases}$$

$$p_{mn} = -i\left(\frac{\hbar m\omega}{2}\right)^{1/2} \begin{cases} -(n+1)^{1/2} & m = n + 1 \\ n^{1/2} & m = n - 1 \\ 0 & \text{otherwise} \end{cases}$$

$$a_{mn} = \begin{cases} n^{1/2} & m = n - 1 \\ 0 & \text{otherwise} \end{cases} \qquad a^{\dagger}_{mn} = \begin{cases} (n+1)^{1/2} & m = n + 1 \\ 0 & \text{otherwise} \end{cases}.$$

The matrices are therefore banded; the matrix x, for example, is

$$\mathbf{x} = \left(\frac{\hbar}{2m\omega}\right)^{1/2} \begin{bmatrix} 0 & \sqrt{1} & 0 & 0 & \cdots \\ \sqrt{1} & 0 & \sqrt{2} & 0 & \cdots \\ 0 & \sqrt{2} & 0 & \sqrt{3} & \cdots \\ 0 & 0 & \sqrt{3} & 0 & \\ \vdots & \vdots & \vdots & & \ddots & 0 \end{bmatrix}$$

2.25.

$$\hat{\ell}_x = i\hbar\left(\sin\phi\frac{\partial}{\partial\theta} + \cot\theta\cos\phi\frac{\partial}{\partial\theta}\right)$$

$$\hat{\ell}^2 = -\hbar^2\left[\frac{1}{\sin\theta}\frac{\partial}{\partial\theta}\left(\sin\theta\frac{\partial}{\partial\theta}\right) + \frac{1}{\sin^2\theta}\frac{\partial^2}{\partial\phi^2}\right]$$

2.26. The clumsy but direct way to do this is to expand each exponential out to the desired order and compare terms.

2.27. Differentiate $S(t)$ under the integral sign and make use of the Hermiticity of $\hat{H}$ and of QM Postulate 7.

2.28.   Heisenberg equations: $\dfrac{\partial\hat{x}}{\partial t} = \dfrac{\hbar^2}{\mu}\hat{p}, \quad \dfrac{\partial\hat{p}}{\partial t} = -\hbar^2\mu\omega^2\hat{x}$

Classical equations: $\dfrac{dx}{dt} = \dfrac{p}{\mu}, \quad \dfrac{dp}{dt} = -\mu\omega^2 x$

2.31. Let $x = \omega t/2$; then extrema occur at the roots of the transcendental equation $\sin x(x \cos x - \sin x) = 0$. The root $x = 0$ gives the central maximum, and application of the theorem of l'Hôpital shows that $F(t, 0) = t^2$ there. The equation $\sin x = 0$ ($x \neq 0$) leads to minima, and thus $x = \pm n\pi$. The equation $x \cos x - \sin x = 0$ leads to maxima, at the first roots of which $F(t, \omega)$ has the values shown:

x	0	± 4.4934	± 7.7253
$F(t, \omega)$	t^2	$0.0472\,F(t, 0)$	$0.0165 F(t, 0)$

CHAPTER 3

3.2. $\tilde{\nu}_0 = 2989.7$ cm^{-1}, $D = -5.1915 \times 10^{21}$ g cm^{-2} s^{-2}; second overtone is calculated to occur at 8346 cm^{-1}.

3.3. If $\varepsilon = h^2/32mL^2$, then

$$\Psi_1^{(1)}(x) = \sqrt{\frac{2}{L}} \sin\frac{\pi x}{L} - \frac{1}{\pi}\sqrt{\frac{2}{L}} \sum_{k=2}^{\infty} \frac{k}{(1-k^2)^2} \sin\frac{k\pi x}{L} \cos\frac{k\pi}{2}$$

and the left half of the box is slightly favored.

3.5. The normalized first-order wave function is

$$\Psi_0^{(1)}(x) = \left(\frac{1}{1 + (q^2 E^2/2\mu\hbar\omega^3)} \right)^{1/2}$$

$$\times \left[\Psi_0^{(0)}(x) - \left(\frac{qE}{\hbar\omega} \right)\left(\frac{\hbar}{2\mu\omega} \right)^{1/2} \Psi_1^{(0)}(x) \right]$$

and the probability that the molecule will be found in the $v = 1$ state is given by the square of the coefficient of $\Psi_1^{(0)}$:

$$p_1 = \frac{q^2 E^2/2\mu\hbar\omega^3}{1 + q^2 E^2/2\mu\hbar\omega^3}$$

$$\simeq 1.37 \times 10^{-9}$$

if $E = 10^7$ V m^{-1}, $\omega = 5.534 \times 10^{14}$ s^{-1}, $\mu = 1.628 \times 10^{-27}$ kg, $\mu_{eq} = 1.08$ D, and $R_0 = 1.275$ Å.

3.9. (a) 0 in all four cases;
   (b) $\frac{1}{4}\hbar^2$ in all four cases;
   (c) $\hat{s}_x\alpha = \frac{1}{2}\hbar\beta$, $\hat{s}_x\beta = \frac{1}{2}\hbar\alpha$;
   (d) $\hat{s}_y\alpha = \frac{1}{2}i\hbar\beta$, $\hat{s}_y\beta = -\frac{1}{2}i\hbar\alpha$;
   (e) $e_x(\frac{1}{2}\hbar) + e_y(\frac{1}{2}i\hbar)$;
   (f) $\frac{1}{2}\hbar^2$.

3.11. No; yes.

3.12. Det $= -192$ as written.

3.13. Integrals have the form (after integrating away the spin coordinates)

$$\iint \chi_\alpha^2(i)\chi_\beta^2(j)r_{ij}^{-1}\,dv_i\,dv_j \qquad (\alpha, \beta = 1s, 2s)$$

3.15.
$$E_1^{(2)} \simeq \frac{h^2}{8mL^2} - \frac{128m\varepsilon^2L^2}{\pi^2h^2}\sum_{k=2}^{\infty}\frac{k^2\cos^2(k\pi/2)}{(k^2-1)^3}$$

$$\simeq 0.9825\frac{h^2}{8mL^2} \quad \text{if } \varepsilon = \frac{h^2}{32mL^2}$$

Use of the variation theorem leads to the equation

$$E(\lambda) = \frac{\hbar^2\pi^2}{2mL^2}\left(\frac{1 - \dfrac{4\lambda}{3\pi} + 4\lambda^2}{1 + \lambda^2}\right)$$

and minimization gives $\lambda = 0.0704$, whence $E = 0.9899(h^2/8mL^2)$.

3.17. $\mathscr{E}_{\text{exc}} = -3.014$ eV, and this state is unstable.

3.18.

R (au)	1.39725	1.79725	2.19725	2.59725	2.99725	3.79725
Z	1.39273	1.28311	1.20141	1.14047	1.09516	1.03768
E (au)	-0.55495	-0.58411	-0.58501	-0.57619	-0.56453	-0.54217

CHAPTER 4

4.2. (a) 6×6;   (b) 5!;   (d), (e) 4!

4.3. (a) -243.84 eV;   (b) $D_e = -0.299$ eV;
   (c) $D_e = 2.64$ eV; $\mathscr{E}$ would have to be -219.72 eV;
   (d) 94.16 eV;   (e) IP(LiH)$_{\text{calc}} = 8.26$ eV.

4.4. (a) $E = -3596.88$ eV.

4.5. Seven others necessarily equal $(ab|cd)$.

4.6. If $\chi_1(j) = c_{A1}1s_A(j) + c_{B1}1s_B(j)$ and $\chi_2(j) = c_{A2}1s_A(j) + c_{B2}1s_B(j)$, then

$$\begin{bmatrix} F_{11} & F_{12} \\ F_{21} & F_{22} \end{bmatrix}\begin{bmatrix} c_{A1} & c_{A2} \\ c_{B1} & c_{B2} \end{bmatrix} = \begin{bmatrix} 1 & S_{12} \\ S_{21} & 1 \end{bmatrix}\begin{bmatrix} c_{A1} & c_{A2} \\ c_{B1} & c_{B2} \end{bmatrix}\begin{bmatrix} \varepsilon_1 & 0 \\ 0 & \varepsilon_2 \end{bmatrix}$$

4.7. $P_{11} = 0.7154$; $P_{12} = 0.0093$; $P_{14} = 0.7624$; $1s_H|2pz_F$ overlap population makes the greater contribution.

4.9. Let $\chi_{3s'} = A\chi_{1s} + B\chi_{2s'} + C\chi_{3s}$; then we have the three equations

$$1 = A^2 + B^2 + C^2 + 2ACS_{13} + 2BCS_{2'3}$$
$$0 = A + CS_{13}$$
$$0 = B + CS_{2'3}$$

4.10.

	$\xi(1s)$	$\xi(2s)$
Bessis	2.6321	0.5869
Ransil	2.6865	0.6372

4.13. $B_n(0) = (n + 1)^{-1} \times \begin{cases} 0 & n = \text{odd} \\ 2 & n = \text{even} \end{cases}$

4.14. $H_{12} =$
$- \frac{1}{2}R^2(\xi_H\xi_L)^{3/2}\left[\frac{1}{4}\xi_L^2 R(A_2B_0 - A_0B_2) + (4 - \xi_L)A_1B_0 + (\xi_L - 2)A_0B_1\right]$
$H_{21} =$
$- \frac{1}{2}R^2(\xi_H\xi_L)^{3/2}\left[\frac{1}{4}\xi_H^2 R(A_2B_0 - A_0B_2) + (4 - \xi_H)A_1B_0 - (\xi_H + 2)A_0B_1\right]$
where the argument of the A_n functions is $\alpha = \frac{1}{2}R(\xi_H + \xi_L)$, and the argument of the B_n functions is $\beta = \frac{1}{2}R(\xi_L - \xi_H)$. Let $R = 3.01542$ au, and let ξ_H and ξ_L vary. Comparisons to six decimals of accuracy are

	$\xi_H = 1, \xi_L = 2.7$	$\xi_H = 1.2, \xi_L = 2.65$	$\xi_H = .97, \xi_L = 2.6$
H_{12}	-0.418512	-0.328384	-0.444527
H_{21}	-0.418512	-0.328384	-0.444527

4.15. $2, 3 \pm \sqrt{7}$.

4.16. If only the positive square roots of the eigenvalues are taken, then

$$\mathbf{A}^{1/2} = \begin{bmatrix} 1.57401 & 0.58473 & 0.42494 \\ 0.58473 & 1.91057 & -0.08838 \\ 0.42494 & -0.08838 & 0.90090 \end{bmatrix}$$

4.19. $\lim_{R \to \infty} H_{22} = -4.455$ au.

4.20.
$$S^{1/2} = \begin{bmatrix} 0.91680 & 0.03890 & 0.25786 & 0.30231 \\ 0.03890 & 0.99589 & 0.07869 & -0.00438 \\ 0.25786 & 0.07869 & 0.96187 & -0.04054 \\ 0.30231 & -0.00438 & -0.04054 & 0.95234 \end{bmatrix}$$

4.21.
$$A'^{-1} = \begin{bmatrix} 0.33333 & -0.16666 & -0.33333 \\ 0.16666 & 0.16666 & -0.16666 \\ -0.33333 & 0.16666 & 1.33333 \end{bmatrix}$$

4.22.

	(12|23)	(23|44)
Exact value	0.0123839	0.0525007
Mulliken approximation	0.01013	0.04655
|% error|	18.2	11.3

4.23. $P_{12}S_{12} = -0.01481$, $P_{13}S_{13} = 0.20174$, $P_{14}S_{14} = 0.19904$, $2N = 3.99999958$.

CHAPTER 5

5.2. $z_{33} = z_{44} = -\frac{1}{2}R$; to find z_{24} one needs the formula for the overlap integral between two $2p_z$ orbitals centered on the same nucleus.

5.3. Use $R_0 = 0.91681$ Å and Slater's rules for the screening factors; also make use of Exercise 5.2. Then, $\langle \mu_{HF} \rangle_{calc} = 1.878$ D.

5.4. $q_F = -0.3747\, e$.

5.6. NaF: $q_F = -0.5996\, e$;   HF: $q_F = -0.2999\, e$.

5.7. (a) $H_{11} = -1.125439$, $H_{12} = -0.970220$, $S_{12} = 0.713533$
(b) $(11|11) = 0.6875$, $(11|22) = 0.533333$, $(12|12) = 0.310781$, $(11|12) = 0.435552$
(c) $c_{11} = c_{12} = 0.540180$, $E = -1.8355916$, $\mathscr{E}(R) = -1.122086$
(d) $D_0 = 0.1128$

5.8. In each experiment the total polarization energy is the sum of two contributions, which are to be computed separately.

5.9. Use the idea that scalars are invariant to rotations of the axes.

5.10. $\alpha' = \begin{bmatrix} 29.70 & 0 & 0 \\ 0 & 25.81 & 3.90 \\ 0 & 3.90 & 25.81 \end{bmatrix}$

5.12. (b) Look at the determinant of **S**;   (c) $\mathrm{Tr}\,\alpha = \mathrm{Tr}\,\alpha' = 81.32$.

5.13. See Exercise 5.12 again.

5.14. $\mathbf{T}' = \begin{bmatrix} -2.1269 & 0 & 0 \\ 0 & 1.7772 & 0 \\ 0 & 0 & 6.3497 \end{bmatrix}$

5.17. $F_{11} = -0.450084$, $F_{12} = -0.599608$, $\varepsilon = (F_{11} \pm F_{12})/(1 \pm S_{12}) = -0.612589$, $+0.521959$; then

$$\alpha_{xx} = \alpha_{yy} = \frac{8}{\varepsilon_2 - \varepsilon_1} c_{11}^2 [X2(1,1) + X2(1,2)] = 3.06 \text{ au}$$

$$\alpha_{zz} = \frac{8}{\varepsilon_2 - \varepsilon_1} c_{11}^2 [Z2(1,1) + Z2(1,2)] = 4.21 \text{ au}$$

5.18. Yes.

5.19. The matrix is orthogonal.

5.20. The problem is asking: If A'_{ij} means $\int \chi_i \hat{A} \chi_j \, dq$, how is A'_{ij} related to the elements of $\mathbf{A}$?

5.21. First show that the matrix $\boldsymbol{\varepsilon}$ is unchanged by the orthogonal transformation (solve the Roothaan matrix equations for $\boldsymbol{\varepsilon}$ and $\boldsymbol{\varepsilon}'$); then show that $\mathbf{Fc} = \mathbf{Sc}\boldsymbol{\varepsilon}$ becomes $\mathbf{F'c'} = \mathbf{S'c'}\boldsymbol{\varepsilon}$, where $\mathbf{F}', \mathbf{c}', \mathbf{S}'$ are the results of orthogonal transformations on $\mathbf{F}, \mathbf{c}, \mathbf{S}$.

5.22. (a) Recall eq. (4.1-12); (c) Exercise 7.46 may be useful; arrive at the conclusion that $E = E'$.

5.23. For F_{11}, only 7 distinct repulsion integrals survive; in the full Hartree–Fock treatment there are 16 distinct such integrals. For F_{13}, only 3 distinct repulsion integrals survive.

CHAPTER 6

6.2. $\Omega(E_n) = -me^4/(8\varepsilon_0^2 h^2 E_n)$, $d\Omega/dE_n = 8n^4\varepsilon_0^2 h^2/me^4 = 10^8$ au^{-1} if $n = 100$.

6.4. $S = C \ln 5$.

6.6. Number of distributions $= 286$; number of ways $\Omega = \Sigma W$ of realizing these distributions $= 1,048,576$. Maximum in W occurs for $(3, 3, 2, 2)$.

6.7. Possible *types* of energy distribution are found by solving the Diophantine equation $3a + 2b + 2c + (10 - a - b - c) = 16$, or $2a + b + c = 6$. Four types of distribution result, and number of distributions $= 16$.

6.8. Total number of ensemble states $= 38,760$. Average populations of the four system states: 4.789, 2.211, 2.211, 0.789.

6.9. Maximum in W occurs for $(5, 2, 2, 1)$.

6.10. For example, choose agreement for states $1, 4$; then $\langle N_2 \rangle = \langle N_3 \rangle$ are calculated as 1.944, to be compared with the exact value of 2.211.

6.13. For 15!, the percentage error in using the expression from footnote 5 is 0.226%; the two-term approximation gives an error of 89.8%.

6.16. $\lim_{x \to 0} \langle S \rangle = 0$.

6.18. For a pure substance, start with the relation $dE = dq_{rev} - dw_{rev} + \mu \, dN$.

6.19. $\sigma_E / \langle E \rangle = 0.8448$; from eq. (6.7-5b) one finds $\sigma_E / \langle E \rangle = 0.3651$.

6.23. $H_{real} = H_{id} + k_B T[T(\partial \ln f / \partial T)_{N,V} + V(\partial \ln f / \partial V)_{N,T}]$

CHAPTER 7

7.2. From $\tilde{G}(T, P) = \tilde{G}°(T)_{id} + RT \ln f$, obtain the relation

$$\tilde{S}(T, P) = \tilde{S}°(T)_{id} - RT\left(\frac{\partial \ln f}{\partial T}\right)_P - R \ln f$$

Next obtain $\tilde{S}(T, P)_{id} - \tilde{S}°(T)_{id} = -R \ln P$, and the result follows.

7.3. $\tilde{V} \simeq RT/P + b - a/P\tilde{V} + ab/P\tilde{V}^2$; $\tilde{S}(T, 1) - \tilde{S}(T, 1)_{id} \simeq -0.0317$ J mol^{-1} K^{-1}.

7.4. $\langle \tilde{S}_{elec} \rangle = 17.52$ J mol^{-1} K^{-1} $\Rightarrow \langle \tilde{S} \rangle = 167.86$ J mol^{-1} K^{-1} (calc).

7.5. $\langle \tilde{S} \rangle = 129.20$ J mol^{-1} K^{-1} (calc).

7.7. Using eq. (7.2-4): $\langle \tilde{S}°_{vib} \rangle = 0.2639 \, R$; using three-term expansion of q_{vib}: $\langle \tilde{S}°_{vib} \rangle = 0.2611 \, R$.

7.9. Obtain $q_{trans} = 2\pi m k_B Tab/h^2 \Rightarrow \langle \tilde{A}_{tr} \rangle = -RT \ln(2\pi m k_B Tabe/N_0 h^2)$ and

$$\langle \tilde{S}_{tr} \rangle = R\left[1 + \ln\left(2\pi m k_B Tabe/N_0 h^2\right)\right]$$

7.11. $J_{mp} = (T/2\theta_r)^{1/2} - \frac{1}{2}$; for HBr at 298 K, $J_{mp} \simeq 3$, and for NO at 298 K, $J_{mp} \simeq 7$.

7.12.

	Trans	Vib	Rot	Total	
$\langle \tilde{S}° \rangle$	172.02	5.37	67.68	245.07	J mol^{-1} K^{-1}
$\langle \tilde{C}_p° \rangle$	$12.47 + R$	6.80	8.31	35.89	J mol^{-1} K^{-1}

7.13. $\langle \Delta \tilde{S}_f° \rangle_{HBr} = 57.32$ J mol^{-1} K^{-1} (calc).

7.14. 53.5% parahydrogen.

7.16. High temperature: $33\frac{1}{3}$% paradeuterium; at 0 K: 0% paradeuterium.

7.18. For $R(r)$ to be normalizable, we require $\lim_{r \to 0} R(r) = 0 \Rightarrow$ $\lim_{x \to -r^{\circ}_{AB}} T(x) = 0$.

7.19. $J \simeq 44$; if J is halved to 22, the centrifugal distortion energy becomes 0.0653 of the original value.

7.20. $\tilde{\nu}_0 = 2887.96$ cm^{-1}, $B = 9.4935$ cm^{-1}, $D = 0.004832$ cm^{-1}.

7.21. $\tilde{\nu}_0 = 2861.60$ cm^{-1}, $B = 14.4576$ cm^{-1}, $D = 0.0014828$ cm^{-1}.

7.22. $D_0 = 103.50$ kJ mol^{-1}.

7.23. Write the energies relative to that at the minimum ($\mathscr{E}_3 = 0$). Then,

$$\text{ESQ} = \sum (\text{squares of deviations}) = \sum_{i=1}^{6} \left[\mathscr{E}_i - 0.60262(1 - e^{-a(r_i - 1.99725)})^2 \right]^2$$

and we desire $d(\text{ESQ})/da = 0$. Obtain, $a = 0.5011 a_0^{-1}$ cm^{-1}, and $\tilde{\nu}_0 = 3983$ cm^{-1}.

7.24. 50 K: $\beta/3 = 0.0801$ 250 K: $\beta/3 = 0.0160$
           $\beta^2/15 = 0.0038$ $\beta^2/15 = 0.00015$.

7.28. For HF, $\Theta_r = 30.149$ K $\Rightarrow \frac{1}{3}R\Theta_r = 0.752$ J mol^{-1}, while $RT = 2.479$ J mol^{-1} at 298 K.

7.29–7.30.

	C$_2$H$_5$OH	NH$_3$	CH$_3$SH	SO$_2$	CH$_3$Br	BrF
$\langle \tilde{C}_p^{\circ} \rangle / R$, calc	8.06	2.99	5.97	5.79	5.29	3.32
$\tilde{C}_p^{\circ}/R$, exp	7.87	4.22	6.04	4.80	5.11	3.97
$\langle \tilde{S}^{\circ} \rangle / R$, calc	34.95	25.56				
$\tilde{S}^{\circ}/R$, exp	33.99	23.14				

7.32. $\mathbf{I'} = \begin{bmatrix} 21.51 & 0 & 0 \\ 0 & 4.84 & 5.53 \\ 0 & 5.53 & 16.67 \end{bmatrix} \times 10^{-46}$ kg m^2

$\text{Tr}\,\mathbf{I'} = 43.02 \times 10^{-46}$ kg m^2; $\text{Tr}\,\mathbf{I} = 18.27 \times 10^{-46}$ kg m^2

7.33.

	Trans	Vib	Rot	Total	
$\langle \tilde{S}^{\circ} \rangle$	174.16	74.25	85.66	334.07	J mol^{-1} K^{-1}
$\langle \tilde{C}_p^{\circ} \rangle$	12.47 + R	62.42	12.47	95.68	J mol^{-1} K^{-1}

$I_x = I_y = I_z = 74.95 \times 10^{-46}$ kg m^2

7.35. If $V_0 \simeq 0$, then for C$_2$H$_6$ at 298 K with $I_r = 2.76 \times 10^{-47}$ kg m^2, one obtains $\langle \tilde{S}_{ir} \rangle = 11.96$ J mol^{-1} K^{-1}.

7.37. $\langle \tilde{S}_{ir} \rangle = 12.5$ J mol^{-1} K^{-1}; since restricting the rotation can only lower $\langle \tilde{S}_{ir} \rangle$ relative to that in the freely rotating case (Exercise 7.35), conclude that $(CH_3)_2Li$ has free rotation.

7.38. $K_{actual}/K_{froz} = 1.9$; $K_{actual}/K_{froz} = 0.552$

7.40. 303 cm^{-1}

7.41. O—Cl $= 1.700$ Å, $\angle$Cl—O—Cl $= 110.9°$
$I_x = 25.0987 \times 10^{-46}$ kg m^2, $I_y = 2.01498 \times 10^{-46}$ kg m^2, $I_z = 23.0837$ $\times 10^{-46}$ kg m^2, $\tilde{G}° = \tilde{A}° + RT = RT - k_B T \ln(Q_{tr}Q_{rot}Q_{vib}Q_{elec})$

	Cl$_2$	O$_2$	OCl$_2$	
$\tilde{G}°$	-20.742	-15.287	-20.873	RT

$\Delta \tilde{G}_f°(OCl)_2 = \tilde{G}°(OCl_2) - \tilde{G}°(Cl_2) - \frac{1}{2}\tilde{G}°(O_2) = 18.62$ kJ mol^{-1}
This value is with respect to the bottoms of the respective potential wells.

7.44. $MM^{-1} = 1 \Rightarrow MM^T = 1$ if M is orthogonal, so $\det|M| \times \det|M^T| = 1$; but exchange of rows with corresponding columns in M leaves the determinant unchanged, so $\det|M^T| = \det|M|$. Therefore, $(\det|M|)^2 = 1$, and $\det|M| = \pm 1$. For the example shown, $\det|M| = +1$.

7.46. For orthogonal matrix M and from Exercise 7.45, we have

$$\det|M| = \sum_i m_{ik} M_{ik} = \sum_i m_{ik}^2 = 1$$

Then, $\det|M'| = \sum_i m_{ik} M'_{ik} = \sum_i m_{ik} M_{i\ell} = \sum_i m_{ik} m_{i\ell} = 0$ since M' contains two identical columns. Hence, $\sum_i m_{ik} m_{i\ell} = \delta_{k\ell}$.

7.47. 920 m s^{-1}.

7.48. Make use of the Leibniz theorem for the differentiation of a definite integral to find $\langle E \rangle$ and $\langle C_v \rangle$; use the theorem of l'Hôpital to find $\lim_{T \to \infty} \langle C_v \rangle$.

7.50. Since $x^3/(e^x - 1)$ is defined at $x = 0$, $\int_0^{13} [x^3/(e^x - 1)] dx$ exists. For $x > 13$, one has

$$e^x - 1 > x^5 \Rightarrow \int_{13}^\infty \frac{x^3 dx}{e^x - 1} < \int_{13}^\infty \frac{x^3 dx}{x^5} = \frac{1}{13}$$

Hence, $\int_0^\infty [x^3/(e^x - 1)] dx$ converges. For lead, $\tilde{E}_0 = 9N_0 h\nu_D/8 = 822$ J mol^{-1}.

7.51. For copper, $\Theta_D = 315$ K, $\alpha = 499.8 \times 10^{-7}$ K^{-1}, $\kappa = 0.763 \times 10^{-11}$ Pa^{-1}, $\tilde{V} = 7.1225 \times 10^{-6}$ m^3. The following comparisons are in J mol^{-1} K^{-1}.

T (K)	50	100	150	200	250	300
$\langle \tilde{C}_p \rangle$, calc	6.11	16.14	20.55	22.57	23.65	24.32
$\tilde{C}_p$, exp	6.15	16.01	20.51	22.63	23.78	24.46

7.53. In the first expression for $\langle S \rangle$, approximate $D(\Theta_D/T)$ by expanding $e^x - 1$ out to third order in x. Approximate $\ln[1 - \exp(-\Theta_D/T)]$ by first expanding the exponential out to third order in T, and then expanding the logarithm out to second order in T. For potassium, $\langle \tilde{S} \rangle = 62.9$ J mol^{-1} K^{-1} (calc).

7.54. $\Theta_D = 166.5$ K, $\gamma = 0.001597$ J mol^{-1} K^{-2}.

7.55. Start with the two-dimensional wave equation, and obtain

$$\langle A \rangle = \tfrac{2}{3}N\hbar\omega_D + \frac{4Nk_BT}{\omega_D^2}\int_0^{\omega_D}\omega \ln(1 - e^{-\hbar\omega/k_BT})\,d\omega$$

from which at low temperatures one then finds

$$\langle \tilde{C}_v \rangle \simeq 24R\left(\frac{T}{\Theta_D}\right)^2\sum_{n=1}^{\infty}\frac{1}{n^3}$$

No analytical form is known for the infinite series, but it is close to $4\pi^3/103 \Rightarrow \tilde{C}_v \simeq 96\ \pi^3RT^2/103\Theta_D^2$. At 9.3 K the value of $\tilde{C}_p$ (assumed here equal to $\tilde{C}_v$) is 0.00228; thus, $\Theta_D = 1047$ K. We then have for $\tilde{C}_p$ at 26 K: (0.0178R, calc; 0.0192R, exp); at 51.4 K: (0.0696R, calc; 0.073R, exp).

7.56. By Gaussian quadrature: 0.5707932; by Simpson's rule ($h = 0.233333$): 0.39958.

7.57. With Compton's equation applied to lead, choose a fit at $T = 80$ K; then $\tau = 28.85$.

T (K)	10	40	60	150	200
$\langle \tilde{C}_p \rangle$, calc	5.42	20.89	22.86	24.59	24.78
$\tilde{C}_p$, exp	2.76	19.57	22.43	25.27	25.87

CHAPTER 8

8.1. $a = 355.6$ L^2 atm K$^{1/2}$ mol^{-2}, $b = 0.1314$ L mol^{-1}.

8.4. $T_B = (a/bR)^{2/3} = 2.8982T_{cr} = 552$ K; exp = 506 K.

8.5. Some calculated values of $\Delta\tilde{E} = \tilde{E}(V) - \tilde{E}(100\ \text{L})$: $\tilde{V} = 33\tfrac{1}{3}$ L, $\Delta\tilde{E} = -6.493$ J; $\tilde{V} = 3.7037$ L, $\Delta\tilde{E} = -84.26$ J; $\tilde{V} = 0.4115$ L, $\Delta\tilde{E} = -773.1$ J.

8.6. An example of a $(2 + 2 + 2)$-partition: $\{N_{22} = 2, N_{31} = 2, N_{43} = 2\}$; an example of a $(1 + 1 + 1 + 3)$-partition: $\{N_{12} = 1, N_{32} = 3, N_{42} = 1, N_{43} = 1\}$; an example of another partition: $\{N_{31} = 1, N_{32} = 1, N_{33} = 4\}$.

8.7. After obtaining $\langle n \rangle$, use the thermodynamic expression $P = (\mu n - E + TS)/V$ for n moles of a pure substance.

8.8. Number of lattice points = 294; area of quarter circle = 314; % error = 7%.

8.9. Number of lattice points = 31205; area of quarter circle = 31415; % error = 0.67%.

8.10. $q_{class} = k_B T \sqrt{m/k}\,/\hbar$

8.11. $$Q_1 = 8\pi^2 I k_B T V/(h^2 \Lambda^3), \text{ where } \Lambda = \left(h^2/2\pi m k_B T\right)^{1/2};$$

$$Q_2 = \frac{Q_1^2}{32\pi^2 V^2} \iint_{-\infty}^{\infty} e^{-U(q_1,\,q_2)/k_B T}\, dq_1\, dq_2$$

8.13. $$b_4 = \frac{1}{V}\left(Q_4 - Q_1 Q_3 - \tfrac{1}{2}Q_2^2 + Q_1^2 Q_2 - \tfrac{1}{4}Q_1^4\right)$$

$$D(T) = -\left(\frac{V}{Q_1}\right)^4\left[20b_2^3\left(\frac{V}{Q_1}\right)^2 - 18b_2 b_3\left(\frac{V}{Q_1}\right) + 3b_4\right]$$

8.15.

T (K)	273	320	373	500	
$B(T)$, calc	−66.0	−41.5	−22.8	+2.9	cm^3 mol^{-1}

8.16.

Gas	H$_2$O	Ar	CH$_4$	CCl$_4$	Butane	
$\tilde{V}_{cr}$	55.43	75.27	98.96	275.66	219.65	cm^3 mol^{-1}
σ_D, calc	2.80	3.10	3.40	4.78	4.43	Å

8.17. $\log_{10} T_B \simeq 2.02 \Rightarrow T_B = 105$ K.

8.18. Find an r such that $dU/dr = 0$ and $d^2U/dr^2 > 0$; $r = 5.579a_0$, $U = -0.07576$ kJ mol^{-1}.

8.19. Square-well and LJ potentials both satisfy the stated restriction.

8.20. Expand the exponential in a Maclaurin series.

8.21.

T (K)	273	340	400	600	
$T(dB/dT)$, calc	173	125	98	53	cm^3 mol^{-1}
$C(T)$, calc	3400	1800	1350	1300	cm^6 mol^{-2}

8.22.

T (K)	75	90	110	125	250	500	700	
$B(T)$, calc	-254	-185	-130	-102	-16.9	17.5	25.8	cm^3 mol^{-1}

8.23.

T (K)	273	340	400	500	
$B(T)$, calc	-55.2	-25.0	-8.2	9.4	cm^3 mol^{-1}

8.24. Integrate both sides of the Maxwell relation $(\partial S/\partial V)_T = (\partial P/\partial T)_V$ with respect to V; for CH$_4$ at 273 K and 1 atm, $\rho \simeq 0.04464$ mol L^{-1}, and using data in Exercises 8.15, 8.21, obtain $\tilde{S}_{real} - \tilde{S}_{id} = -0.0053\, R$.

8.25. Introduce the virial expansion for $\tilde{V}$ in powers of P, and then make use of a result in Exercise 8.14.

8.26. To calculate μ_{JT}° from Table 8.6, use the program in Exercise 8.22, and calculate $T(dB/dT)$ by numerical differentiation; μ_{JT}°(calc) $= 0.467$ K atm^{-1}. To calculate μ_{JT}° from the Dieterici equation, refer to Exercise 8.15; μ_{JT}°(calc) $= 0.569$ K atm^{-1}.

8.27. Make use of the law of cosines.

8.28. Plot $PV^{-1}[P^\circ/(P^\circ - P)]$ versus P; $V_m \simeq 0.341$ cm^3, $P_h \simeq 0.027$ Torr.

8.29. $g^{(k)} = [\rho_0/(n-k)]\int g^{(k+1)}\, d^3 r_{k+1}$.

8.30. $C = \rho_0^{-3} \Rightarrow g^{(3)}(\mathbf{r}_1, \mathbf{r}_2, \mathbf{r}_3) = g^{(2)}(\mathbf{r}_1, \mathbf{r}_2)g^{(2)}(\mathbf{r}_1, \mathbf{r}_3)g^{(2)}(\mathbf{r}_2, \mathbf{r}_3)$

8.31. Integrate $4\pi\rho_0 r^2 g(r)$ by the trapezoidal rule from $r = 2.82$ Å [where $g(r) \simeq 0$] to $r = 5.22$ Å; number of nearest neighbors $= 12.8$.

8.32.

Distance from Central Atom	Number of Neighbors
$\frac{1}{2}L\sqrt{3}$	8
L	6
$L\sqrt{2}$	12
$2L$	6
$\frac{1}{2}L\sqrt{17}$	24
$\frac{1}{2}L\sqrt{19}$	24

$$L = 2.8644 \text{ Å}$$
$$\rho_0 = 0.08492 \text{ atom Å}^{-3}$$

8.33. Make the substitution $x = is\cos\alpha$.

8.35. Arrive at $\langle T \rangle = (3N/2m)\langle p_{xi}^2 \rangle$, then use eq. (8.6-1).

8.37. Write

$$3Nk_BT = 3PV + \tfrac{1}{2}N(N-1)\int\int f^{(2)}(\mathbf{r}_i,\mathbf{r}_j)\left(\mathbf{r}_i \cdot \frac{\partial u(\mathbf{r}_i,\mathbf{r}_j)}{\partial \mathbf{r}_i}\right)d^3r_i d^3r_j \text{ from}$$

eq. (8.6-3), and now simplify.

8.39. (b) Since $\rho = \rho(n,V) \equiv \dfrac{n}{V}$, then ρ is a homogeneous function of order 0. Apply Euler's theorem, followed by the chain rule, to get

$$-n\left(\frac{\partial \rho}{\partial P}\right)_{V,T}\left(\frac{\partial P}{\partial n}\right)_{V,T} = V\left(\frac{\partial \rho}{\partial P}\right)_{n,T}\left(\frac{\partial P}{\partial V}\right)_{n,T}$$

But pressure is also homogeneous of order 0, and the final result follows.

8.40. (b) For very large n, truncate the series for $p_G^{(k)}$ to just one term, and for $k = 2$ write $p_G^{(2)} \simeq e^{n\mu/k_BT}Q_n p_n^{(2)}/\Xi \simeq p_n^{(2)}$; (c) eliminate $\langle n^2 \rangle$ from the equations in Exercises 8.39(d), 8.40(b).

8.41. $\Phi(x) = \tfrac{1}{2}e^{x^2}$

8.42. $\Phi(x) = x - \pi\lambda(-1)^n \sin nx/[n(1 - \tfrac{1}{2}\pi\lambda)]$

8.43. $k_BT\dfrac{\partial \ln p^{(3)}}{\partial \mathbf{r}_1} = -\dfrac{\partial u(r_{12})}{\partial \mathbf{r}_1} - \dfrac{\partial u(r_{13})}{\partial \mathbf{r}_1}$

$- \displaystyle\int \frac{[\partial u(r_{14})/\partial \mathbf{r}_1]\, p^{(4)}(\mathbf{r}_1,\mathbf{r}_2,\mathbf{r}_3,\mathbf{r}_4)}{p^{(3)}(\mathbf{r}_1,\mathbf{r}_2,\mathbf{r}_3)} d^3r_4;$ simplification could be

achieved by some kind of superposition approximation, for example, $p^{(4)}(1,2,3,4) = \rho_0^{-8}\{ p^{(3)}(1,2,3) \cdot p^{(3)}(1,2,4)p^{(3)}(1,3,4)p^{(3)}(2,3,4)\}$.

8.44.

ρ^*	0	0.1	0.2	0.3	0.4	0.5	0.7	0.88
Z	1	1.240	1.553	1.967	2.518	3.262	5.710	10.05

8.45. Model $U(r)$ in terms of a Heaviside step function (see Exercise 2.17): $U(r) = k_BT[U_0 - H(r/d - 1)]$, where U_0 is very large and is in units of k_BT, and $H(x - a) = 0$ for $x < a$ and 1 for $x > a$. Then make use of the fact that the derivative of a Heaviside step function is a Dirac delta function.

8.46. $\tilde{V}_\ell/\tilde{V}_s = 1.07_3$ (calc).

8.49. The idea is to show that if $g(r)$ is set equal to 1 for large r in the integrand, then the right side of the equation becomes unity.

Author Index

331

Subject Index

Acetylene, electron population analysis, 125
Algorithm, 15
Ammonia:
  compressibility factor at critical point, 239
  computer simulation, 295
  hard-sphere diameter, 263
  pair correlation function for liquid, 285
Amplitude, 218, 275
Angular momentum:
  eigenfunctions for square, 47, 195
  in molecules, 194
  relative, for classical two-body system, 11
  spin, see Spin angular momentum
Anharmonic oscillator, 65, 205
Annihilation operator, 38, 75
Argon:
  Boyle temperature, 240
  critical constants from solution of various
    integral equations, 291
  hard-sphere diameter, 263
  Joule-Thomson coefficient, 268
  Lennard-Jones parameters, 265, 291
  model for melting, 304
  pair correlation function for liquid, 280
  pair potentials, 284
  pressure, according to Monte Carlo method,
    296
Associated Laguerre polynomial, 48, 195
Atomic charge, 123
  calculations, for selected molecules, 125
  criticism of Mulliken's treatment, 124

Politzer's treatment, 125
Atomic scattering factor, 278
Atomic spin-orbital, 71, 99–101, 102–103
Atomic term symbol, 185
Atomic units, 104
Average energy approximation, 133–135
Average value of dynamical variable, see
    Expected value

Banded matrix, 302
Barriers to internal rotation, 217
Basis (set), 31, 95, 99–101, 147, 155
  universal atomic, 118
Benzene:
  compressibility factor at critical point, 239
  computer simulation, 295
  Lennard-Jones parameters, 265
Bernoulli numbers, 206
Bipolar coordinates, 298
Bohr Correspondence Principle, 250
Boltzons, 182
Born-Green (BGBY) equation, 290
  application to density profile at liquid-gas
    interface, 292
  stimulus for research on phase transitions,
    292–293
Born-Oppenheimer approximation, 76–78,
    89, 119, 128, 198
Bosons, 69, 194–196
Boyle temperature, 238
  of Dieterici gas, 239